123 PIC®
Microcontroller
Experiments for
the Evil Genius

Evil Genius Series

123 Robotics Experiments for the Evil Genius

Electronic Gadgets for the Evil Genius: 28 Build-it-Yourself Projects

Electronic Circuits for the Evil Genius: 57 Lessons with Projects

123 PIC® Microcontroller Experiments for the Evil Genius

Mechatronics for the Evil Genius: 25 Build-it-Yourself Projects

50 Awesome Automotive Projects for the Evil Genius

Solar Energy Projects for the Evil Genius: 16 Build-it-Yourself Thermoelectric and Mechanical Projects

Bionics for the Evil Genius: 25 Build-it-Yourself Projects

MORE Electronic Gadgets for the Evil Genius: 28 MORE Build-it-Yourself Projects

123 PIC®
Microcontroller
Experiments for
the Evil Genius

MYKE PREDKO

McGraw-Hill

New York   Chicago   San Francisco   Lisbon
London   Madrid   Mexico City   Milan   New Delhi
San Juan   Seoul   Singapore   Sydney   Toronto

The McGraw·Hill Companies

Cataloging-in-Publication Data is on file with the Library of Congress

2 3 4 5 6 7 8 9 0   QPD/QPD   0 1 0 9 8 7 6 5

ISBN 0-07-145142-0

The sponsoring editor for this book was Judy Bass and the production supervisor was Pamela A. Pelton. It was set in Times Ten by MacAllister Publishing Services, LLC. The art director for the cover was Anthony Landi.

Printed and bound by Quebecor/Dubuque.

 This book is printed on recycled, acid-free paper containing a minimum of 50 percent recycled de-inked fiber.

McGraw-Hill books are available at special quantity discounts to use as premiums and sales promotions, or for use in corporate training programs. For more information, please write to the Director of Special Sales, McGraw-Hill Professional, Two Penn Plaza, New York, NY 10121-2298. Or contact your local bookstore.

Contents

Acknowledgments

This book would not have been possible without the help, suggestions, and time from the following individuals:

- Carol Popovich, Greg Anderson, Joe Drzewiecki, Andre Nemat, and Fanie Duvenhage of Microchip who have helped me understand which products would be best suited for this book, and have been willing to spend time with me to understand my requirements, answer my legions of questions, and suggest avenues to follow that hadn't occurred to me.

- HT-Soft and their technical support staff for answering my questions quickly and helping to explain the inner workings of the PICC Lite compiler. The PICC™ line of compilers is a tremendous tool for the beginner and professional alike, and one I never hesitate to recommend.

- Brad North, Richard Bonafede, and the students of Rick Hansen Secondary School in Missassauga, Ontario, for helping me to learn more about how students (and teachers) learn about programming, electronics, and the PIC MCU.

- Blair Clarkson and Dave Pilote at the Ontario Science Centre in Toronto, for pushing me to explain basic microcontroller concepts to people that have built the TAB Electronics SumoBot at a workshop and want to do more with it. The BS2 interface and Robot IR Tag experiments are a direct result of this work.

- My editor at McGraw-Hill, Judy Bass, who consistently responds to my questions and suggestions, regardless of how dumb they are, with good humor and thoughtfulness.

- The PIC MCU continues to be one of the best supported devices on line, and I would like to thank the many individuals who have taken the time to put information and projects on the Internet as well as support and help others trying to better understand the PIC MCU or get their applications up and running.

- Celestica and its employees for ideas, answers to strange questions, and opportunities to expand my technical horizons.

- My daughter Marya, who has grown up with a father that is always trying out new projects on her. She tries them out with enthusiasm despite the fact they generally need a bit of tuning before they work perfectly.

- My wife Patience, for keeping everything together, even when our youngest daughter Talitha was sick, and for having dinner on the stove, even with me asking her to do a "quick read" of a section or two. I couldn't do any of it without you.

To all of you, thank you for all you unselfish help and willingness to share your ideas, experiences, and enthusiasm for this book.

myke
http://www.myke.com

Myke Predko is Test Architect at Celestica, in Toronto, Canada, a supplier of printed circuit boards to the computer industry. An experienced author, Myke wrote McGraw-Hill's best-selling *123 Robotics Projects for the Evil Genius*; *PICmicro Microcontroller Pocket Reference*; *Programming and Customizing PICMicro Microcontrollers, Second Edition*; *Programming Robot Controllers*; and other books, and is the principal designer of both TAB Electronics *Build Your Own Robot* kits.

Introduction

When I wrote my first book on the Microchip PIC®
*microcontroller (*commonly abbreviated to "MCU")
almost 10 years ago, the most common criticism I
received about the book was that it took too long to
get to the projects. This is quite foreign to me because I
tend to learn a new device, like a microcontroller, by
first reviewing the datasheets for the part's electrical
information, working at understanding the architecture and how it is programmed, and ending with
understanding what kind of development tools are
available for the part. Looking over this list of tasks, it
is quite obvious that they came about with my background and training. Being a teenager in the 1970s and
going to university in the early 1980s, there wasn't the
variety of easy-to-work-with devices that are available
today, and the sophisticated personal computer-based
development tools that we take for granted were not
even being considered, let alone being developed or
sold. My method for learning about a new part is effective for me and a result of the situation I found myself
in when I first started working with electronics. Today
you can set up a development "lab" and create a basic
application from scratch for the PIC MCU in less than
20 minutes using Microchip's MPLAB® *integrated
development environment* (IDE) and PICkit™ 1
starter kit with HT-Soft's PICC Lite™ C compiler.

The purpose of this book is to introduce you to the
Microchip PIC MCU and help you understand how
to create your own applications. In this introduction,
using the PICkit 1 starter kit *printed circuit board*
(PCB) and free development tools from Microchip
and HT-Soft, I will show you how easy it is to create a
simple PIC MCU program that will flash one of the
light-emitting diodes (LEDs) on the PICkit 1 starter
kit. As you work through the book, your understanding of the PIC device will increase to the point where
you should be comfortable creating your own complex applications in both the C programming language as well as assembly language.

The PICkit 1 starter kit (see Figure i-1) contains
everything you will need to learn how to create and
test your own PIC MCU applications. This includes a
programmer PCB (see Figure i-2), a *universal serial
bus* (USB) cable to connect the PICkit 1 starter kit to
your PC, a CD-ROM containing the source code for
the applications presented in this book, two PIC
MCUs, an eight-pin PIC12F675, and a 14-pin
PIC16F684. In this book, I will be focusing on the
PIC16F684 because its 14 pins allow a greater variety
of different applications to be built from it, but you will
also gain experience with the eight-pin PIC12F675.

On the back cover of the book is a web link that you
can use to order a PICkit 1 starter kit for use with this
book. If you do not buy a PICkit 1 starter kit, the
source code can be downloaded from my web site
(www.myke.com).

In this book, I will be working exclusively with
Microsoft Windows. I recommend that you use the latest version available (at this writing it is Windows/XP
SP2) when working with the PICC Lite compiler and
MPLAB IDE tools used in the book. Development
tools are available for Linux, although not for Apple
Macintosh OS/X (but you should be able to get the
Windows software to work from an emulator). You
will find that the software works well under Windows.

If you look at the CD-ROMs that come with the
PICkit 1 starter kit, you will find they have the
Microchip MPLAB IDE and HT-Soft PICC Lite compiler development tools that are used in the book.
Although you could load these programs onto your
PC, I recommend that you download the latest versions from the Microchip and HT-Soft web sites. These
tools are continuously updated (during the period that
this book was written, MPLAB IDE had five upgrades,
two of them major and changed how some of the operations are performed) to include new features and PIC
MCU part numbers, and to fix any outstanding problems. In this book, I used MPLAB IDE version 7.01

Figure i-1 *The Microchip PICkit 1 starter kit
enables you to create your own PIC MCU
applications and to test them out easily and
inexpensively.*

1

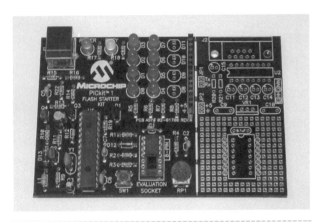

Figure i-2   *The PICkit 1 starter kit's PCB consists of programmer circuits along with eight LEDs, a switch, and a potentiometer that enable you to easily learn how to program and access the peripheral features of the PIC MCU.*

Figure i-3   *Step 1 - Go to www.htsoft.com*

and PICC Lite compiler version 8.05. With the versions that you use, you may see some differences in look or operation, but the features presented in this book will all be present. If you are confused as to how to perform some operation, you can consult the Tutorial section of MPLAB IDE, which can be found under the Help pull-down.

To start setting up the software needed to start developing your own PIC MCU applications, go to www.htsoft.com, as shown in Figure i-3. PICC Lite is a free, full-featured C compiler that supports quite a few of the different PIC MCU part numbers (including the PIC12F675 and PIC16F684 included in the PICkit 1 starter kit package). Next click "Downloads" and then select "PICC Lite (Windows)" (see Figure i-4). To download PICC Lite compiler, you will have to register with HT-Soft at no charge (see Figure i-5). To do this, follow the instructions on the page shown in Figure i-4. It should go without saying that the page will probably not look exactly like the figures here, due to the delay between when I have written this and when you actually access the web site.

The retail PICC compiler is capable of building code for literally all the PIC MCU part numbers and does not have any of the restrictions of the PICC Lite compiler.

After registering, the PICC Lite compiler installation software should start downloading automatically. Depending on your security settings (especially if you use Microsoft Windows/XP Service Pack 2 or later), the download may be blocked (as in Figure i-6). If this is the case, you may have to turn off the security to allow the download to take place. Once the program has downloaded, I recommend selecting "Run" instead of "Save" (see Figure i-7). This will install the PICC

Figure i-4   *Step 2 - PICC Lite Compiler download*

Figure i-5   *Step 3 - Registering a an HT-Soft customer*

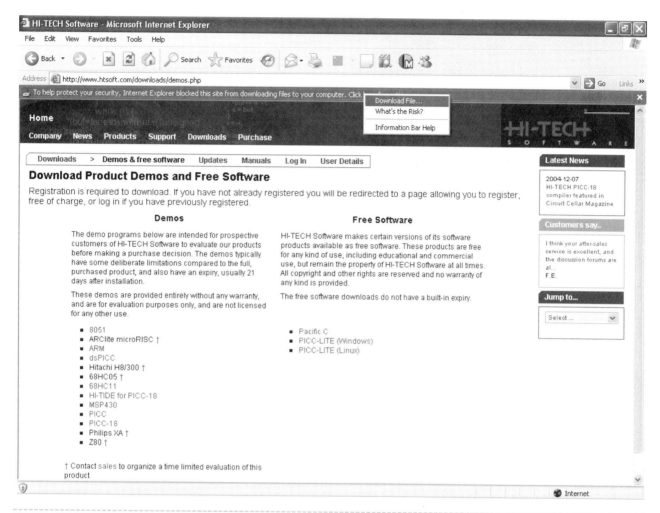

Figure i-6   *Step 4 - Allowing PICC Lite Compiler file download in Windows/XP SP2*

Lite compiler without leaving you with an .exe or .zip file to delete later.

When the PICC Lite compiler installation screen appears (see Figure i-8), click "Next" and follow the defaults. If you are prompted to load the software/drivers for MPLAB IDE, do so, as they will be required to use PICC Lite compiler with the Microchip tools and provide you with a truly integrated development environment, or IDE. The MPLAB IDE is a single program containing an editor, assembler, linker, simulator, and PICkit 1 starter kit's programmer interface, and it will be the only program you have to run to create PIC MCU applications. PICC Lite compiler will integrate with MPLAB IDE when the latter is installed so you will have a single Windows program for developing C and assembly language programs for the PIC MCU.

After PICC Lite compiler is installed, you will be asked if you want to restart your computer. Click "No," then power down your computer, and power back up. I find that soft resets (ones in which power is

Figure i-7   *Step 5 - PICC Lite downloading*

not removed) may not reset all the PC's parameters, and the software installed on boot may not work properly. Powering down and then back up eliminates these potential problems.

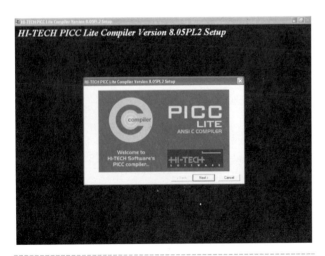

Figure i-8   *Step 6 - PICC Lite Installer*

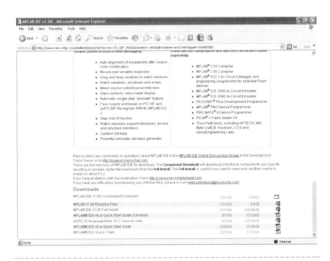

Figure i-10   *Step 8 - MPLAB IDE Download Page on www.microchip.com*

With the PC back up, go to Microchip's web site (www.microchip.com shown in Figure i-9) and click "MPLAB IDE" or "Development Tools," followed by "Full Install" (see Figure i-10). The MPLAB IDE software is quite large (30 MB) and will take some time to download if you use a dial-up connection. This time click "Save" instead of "Run" and store the .zip file into a temporary folder on your PC (see Figure i-11). You should be able to unzip the file by double-clicking on it, and the file management software on your PC will expand the file into the directory of your choosing (ideally the same one you started with).

After the MPLAB IDE install files are unzipped, double-click on "Setup" and follow the instructions to install the MPLAB IDE (see Figure i-12). If asked, make sure the programming interface for the PICkit 1 starter kit is included and you will not have to look at

Figure i-11   *Step 9 - MPLAB IDE Install files stored in a temporary folder*

any readme files (unless you want to). If you are prompted to reboot your computer, click "No" and then power down and power back up, as you did after installing the PICC Lite compiler software. *Do not connect the PICkit 1 starter kit to your PC using the USB cable until you are told.*

That's it; you've just installed a set of integrated development tools that are just as powerful as some software development products that cost many thousands of dollars. With the tools installed, you can copy the source code files for the application code used in this book from the PICkit 1 starter kit's CD-ROM code folder into a similar "code" or "Evil Genius" folder under the C drive of your PC. Another source for these files can be found on my web site at www.myke.com.

Figure i-9   *Step 7 - Set your browser to www.microchip.com to download the MPLAB IDE as well as PIC MCU Datasheets*

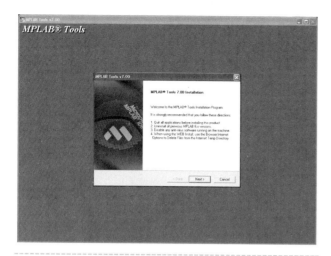

Figure i-12 *Step 10 - MPLAB IDE Installer window*

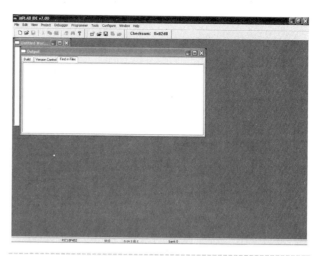

Figure i-13 *Step 11 - MPLAB IDE Start Up desktop*

Now let's try to create a simple program that flashes an LED on the PICkit 1 starter kit. To do this, double-click on the MPLAB IDE icon that has been placed on your PC's desktop. When it first boots up, the MPLAB IDE desktop looks like Figure i-13, and is ready for you to start entering your own application. Click "New" and enter the following code into the window that comes up:

```c
#include <pic.h>
    __CONFIG(FCMDIS & IESODIS & BORDIS & UNPROTECT
&
    MCLRDIS & PWRTEN & WDTDIS & INTIO);
int i;
main()
{
    PORTA = 0;
    CMCON0 = 7;
    ANSEL = 0;
    TRISA4 = 0;
    TRISA5 = 0;
    while (1 == 1)
    {
        for (i = 0; i < 25000; i++);
        RA4 = RA4 ^ 1;
    }
}
```

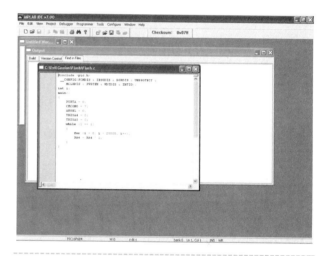

Figure i-14 *Step 12 - First Application entered into MPLAB IDE Editor window and saved*

Different parts of this program will be displayed using different colors; don't worry if it looks a bit strange. The program should be saved as "c:\Evil Genius\Flash\Flash.c" (see Figure i-14). Once the program is saved, close the window that contains it.

All programs should be run as part of an MPLAB IDE project that saves options and features selections specific to the application without requiring you to reload them each time you start up MPLAB IDE. The first step is to specify the project name and where it is going to be stored (see Figure i-15).

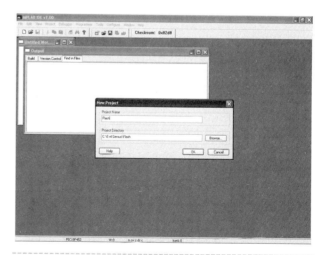

Figure i-15 *Step 13 - Creating an MPLAB IDE project*

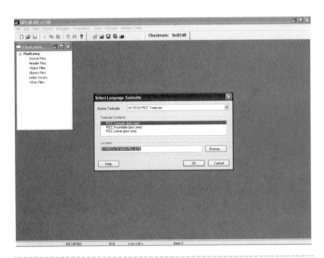

Figure i-16 *Step 14 - Selecting the PICC Lite C Compiler as the MPLAB IDE Build tool*

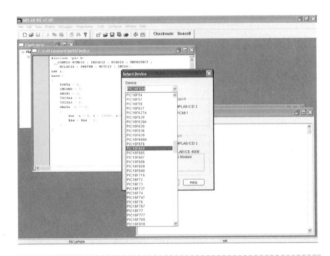

Figure i-18 *Step 16 - Selecting the PIC MCU part number to work with (note the large number of PIC MCUs to choose from).*

The Flash application is written in the C programming language for the PICC Lite compiler. To specify the PICC Lite compiler, click "Project" and then "Select Language Toolsuite." You may have to scroll through the Active Toolsuite to find PICC Lite compiler (see Figure i-16). When you have selected it, make sure that the location of PICL.exe is correct. If it is not, look for the C:\PICCLITE folder on your PC, and point the Toolsuite Contents to picl.exe in the BIN subfolder.

When you are working with assembly language programs, you may have to perform the same operation there as well. In this case, the assembly language programs (such as mpasmwin.exe) can be found in the Program Files\MPLAB folder or in its subfolders.

Right-click on "Source Files" in the Flash.mcw window (see Figure i-17) and select "Flash.c" from the

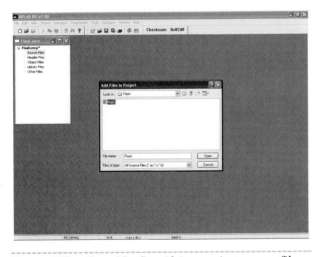

Figure i-17 *Step 15 - Specifying project source file*

c:\Evil Genius\Flash\Flash.c folder (where you stored the program earlier). To load Flash.c onto the desktop, double-click on "Flash.c" in the Flash.mcw window. This will associate Flash.c with the project that was just created. Each time you work with a new program, it should have a new project associated with it.

Next, you will have to make sure the proper PIC MCU is selected for the application. Click "Configure" and then "Select Device," and find PIC16F684 in the list (see Figure i-18). I'm sure you will be amazed at the number of different PIC microcontroller part numbers that come up. After working through this book, you will discover you can program and use the vast majority of these chips in your applications. The difference in the part numbers is the number of pins and interfacing features built into the PIC MCU. Programming and interfacing are identical to what has been presented in this book.

Now you are ready to try and "build" the application. You can click "Project" and then "Build All," or press Ctrl+F10 to compile the application and store the result in a .hex file that will be programmed into the PIC MCU later. If any errors occur, go back over the code you keyed in and compare it to the previous listing. This is the most likely source of the problem. Once the program has compiled correctly, you will get the summary information shown in Figure i-19, listing the amount of space required to store and run the program in a PIC MCU.

With the program compiled, plug your PICkit 1 starter kit into a USB cable plugged into your PC. Afterwards, the MPLAB IDE screen will look like Figure i-20 with the status window changing to list the Firmware version of the PICkit 1 starter kit. If it does

Introduction

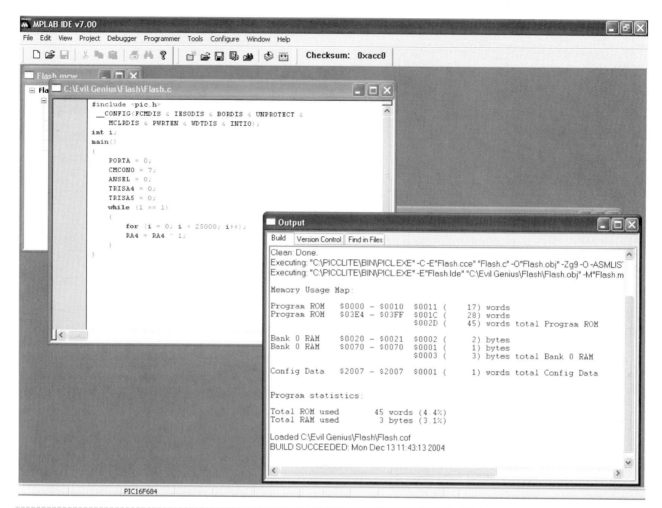

```
MPLAB IDE v7.00
File  Edit  View  Project  Debugger  Programmer  Tools  Configure  Window  Help

                                              Checksum:  0xacc0

Flash.mcw

C:\Evil Genius\Flash\Flash.c

    #include <pic.h>
    __CONFIG(FCMDIS & IESODIS & BORDIS & UNPROTECT &
        MCLRDIS & PWRTEN & WDTDIS & INTIO);
    int i;
    main()
    {
        PORTA = 0;
        CMCON0 = 7;
        ANSEL = 0;
        TRISA4 = 0;
        TRISA5 = 0;
        while (1 == 1)
        {
            for (i = 0; i < 25000; i++);
            RA4 = RA4 ^ 1;
        }
    }

Output

Build   Version Control   Find in Files

Clean: Done.
Executing: "C:\PICCLITE\BIN\PICL.EXE" -C -E"Flash.cce" "Flash.c" -O"Flash.obj" -Zg9 -O -ASMLIS
Executing: "C:\PICCLITE\BIN\PICL.EXE" -E"Flash.lde" "C:\Evil Genius\Flash\Flash.obj" -M"Flash.m

Memory Usage Map:

Program ROM   $0000 - $0010   $0011 (     17) words
Program ROM   $03E4 - $03FF   $001C (     28) words
                              $002D (     45) words total Program ROM

Bank 0 RAM    $0020 - $0021   $0002 (      2) bytes
Bank 0 RAM    $0070 - $0070   $0001 (      1) bytes
                              $0003 (      3) bytes total Bank 0 RAM

Config Data   $2007 - $2007   $0001 (      1) words total Config Data

Program statistics:

Total ROM used        45 words (4.4%)
Total RAM used         3 bytes (3.1%)

Loaded C:\Evil Genius\Flash\Flash.cof
BUILD SUCCEEDED: Mon Dec 13 11:43:13 2004

PIC16F684
```

Figure i-19   *Step 17 - Build information for "Flash.c" application*

not change, then click on the *programmer toolbar* followed by "Select Programmer" and then the PICkit 1 icon. A four-button programmer toolbar will also be displayed on the desktop. Remove the PIC12F675 that came installed in the PICkit 1 starter kit and put in the PIC16F684 that was in the PICkit box. Remember to store the PIC12F675 in a safe place (using the piece of foam the PIC16F684 was on is a good choice). Click "Programmer," "Program Device," or the Program icon (place your mouse over the programmer icons to make a button legend appear) to download the application into the PIC16F684. The programming operation will take a few moments (the operation is indicated by a growing bar on the bottom-left corner of the desktop, and a "Program Succeeded" message will be shown on the status window, Figure i-21).

Once the build/compile operations, followed by the device programming steps, are complete, the D0 LED of the PICkit 1 starter kit will start to flash. If it doesn't, you should review the source code and the process of creating the project.

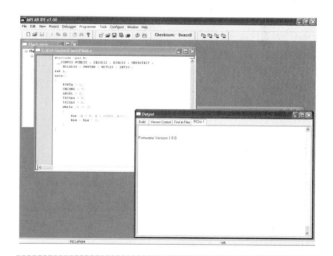

Figure i-20   *Step 18 - Information provided when the PICkit 1 starter kit is plugged into the development PC's USB Port. Note: Four-button Programmer toolbar at the top center of MPLAB IDE desktop.*

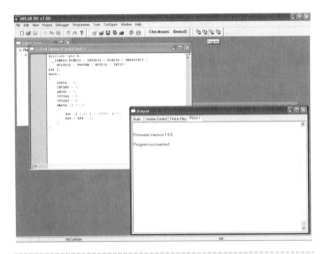

Figure i-21 *Step 19 - The first application has been programmed into a PIC16F684 and inserted into the PICkit 1 started kit connected to the development PC. The D0 LED should now be flashing!*

That's it; you have just set up a very sophisticated microcontroller application code development lab and created your first application. I realize very little of the program or the process you went through to get to this point makes sense, but as you work through the different experiments in this book, their functions will become more obvious and easier for you to use on your own. Just as if you were to review the datasheets for the part's electrical information, working at understanding the architecture and how it is programmed will help you understand what kind of development tools are available for the part.

Throughout the rest of this book, I will help you learn about and work with the PIC MCU. The experiments range from the very trivial to some very complex interfacing applications that are really quite a bit of fun. There is a lot of material in this book and a lot to learn; try to work through each experiment in a section before taking a break—I would be surprised if you were able to get through the entire book in less than a year.

As part of the learning exercise, try to develop your own circuits and code—this will cement the knowledge you gain from the book and help you build the skills needed to create your own applications. Don't be afraid to use my designs and code as a base or as a part of yours (using cut and paste). Playing "what if" can be a lot of fun and very instructive. I believe that before somebody is comfortable working on a new device and development system, he or she should have 50 or so of his own applications under his or her belt.

Don't be discouraged when, at first, your applications don't work. It isn't unusual for it to take a week or two to get a person's very first application working (even if it's flashing an LED, as I do in this introduction). The second time it will take half as long, the third a quarter, and so on. Over a fairly short period of time, you will be able to create applications efficiently that are as sophisticated as those created by professionals.

Seeing your own PIC microcontroller applications working will be an amazing experience for you, and it will give you a sense of pride to know you can do something that only a small percentage of the world's population can do. Unfortunately, the process of getting there is full of frustration, confusion, and hard work. Along with teaching you about the PIC MCU, the purpose of this book is to help you gain the skills necessary to develop your own applications with a minimum of the inevitable frustration, confusion, and hard work.

Prerequisites

This book was written to be the second in a sequence (*123 Robotics Experiments for the Evil Genius* was the first book) and, as such, many of the basic electrical, mechanical, and programming concepts used in this book were presented in the first. To understand fully the experiments presented in this book as well to be able to create your own applications, you will have to be familiar with the concepts listed below:

- Basic electrical laws
  - Parts of a circuit
  - Ohm's law
  - Series resistances
  - Parallel resistances
  - Kirchoff's voltage law
  - Kirchoff's current law
  - Thevinin's equivalency
  - Resistor markings
- Semiconductor basics
  - Diode operation
  - LEDs (including 7 Segment Displays)
  - Bipolar transistor operation and pinouts
  - MOSFET operation
- Binary electronic logic
  - The six basic gates
  - Different logic technologies
  - The Boolean arithmetic laws
  - Types of flip flops
  - Common circuits

- Adders
- Muxes/Demuxes
- Counters
- Shift registers and *linear feedback shift registers* (LFSRs)
- Oscillators
  - Basic relaxation oscillator
  - Reflex oscillator
  - Crystals and ceramic resonators
  - 555 timer chip
- Common electronic devices
  - 741 op amp
  - 386 audio amplifier
  - IR object sensors
- Power supplies
  - Batteries
  - 780x/78L0x voltage regulators
  - Switch mode power supplies
- Numbering systems
  - Scientific notation
  - Metric prefixes
  - Capacitor markings
  - Binary numbers and conversions
  - Hexadecimal numbers and conversions
- Programming concepts
  - Data types
  - Variable declaration
  - Assignment statement
  - Variable arrays
  - If/else/endif statement
  - While statement
  - Subroutines
- Microcontroller concepts
  - Memory organization
  - Input and output pins
  - Special pin functions
  - Power supplies

Comments for Teachers and Students

While the target audience for this book is Grade 11 (junior) and Grade 12 (senior) high school students preparing for post secondary education in engineering,

computer science, mathematics, or physical sciences, it should be appropriate for university students or technical hobbyists looking for more information on programming and interfacing PIC microcontrollers into an application. To ensure that the material presented here would be relevant to high school students, the Ontario Ministry of Education curriculum guidelines for Computer Engineering (found at www.edu.gov.on .ca /eng/document/curricul/secondary/grade1112 /tech/tech.html#engineering) have been used as a reference when topics, experiments, and materials were selected. This book should be useful both as a course text and as a reference for both teachers and students. After working through these experiments, you will not only have a good understanding of how PIC16F684 and PIC12F675 microcontrollers (and indeed the entire PIC family of microcontrollers) are programmed and interface to other devices, but you will also be well on your way to being capable of creating your own sophisticated PIC MCU applications.

Reading through the experiments in this book will not make you proficient in creating your own PIC MCU application. I am a firm believer in doing, and I would expect that in any course, there would be many assignments that consist of modifications of the experiments presented in this book. These assignments should give students the task of creating their own applications and, as part of the process, planning, wiring, and debugging them. For the students to become reasonably proficient in developing applications, they should be given 10 to 20 applications as assignments over the course of a term in order to familiarize themselves with the MPLAB IDE software development environment, the PICkit, and the PIC MCU operation.

Along with providing information on Microchip PIC microcontrollers (the PIC16F684 in particular), electronics interfacing programming in C, and assembler programming, this book attempts to develop and engender the important thinking and problem-solving skills that are expected from a graduate engineer. These skills include the ability to work independently, perform basic technical research, create a development plan, and effectively solve problems. Along with being useful in professional careers, these skills are critical to your success in college and university.

The experiments in this book do not lend themselves well to group activities—you'll find that it is difficult to divide small microcontroller applications into different tasks that can be carried out by different members of a group. For this reason, I recommend assignments that are limited as one-student projects. Projects given during the term should be solvable in less than 100 lines of code (two printed pages) and should not require a lot of research on the part of the student. Instead time should be spent trying to decide the best way to solve the program and structure the software. All of these projects, except for a *summative project*, should be easily completed in a week or less.

A Note to Students:

- Teachers are very good at figuring out when an assignment is copied or plagiarized, and getting caught will land you and any other students who are involved in trouble. Copying code or circuitry from the Internet or other books will be identified quickly as it is difficult to rework them to fit into your design system or style of programming. In any case, cheating will not help you gain the necessary skills for developing your own application (and ultimately passing the course). In short, hand in only your own work and make sure that you can explain the how and why of your assignments. It will pay off in the long term.

- When given an assignment, spend a few moments a day on it, no matter what. Don't leave it until the night before it's due. By doing a little bit of work on it each day, your subconscious will work on the assignment, and when it comes time to finally create it and write it up, it will seem a lot easier.

- Don't be afraid to try something different. The worst thing that can happen is that it won't work—the upside is you will discover a way of approaching the problem that is very efficient and easy to implement. In this book, there are several examples where I just tried different things and discovered a better solution to a problem than I would have expected (and cases where the alternative solution was worse than the original).

A Note to Teachers:

- For teachers considering using this book as a text for their courses, I would like to emphasize that the material is designed for the PIC16F684, PICkit 1, and, to a lesser extent, the PIC12F675. I realize that investments have been made in other PIC (and other) MCUs, in programmers, in and other equipment, but the MPLAB IDE/PICC Lite compiler and PIC16F684/PICkit combination is an extremely flexible and cost-effective tool for application development.

- Try to use real-world situations for your assignments and try to vary them, both to interest your students and to make it harder for them to copy from preexisting materials. A good way to come up with these assignments is to keep a notebook handy and record the different things you see in your travels. Toys 'R Us, Radio Shack, The Sharper Image, and other retailers can be wonderful sources of inspiration.

- Lead by example. You should work through as many different applications until you are comfortable doing so and can debug most of the problems you experience. An important tool to be familiar with is the MPLAB IDE simulator. A couple of teachers have told me that they don't bother with the simulator when they are teaching because students prefer to see flashing lights or something happening. Please try to break this habit and encourage the use of the simulator as a tool for verifying the operation of the code as well as for debugging execution problems. Using the simulator will give students the ability to see the program working or not working, and why it is not working before making the effort to wire the application. For virtually all applications that run the first time power is applied, the developer should have demonstrated the correct operation of code on the simulator before he or she attempts to burn it into a PIC MCU.

Icons and Conventions

At the start of each section and experiment, you will find one or more icons indicating the parts and tools you need to have available to complete the experiment(s). I have chosen to do this rather than providing one central list of parts required for all the experiments in this book. The reason for doing this is to help you efficiently plan for the section's experiments without having to buy many hundreds of dollars of parts (many of which you will not need for months). I have tried to keep the number of parts to a reasonable minimum by designing as many experiments as possible to execute within the PICkit 1 starter kit—once the PIC MCU is programmed, the application programmed into it can execute on the PICkit 1 starter kit without modification.

The icons used to specify parts and operations are as follows:

At the start of each section, I will list the required parts for the section under this icon (see Figure i-22). As I indicated, this is a summation of the parts used in all the experiments of the section.

Introduction

Figure i-22   *Required Parts icon*

Figure i-23   *PC/Simulator icon*

Figure i-24   *PICkit 1 starter kit icon*

Figure i-25   *Parts Bin icon*

Figure i-26   *Toolbox icon*

Virtually all of the experiments will require a PC (indicated by Figure i-23) running Windows and loaded with the MPLAB IDE and PICC Lite compiler. This PC should be running one of the following versions of Microsoft Windows:

- Windows 98 SE
- Windows ME
- Windows NT 4.0 SP6a Workstations (NOT Servers)
- Windows 2000 SP2
- Windows XP

And the PC should have 32 MB memory (128 MB recommended), 95 MB of hard disk space (1 GB recommended), CD-ROM drive, along with Internet Explorer 5.0 or greater with an Internet connection for installation and online Help along with one free USB port.

When the PICkit 1 starter kit icon (see Figure i-24) appears at the start of an experiment, it means that the PICkit is required either as a platform for the experiment or as a PIC MCU programmer.

The Parts Bin icon (see Figure i-25) specifies which parts are required for the application. Parts may be reused between experiments.

The last icon (see Figure i-26) lists the tools required to create the application's circuit. You should not require any specialized tools for the experiments presented in this book. Remember to wear safety equipment when doing any cutting or drilling.

In terms of conventions, I will use both *System Internationale* (SI), better known as the metric system, and English measurements together where possible. I recognize that there are still a number of issues with specifying the correct measurement system, so by listing both I hope there will be less confusion. Standard electrical prefixes will be used, and I assume the reader is familiar with them. They include the following:

k for thousands

M for millions

m for thousandths

μ for millionths

p for trillionths

I restrain from specifying part numbers except when I believe that only one manufacturer's part should be used. Often, equivalents to various parts can be found in surplus and general electronics stores at a very low cost.

For the screen shots shown in the book, I have used MPLAB IDE version 7.01. I know that after this book goes to press, later and more capable versions of MPLAB IDE will be available for download from Microchip's web site, www.microchip.com

The MPLAB IDE functions presented in this book will not change from version to version (except for fixes to discovered problems), so despite some cosmetic changes in appearance, their operation will not change. I recommend that you always use the latest version available from the Microchip web site because if you have problems, the first recommendation that you will be given will be to try again with the latest version of the software.

Finding Parts

When you are first starting out in electronics, it can be difficult to find retailers to provide you with the parts and tools to create your own circuits. Over time, you will develop a network of stores that have the parts you need, but if you are starting out or are looking for better suppliers, here are some suggestions:

- Digi-Key (www.digikey.com). I have not found a source of parts anywhere in the world that matches the selection, price, and service of Digi-Key.

- Jameco (www.jameco.com). Another excellent supplier that carries Microchip parts. Also carries a good selection of robot parts (including gears and motors).

- Mouser Electronics (www.mouser.com). Along with Microchip products, Mouser also has an excellent selection of opto-electronic parts.

- Radio Shack (www.radioshack.com). I find Radio Shack to have a number of components that I can't find anywhere else (thermistors, TRIACs) as well as good-quality wire at a reasonable price. Recently, Radio Shack has included some Microchip PIC MCU chips and the Parallax BASIC Stamp to their catalog.

- Local electronics stores. In Toronto, I recommend Supremetronic (www.supremetronic.com) as it has all the passive and discrete parts that I need, prototyping PCBs, and other useful parts that are nice to handle and choose from instead of deciphering PDFs on line.

- Local surplus shops. If you are in the Toronto area, I'm sure you will recognize Active Surplus (www.activesurplus.com) by the large stuffed gorilla out front. While having a good selection of electronic parts, surplus stores often have a variety of other parts and subassemblies that are perfect for hacking or designing new controls.

Introduction

Under the Covers of the PIC16F684

PC/Simulator

PICkit™ 1
starter kit

Required Parts

1 PIC16F684

Before astronomers begin to investigate and learn more about a star, they make sure they fully understand the tools they are going to use. The tools to be used are chosen for their ability to investigate the specific aspects of the star that is to be studied. Although a few discoveries have been made with poorly understood equipment, the vast majority of observations have resulted in failure. Just like astronomers, before we investigate and learn about the Microchip PIC® microcontroller (the PIC16F684 specifically), we want to know as much as possible about the tools we are going to be using.

The PICkit™ 1 starter kit is an excellent tool for learning about the PIC microcontroller as it includes, along with its programming capability, a basic test circuit that you can use with a PIC MCU. As shown in Figure 1-1, the PICkit starter kit provides from 8 up to 12 individually addressable LEDs, along with a button input and a potentiometer for variable-voltage inputs.

The built-in programmer interfaces to the *development PC* via a USB port, which is preferable to serial or parallel ports. Overall, the PICkit 1 starter kit is almost a perfect tool for learning how to program the PIC MCU and interface to hardware.

When I described the PICkit 1 starter kit as being almost perfect, it probably set off some alarm bells in your head—this type of qualification is normally used to describe things like a used car or a blind date. In this case, I am being as literal as possible. The PICkit 1 starter kit is an extremely good product and an excellent first tool (probably the best that I know of) with which to learn to program and interface to the PIC MCU. The three potentially negative issues I would like to bring to your attention regarding the PICkit 1 starter kit are actually quite minor and, to some extent, can be exploited to help you better understand how the PIC MCU works.

The biggest issue that will have to be addressed in this book is the organization of the PICkit 1 starter kit's LEDs. If you were to follow the wiring of the eight LEDs and the four I/O pins they are connected to, you would notice that only two of the pins can be outputs (as shown in Figure 1-2). And if all the I/O pins to which the LEDs are connected (RA5, RA4, RA2, and RA1) were made outputs, you would have multiple LEDs lit if you make any of the I/O ports high. The solution to this problem is to enable only two pins as outputs at a time (one high and one low). This allows eight individually addressed LEDs, but not an arbitrary number of LEDS to be on at any given time.

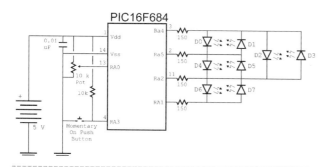

Figure 1-1 *Equivalent PICkit 1 starter kit circuit*

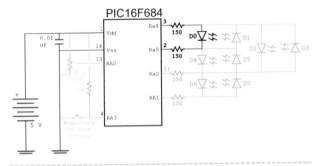

Figure 1-2 *PICkit 1 starter kit LED on*

Arbitrary numbers of LEDs can be turned on by scanning through the LEDs, just as a TV's electron beam scans across the cathode ray tube, turning on phosphors one at a time. This trick will be demonstrated later in the book and will be used to display eight bits of data at a given time. The organization of the eight LEDs (along with the button and potentiometer) was made to support eight- and 14-pin parts in the PICkit 1 starter kit's socket. That organization and the choice of pins (and how they are organized on the PIC MCUs that fit in the PICkit 1 starter kit's socket) is actually quite inspired, as it leaves the six pins of the 14-pin microcontroller's PORTC available for other uses.

The second problem is that it takes a few mouse clicks to turn off the power going to the programmed part. I seldom remember to turn off the power, and I doubt you will either. Although the different PIC microcontrollers are very robust devices from the electrical overload perspective, you should never pull them from a socket while power could still be applied to some of the pins. Although the *zero insertion force* (ZIF) socket, which I show you how to install later in the book, goes a long way in mitigating this problem, you should still be cognizant that you could potentially be damaging the PIC16F684 every time you plug it into and unplug it from the PICkit 1 starter kit while the PICkit is still connected.

The final issue to be aware of is the potential liability of the USB port used to connect the PICkit 1 starter kit to the development PC. Although most commercial and home PCs have built-in USB ports and versions of the Microsoft Windows operating system that can access the PICkit 1 starter kit very simply, there are a number of PCs in educational and institutional settings that do not have the required ports or software. There is no easy fix to this problem other than trying to find the fastest, biggest, and most modern memory PC to be used as a programming station.

None of these three issues are major show stoppers —they are really just speed bumps, and they can be overcome fairly easily.

Experiment 1—I/O Pins

Arguably the most important feature of the PIC MCU is its set of *input/output* (I/O) pins. The 12 pins available to the application developer allow the microcontroller to sense the outside world and output in different ways. The PIC16F684's I/O pins are capable of many different functions including analog and different digital signal processing. But as I first start introducing the chip, I will treat the I/O pins as simple digital I/O, capable of sensing or outputting simple binary signals. The more advanced features will be addressed later in the book when you become more comfortable with programming and working with the PIC16F684.

The basic PIC MCU digital I/O pin design is shown in Figure 1-3. The *TRIS* bit controls whether or not the pin can output the value saved in the *PORT* bit. The term *TRIS* is an abbreviation of Tri-State and references the tri-state driver that can drive the PIC MCU's pin. When the TRIS bit is low (0), the value in PORT is driven onto the pin, and the pin is said to be in *output mode*. When the TRIS bit is high (1), the PIC MCU pin is held in a high-impedance state, and the data level at the pin can be read without being affected by the contents of the PORT bit. This is known as *input mode*. Remembering which TRIS state accounts for which mode is quite easy to remember: A TRIS value of 1

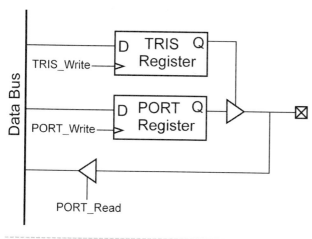

Figure 1-3 *I/O pin*

puts the pin in input mode, and a TRIS value of 0 puts the pin in output mode. The TRIS bit value approximates the first letter of *input* or of *output*.

By convention, you will see I/O pins referred to using the format

R&#

where the ampersand (&) represents the port, and the number sign (#) represents the port's pin. So PORTA, pin 4 is known as *RA4*. This shorthand can be a bit confusing because it refers to both the pin and the PORT register. The TRIS bits are usually written in the format

TRIS&#

with & and # being used in the same way as they are in the PORT bit/pin definition. These naming conventions are used for both C and assembly language programming.

When I first presented you with the cFlash.c in the Introduction, the manner in which the LED was turned on and off is not easy to see. To help you see the operation of the I/O port and how it affects the PICkit 1 starter kit's D0 LED, I created cPins.c, which turns on and off the LED using the PORT and TRIS bits more explicitly.

```
#include <pic.h>
/* cPins.c - Examine Operation of PIC MCU Pins

This Program is a modification of "cFlash.c"
with more explicit writes to the TRIS and PORT
bits.

RA4 - LED Positive Connection
RA5 - LED Negative Connection

myke predko
04.09.15

*/
```

```
__CONFIG(INTIO & WDTDIS & PWRTEN & MCLRDIS &
UNPROTECT \
 & UNPROTECT & BORDIS & IESODIS & FCMDIS);

int i, j;

main()
{

  PORTA = 0;
  CMCON0 = 7; // Turn off Comparators
  ANSEL = 0; // Turn off ADC
  TRISA4 = 0; // Make RA4/RA5 Outputs
  TRISA5 = 0;

  while(1 == 1) // Loop Forever
  {
  for (i = 0; i < 255; i11) // Simple Delay Loop
  for (j = 0; j < 129; j11);

  RA4 = 1; // D0 LED On

  for (i = 0; i < 255; i11) // Simple Delay Loop
  for (j = 0; j < 129; j11);

  RA4 = 0; // D0 LED Off

  for (i = 0; i < 255; i11) // Simple Delay Loop
  for (j = 0; j < 129; j11);

  RA4 = 1; // D0 LED On Again

  for (i = 0; i < 255; i11) // Simple Delay Loop
  for (j = 0; j < 129; j11);

  TRISA4 = 1; // Put RA4 into Input Mode

// LED Turned Off Due to RA4 Not Driving Current
Out

  for (i = 0; i < 255; i11) // Simple Delay Loop
  for (j = 0; j < 129; j11);

  RA4 = 0; // Restore Original Operating
  TRISA4 = 0; // Conditions
  } // elihw
} // End cPins
```

Before entering the "while (1 == 1)" statement, the code puts the RA4 and RA5 pins in output mode (writes a 0 to them) after clearing them (setting all bits to zero). After the while statement, the code delays for a half second and then loads RA4 with a 1, which drives or sources current from RA4, through D0, and is taken in, or *sinked*, by RA5. The code then waits another half second before loading RA4 with a zero, turning off D0. This should be fairly easy to understand; current flows from RA4 through D0 and into RA5.

Next, the program waits another half second before loading RA4 with a zero, turning off D0. After another half-second delay, RA4 is loaded with a one, and D0 is turned on again. After another delay, the LED is turned off by putting RA4 into input mode (the PORT value of RA4 does not change). With RA4 in input mode, no current can flow through D0. After another delay, RA4 bit is put back into output mode with the PORT bit being low so the LED will be off. At this point, the program repeats.

This program is a very simple example of how the PORT bits work. With the basic PICkit 1 starter kit circuit, only one LED can be turned on at any one time, which gives you an opportunity to try and decode the schematic of the PICkit 1 starter kit and try to turn on different LEDs or a series of LEDs in sequence. Later in the book, I will present you with code that will do this. But for now, you might want to try and modify cPins.c to turn on other LEDs by putting RA4 and RA5 into input mode, and then selecting two other pins to put into output mode and turn on another LED. When you do this, just use the explicit writes to the PORT and TRIS bits that I do in cPins.c and save the more sophisticated methods for controlling the LEDs for later.

Experiment 2—Configuration Word

One way the various PIC microcontrollers differ from the other chips out there is in their ability to have certain operating configuration parameters set when they power up. The configuration parameters are specified by writing to a special word in program memory, known appropriately enough as the *Configuration Word*. When most users start working with the PIC MCU, understanding how this word is configured is ignored until it is time to program a chip and test it in an application, at which point they try to figure out the correct values for the configuration word, often getting an incorrect value. Most incorrect values will cause the PIC MCU not to power up or apparently reset itself every few moments.

To avoid this problem, when I presented you with the initial cFlash.c program, I included the proper configuration word specification in the program, which is automatically recognized by the PICkit 1 starter kit programmer and stored in the PIC MCU, so that it will power up correctly without any intervention from you. Setting the configuration fuses manually leaves too much of an opportunity for error.

The configuration word specification is the following statement and starts with two underscores in the leftmost column of the source code:

```
__CONFIG(INTIO & WDTDIS & PWRTEN & MCLRDIS & UNPROTECT \
  & BORDIS & IESODIS & FCMDIS);
```

This statement enables the internal oscillator of the PIC16F684, disables the *Watchdog Timer*, the external reset pin, the low-voltage detect circuitry, and the advanced clocking options. Along with this, code protection is disabled, meaning that the contents of the chip can be read out. Note that the parameter words (called *labels*) have different values for different PIC MCUs, and different PIC MCUs may have different parameter words all together.

Each of the parameters of the __CONFIG statement is ANDed together to form a value that is saved in the configuration word when the PIC MCU is programmed. This value could be calculated manually by reading the "Special Features of the CPU" section of the PIC16F684 datasheet and creating the correct 14-bit value, or you can take the values built into the PICC Lite™ compiler or MPASM™ assembler include files listed in Table 1-1 and AND them together as I have done in the previous statement. ANDing together the include file values forces the compiler or assembler to calculate the configuration word value for you, saving some time and ensuring that the values are correct.

When specifying the values for the configuration word, make sure that every bit of the configuration word is represented. If a bit is forgotten, then chances are the other values will make it a 1, which may or may not be the value that you want for the configuration word bit. To emphasize the importance of having specified a label for every configuration word bit, I want to point out that trying to figure out why a PIC MCU won't run properly when a configuration word bit is misprogrammed is incredibly difficult.

As I will discuss later in the book, the PIC16F684 has a high-accuracy internal clock, negating much of the need for an external clock (although one can be added), the *INTIO*, in the __CONFIG, specification enables this clock and saves you from having to add your own clock circuitry. The other options that I have selected were chosen because they tend to make your life easier and alleviate the need for you to come up with any special external circuitry to the PIC microcontroller.

As you work on your own applications, you may want to change some of these values and experiment with configuration word options. Chances are you will end up with a situation where the selected options put the PIC MCU into a state where it cannot run properly. If you end up in this situation, remember to

Table 1-1

Configuration Fuse Parameter Specifications with Affected Bit(s) Listed First

Bit #	PICC Lite Label	MPASM Assembler Label	Comments
13–12	N/A	N/A	Unimplemented; Read as 1
11	FCMEN	_FCMEN_ON	Fail-Safe Clock Enabled
11	PCMDIS	_FCMEN_OFF	Fail-Safe Clock Disabled
10	IESOEN	_IESO_ON	Internal/External Switchover Mode Enabled
10	IESODIS	_IESO_OFF	Internal/External Switchover Mode Disabled
09–08	BOREN	_BOD_ON	Brownout Detect/Reset Enabled
09–08	BOREN_XSLP	_BOD_NSLEEP	Brownout Detect/Reset Disabled in Sleep
09–08	SBOREN	_BOD_SBODEN	Brownout Detect/Reset Control by SBOREN
09–08	BORDIS	_BOD_OFF	Brownout Detect/Reset Disabled
07	UNPROTECT	_CPD_OFF	EEPROM Data Memory Protect Disabled
07	CPD	_CPD_ON	EEPROM Data Memory Protect Enabled
06	UNPROTECT	_CP_OFF	Program Memory Protect Enabled
06	PROTECT	_CP_ON	Program Memory Protect Disabled
05	MCLREN	_MCLRE_ON	_MCLR Pin Function Active
05	MCLRDIS	_MCLRE_OFF	_MCLR Pin Function Inactive/Pin is Input RA3
04	PWRTDIS	_PWRTE_OFF	70 ms Power Up Delay Timer Disabled
04	PWRTEN	_PWRTE_ON	70 ms Power Up Delay Timer Enabled
03	WDTEN	_WDT_ON	Enable Watchdog Timer
03	WDTDIS	_WDT_OFF	Disable Watchdog Timer
2–0	RCCLK	_EXTRC_OSC_CLKOUT | _EXTRC	RC Clock, RA5 Clockout
2–0	RCIO	_EXTRC_OSC_NOCLKOUT | _EXTRCIO	RC Clock, RA5 I/O Pin
2–0	INTCLK	_INTRC_OSC_CLKOUT | _INTOSC	Internal Oscillator, RA5 Clockout
2–0	INTIO	_INTRC_OSC_NOCLKOUT | _INTOSCIO	Internal Oscillator, RA4 I/O Pin
2–0	EC	_EC_OSC	External Clock on RA5, RA4 I/O Pin
2–0	HS	_HS_OSC	High Speed (4–20 MHz) Crystal on RA4 & RA5
2–0	XT	_XT_OSC	Nominal Speed (1–4 MHz) Crystal on RA4 & RA5
2–0	LP	_LP_OSC	Low Speed (32 kHz–1 MHz) Crystal on RA4 & RA5

change the configuration fuses back to the default value that I have listed here, and use for most of the experiments. Once returned to this value, you can change each option individually, trying to find the one that caused the application to stop working.

Experiment 3—PIC MCU Variable Memory, Registers, and Program Memory

In the introduction, I outlined some of the differences between a microcontroller (such as the PIC microcontroller) and a PC, and I noted that one of the big differences in the application code was the need for providing code to initialize variables. Another difference is how the application code loaded into a PC or a PIC microcontroller and how it is organized. In this experiment, I will explain these differences (and the need for them) and help you understand how information is organized in the PIC MCU and some of the issues that you must be aware of when you are programming the PIC16F684.

When computer architectures are presented for the first time, something like Figure 1-4 is shown, which is the standard *Princeton* or *von Neumann architecture*. The important feature of this architecture is that the *memory space* is a single address area for the *program memory*, *variable memory*, and *stack memories*; the addresses for these application features are presented as being arbitrary and could be placed anywhere in this address space. An interesting side effect of this computer architecture is that an errant program could escape from the program memory area, start executing through the variable memory, stack RAM, register areas, and treat the data as program statements.

When a program is built for loading into a target Princeton computer system, the file that is loaded into the computer is organized to reflect the memory organization of the memory space. As shown in Table 1-2, the program file is broken up into *segments*, each at a different location, with a different size, and with different information. This is a simple example; large PC applications can have literally dozens of different, code, variable, data, and stack segments whereas a

microcontroller application often only requires only four (reset, code, data, and stack). The compiler and other application build tools are responsible for specifying and allocating memory in the application.

Table 1-2
Princeton Computer Architecture Application File Segment and Function

Segment	Function
Reset	Address application starts executing at. Usually a *goto* at the start of the application code.
Code	Application code. Starting address and size specified.
Data	Variable space for application. Starting address and size specified. Initial values for variables.
Stack	Program counter and data stack.

The PIC MCU is designed to use the *Harvard computer architecture* (see Figure 1-5) in which the program memory and variable memory/register spaces are kept separate from each other. Along with this, the program counter's stack is also kept in a separate memory space. The advantage of this method during program execution is the ability of the processor to fetch new instructions to execute while accessing the program memory/registers. Bad programs can still execute, but at least they won't try to execute data as instructions. The disadvantage of this method is a loss of flexibility in application organization (i.e., changing data or stack segment sizes to accommodate different applications).

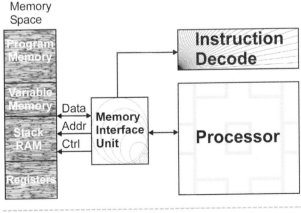

Figure 1-4   *Princeton architecture*

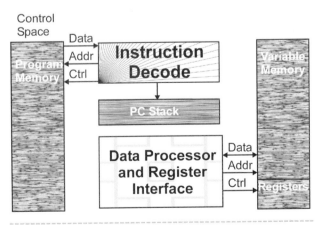

Figure 1-5   *Harvard architecture*

The program files loaded for PIC MCU and other Harvard computer systems are much simpler than the files loaded into the Princeton computer systems. These are produced by the MPLAB IDE, end in the extension .hex, and are usually referred to as "hex files." A typical PIC MCU hex file consists of two or three segments, one for application code (starting at the reset address), one for the configuration fuses, and an optional one for *electrically erasable programmable read-only memory* (EEPROM) data. For the applications presented in this book, the hex files will consist only of the code and configuration fuse segments. There is no need for defining different areas in the memory space.

This last point is subtle, but extremely important. When you *declare* a variable in a Princeton computer system, you are specifying the label to be used with the variable as well as reserving a space in memory for the variable. When you are declare a variable in a Harvard computer system (like the PIC MCU), you are simply specifying the label to be used with the variable; there is no need (or any mechanism) to reserve space for the variable. This means that the location for variables in a Harvard computer system can be chosen with much less rigor than in a Princeton computer system, and the Harvard application can still be expected to run.

The separate memory areas also mean that initial variable values cannot be loaded into the *data segment*; in the Harvard computer system variables are initialized from program memory. In terms of hex file space, you save on the need for a data segment, but you do require more for the initialization space. For microcontrollers, this is not an important point because in either computer architecture, variables will have to be initialized by code, because the applications are not loaded into memory as they are with a PC.

Although I've probably given you the perception that variables can be placed anywhere willy-nilly in the PIC MCU, the truth is a bit more complex. As I work through the book, I will explain the variable memory and register space in more detail, but for now I just want to note that they share the same space and explain a bit about the function of the registers.

What I am casually calling the registers in this experiment are referred to as the *special function registers* by Microchip (and, similarly, what I am calling variable memory is more properly called the *file registers*), the special function registers (SFRs) listed in Table 1-3 are used to monitor the status of program execution as well as provide an interface to the hardware peripheral functions of the PIC MCU. As you read through the book, the function and addressing of these registers will be explained to you.

Table 1-3

PIC16F684 Microcontroller Special Function Registers

Name	Address	Function
INDF	0x00 & 0x80	Index Data Register
TMR0	0x01	TMR0 Value
PCL	0x02 & 0x82	Low 8 Bits of the Program Counter
STATUS	0x03 & 0x83	PIC MCU Processor Status Register
FSR	0x04 & 0x84	PIC MCU Index Register
PORTA	0x05	PORTA I/O Pin Value
PORTC	0x07	PORTC I/O Pin Value
PCLATH	0x0A & 0x8A	Upper 5 Bits of the Program Counter
INTCON	0x0B & 0x8B	Interrupt Control Register
PIR1	0x0C	Peripheral Interrupt Request Register
TMR1L	0x0E	Low Byte of TMR1 Value
TMR1H	0x0F	High Byte of TMR1 Value
T1CON	0x10	TMR1 Control Register
TMR2	0x11	TMR2 Value
TMR2CON	0x12	TMR2 Control Register
CCPR1L	0x13	Low Byte of CCP Register
CCPR1H	0x14	High Byte of CCP Register
PWM1CON	0x15	CCP PWM Control Register
ECCPAS	0x16	CCP Auto-Shutdown Control Register
WDTCON	0x17	Watchdog Timer Control Register
CMCON0	0x18	Comparator Control Register
CMCON1	0x19	Comparator Control Register
ADRESH	0x1E	High Bits of ADC Result
ADCON0	0x1F	ADC Control Register
OPTION	0x81	PIC Operation Control Register
TRISA	0x85	PORTA Data Direction Pins
TRISC	0x87	PORTC Data Direction Pins
PIE1	0x8C	Peripheral Interrupt Enable Register
PCON	0x8E	Power Control Register
OSSCON	0x8F	Oscillator Control Register
OSCTUNE	0x90	Oscillator Tuning Register
ANSEL	0x91	ADC Pin Enable Register
PR2	0x92	TMR2 Period Register
WPUA	0x95	Weak Pull-Up Enable Register
IOCA	0x96	Interrupt on Port Change Select Register
VRCON	0x99	Comparator Vref Control Register
EEDAT	0x9A	EEPROM Data Register
EEADR	0x9B	EEPROM Address Register
EECON1	0x9C	EEPROM Control Register
EECON2	0x9D	EEPROM Write Enable Register
ADRESL	0x9E	Low Bits of ADC Result
ADCON1	0x9F	ADC Control Register

The simulator built into the MPLAB IDE is probably the least used and understood tool available to you, which is unfortunate because it is the most effective tool that you have to find and debug problems. The reasons why applications are not simulated is due to the perceived notions that it is too much work and that the program is either very simple or based on simple changes to a working program. Old hands will cringe at these excuses and remember how they learned the hard way. Personally, I *never* attempt to burn a program into a PIC MCU without first simulating it and making sure that it works properly. I *always* do this before seeing if it will work in the application circuit; by simulating it first you have confidence that the program should run.

In this experiment, I will work through the basics of setting up the simulator for an MPLAB IDE project. I have a few comments regarding how it should be used but, for the most part, I recommend that you use it for all the experiments in this book. Some of the initial programming (C and assembler) experiments are designed to work only in the simulator, but once again, you should simulate the experiments that access hardware before burning them so you can understand how they work and what they are expected to do. As you gain experience with the simulator, you should also develop personal standards for displaying data and understanding what the simulator is telling you in order to find problems and gain confidence in your program.

After you have created a project and have built the source code to ensure there aren't any syntax errors (language formatting errors), you should enable the debugger by clicking on "Debugger," then "Select Tool," and "MPLAB Sim" as shown in Figure 1-6. After enabling the simulator, the simulator toolbar (see Figure 1-7) will appear. The toolbar will allow you to do the following:

- Run the simulated program at full speed. On a 2.4 GHz Pentium running Windows XP, I find that 1 simulated second executes in about 3 seconds.

- Stop execution at any time. A *breakpoint*, which forces the application to stop at a specific location, can also be put in the application code as I will show presently.

- Clicking "Animate" causes the program to run relatively slowly, so you can watch the flow of the program. I find that this feature is best used to illustrate the operation of basic operating concepts.

- The program (subroutine or function) can be executed one step at a time by clicking the *Step In* icon.

- If the operation of a subroutine or function is well understood and felt to be correct or if it takes a long time to execute, clicking the *Step Over* icon will cause the simulator to execute the code in the subroutine or function at full speed and stop at the statement following the call to the subroutine or function.

- If you find yourself in a subroutine or function that is going to execute for a while, you can *Step Out* of it. After clicking this icon, execution will run at full speed, stopping at the instruction after the subroutine or function call statement.

- To restart the simulation from power up, click on the Reset icon. This icon will cause the simulated PIC MCU to return to power up conditions and reset the *Stopwatch* and *Stimulus* functions as well.

To set a *breakpoint* (which causes execution to stop when it is encountered) in your program, double-click on the statement at which you wish the breakpoint to be placed. Execution will stop when the simulator is running or animating an application, when a breakpoint is encountered. The breakpoint's statement is not executed but will execute if the Run, Animate, or one of the Step icons is clicked. Figure 1-8 shows cFlash.c with three breakpoints set and the execution *run arrow* pointing to the first statement in the program.

At this time, you may want to start up your cFlash project, enable the simulator, and put breakpoints at the three locations I have in Figure 1-8. Once you have done this, click on the Reset icon and you'll see that there is no run arrow. If you click on one of the Step icons repeatedly, the arrow won't appear and you may think that you have not enabled the simulator properly.

If you click on the Run icon, the program will stop and the run arrow will appear at the first breakpoint (at "PORTA = 0"). Before the C program starts, there

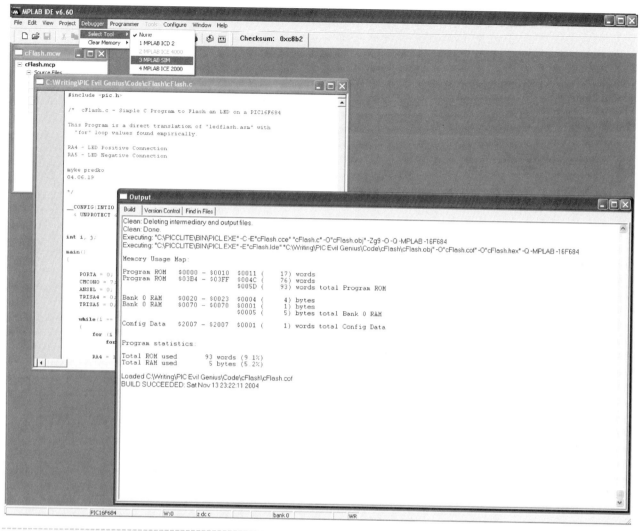

Figure 1-6   *Enabling MPLAB simulator*

is some code that sets up the execution environment and calls the C program. Placing a breakpoint at the first statement in the program and then clicking on the Run icon after resetting the simulator, will provide you with an application that is ready to go from the first statement.

Now would be a good time to start single-stepping through the program.

When you get to the *for statements*, you'll see that the arrow doesn't move for each click of the Step icons or even seemingly for the Animate icons. These statements provide a half-second delay for the program—you can execute through them quickly and stop at the statements after them by clicking the Run icon.

To understand better how long an application is taking to execute, click "Debugger" and then "Stopwatch" to display the Stopwatch window shown in Figure 1-9. This window will show you how many *instruction cycles* have executed and how long it would

have taken if you were working with a PIC microcontroller running at 4 MHz (which is the default execution speed of the PIC16F684). The simulated execution speed can be changed by clicking on "Debugger" and then on "Settings;" this pop-up also allows you to change other operating parameters of the simulator.

You can also monitor the values of registers and variables in your program by adding a *Watch window* (see Figure 1-10). To add a register, select it in the pull-down to the right of "Add SFR" and click on "Add SFR;" to add a variable, select the one you want from the pull-down to the right of "Add Symbol" and click on "Add Symbol." For your initial programs, I recommend that you always put in the I/O (TRIS and PORT) registers as well as all the variables in your program. When you are putting registers and variables into the Watch window, you may feel that the method in which they are displayed is suboptimal. You can change how they are displayed by right-clicking on the

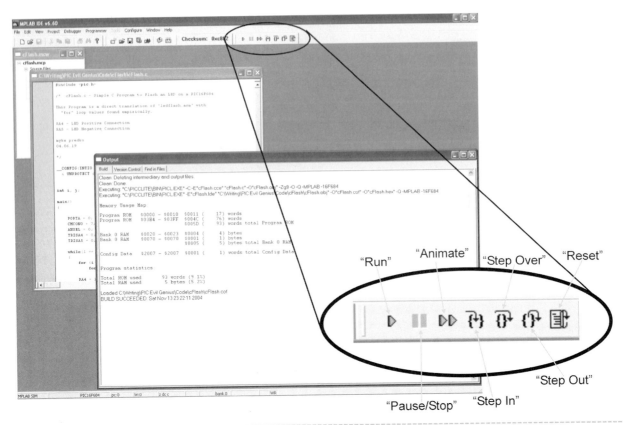

Figure 1-7 *Simulator active*

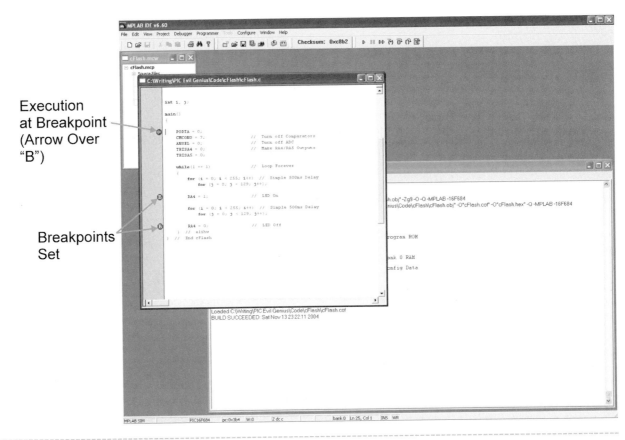

Figure 1-8 *Setting breakpoints*

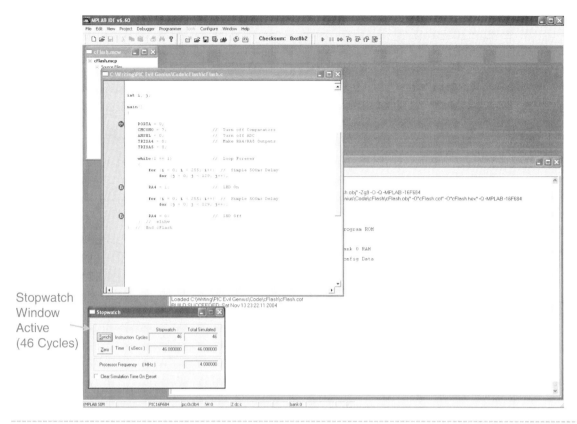

Stopwatch
Window
Active
(46 Cycles)

Figure 1-9   *Adding Stopwatch*

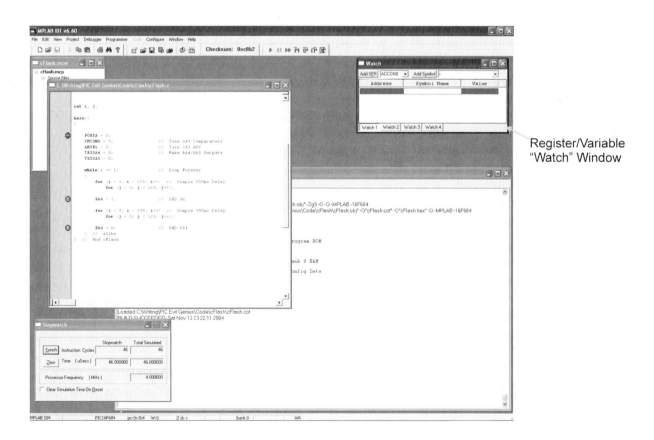

Register/Variable
"Watch" Window

Figure 1-10   *Adding a Watch Window*

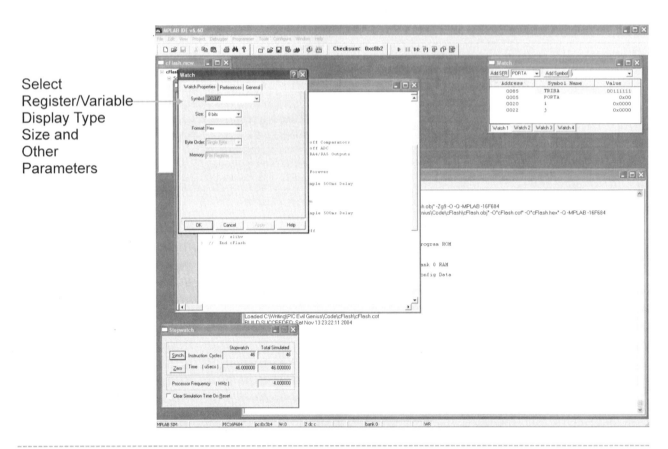

Select
Register/Variable
Display Type
Size and
Other
Parameters

Figure 1-11   *Changing Watch data types*

variable, selecting "Properties," and then changing the data format as shown in Figure 1-11.

With the information I have presented here, you can simulate most of the functions of the code written for the experiments in this book. Right now, you can just monitor the operation of the code, but in a later experiment, I will show you how to set specific input and register stimulus, which will help you see how your application runs under different conditions.
For now, you can change register values by double-clicking on their value in the Watch window and putting in new values manually. Adding stimulus to a simulated program takes a bit more work than the execution, breakpoint, stopwatch, and variable display steps I've shown here, but it provides you with a way to test your application under constant conditions, rather than depending on remembering to set certain values manually.

One of the nice features of the MPLAB IDE simulator interface is that it is the same as the ICD 2 debugger and ICE 2000 emulator interfaces. This means that when you start working with these tools, there should be only a minimal learning curve going from the simulator. The simulator features I've presented here are also present in the debugger and emu-

lator. Debuggers and emulators provide hardware interfaces to the application circuit (actually replacing the PIC MCU in the circuit) that you can monitor, set breakpoints for, and repeat over different sections of code, just as you would in the simulator.

At the start of this experiment, I suggested that you use the MPLAB IDE simulator to test all the application code in this book (and any that you encounter) before burning it into a PIC MCU. By doing so you will gain experience with the simulator and learn to come up with strategies that work best for you. Additionally, you can experiment with placing your editing, watch, and stopwatch windows on the MPLAB IDE desktop in positions that make the most sense to you and that allow you to debug your applications efficiently.

For Consideration

When people start investigating Microchip PIC microcontrollers for the first time, they are generally overwhelmed by the number of different part numbers (microcontrollers with different features) that are available. As I write this, there are over 250 active PIC MCU part numbers to choose from, not including packaging options. Most PIC MCUs are available with at least three different packages; each part number is available with two operating temperature ranges, and the older parts are available in different voltage ranges. This means, when all is told, there are about 2,000 different PIC MCUs and options to choose from. Over the next two pages, I would like to make your choice a bit simpler.

There are six major PIC MCU families to choose from as I have outlined in Table 1-4. The PIC16F684 that is featured in this book may seem like one of the lower-end PIC MCUs, but it is actually a quite flexible member of the mid-range.

The low-end PIC MCU processor architecture is available on many of the entry-level parts. The processor is very similar to the mid-range architecture (which is used in the PIC16F684) but does not have some immediate instructions, does not support interrupt requests, cannot support large amounts of variable memory, and does not have any advanced peripherals. The mid-range PIC MCU architecture is the most popular architecture (and used in the PIC16F684), as it supports moderate amounts of variable memory, advanced peripherals, and interrupts. The PIC17 high-end architecture is somewhat unique to this family of microcontrollers although it does have some similarities to the low-end and mid-range. The PIC18 architecture is really a superset of the mid-range architecture and it offers a number of features that allow it to access more program memory, variable memory, and peripheral registers as well as instructions that will simplify and speed up traditional applications. Once you are comfortable with programming the mid-range chips and have worked through the experiments in this book, you shouldn't have any problems learning how to program and working with the other PIC MCU architectures.

Most new users and hobbyists limit themselves to the parts listed in Table 1-5. These devices are all easily programmed either by the PICkit 1 starter kit, PIC-START® Plus, ICD 2, or a homegrown programmer (of which there are many different designs available for download from the Internet). The Microchip ICD 2 is a debugger interface that gives you many of the capabilities of an *in-circuit emulator* (ICE) without the extreme cost of an ICE.

I want to make a few comments on each of the devices listed in Table 1-5. The PICSTART Plus is a development programmer produced by Microchip, which can be used for all PIC microcontroller part

Table 1-4
Microchip PIC Microcontroller Families

PIC MCU Family	Processor Architecture	Programming Algorithms and ICD	Program Memory Type	Program Memory Size	File Register Size	I/O Pins	Features
PIC10	Low-End (12 Bit)	ICSP and ICD	Flash	256 to 512 instructions	16 to 24 bytes	4	Comparator
PIC12	Low-End (12 Bit)	ICSP	EPROM (OTP)	512 to 1,024 instructions	25 to 41 bytes	6	
PIC12	Mid-Range (14 Bit)	ICSP and ICD	EPROM (OTP) and Flash	512 to 2,048 instructions	25 to 128 bytes	6	Comparator and ADC, Advanced Timers
PIC14	Mid-Range (14 Bit)	ICSP	EPROM (OTP)	4096 instructions	192 bytes	20	Comparator and ADC, Advanced Timers
PIC16	Low-End (12 Bit)	Parallel, ICSP on Flash Parts	EPROM (OTP) and Flash	512 to 2,048 instructions	25 to 73 bytes	12 to 20	
PIC16	Mid-Range (14 Bit)	ICSP and some ICD	EPROM (OTP) and Flash	512 to 8,192 instructions	72 to 368 bytes	12 to 33/36 to 52 for LCD	ADC, Serial I/O, Advanced Timers, LCD, USB
PIC17	"High-End" (16 Bit)	Parallel	EPROM (OTP)	2k to 16k instructions	272 to 902 bytes	33 to 66	External Bus, Serial I/O, Advanced Timers
PIC18	"PIC18" (16 bit)	ICSP and ICD	EPROM (OTP) and Flash	8k to 64k instructions	640 to 4,096 bytes	16 to 72	ADC, Serial I/O, Advanced Timers

numbers—in some cases adapters are required to program the parts. In some cases, another PIC MCU part number has to be used to implement the ICD 2 function for a specific PIC MCU. Although the PICC Lite compiler is either restricted or not available on some of the PIC MCUs listed in Table 1-5, the full product supports each of these PIC MCUs and does not limit the program size or number of variables required for the application. Each of the listed PIC MCUs has flash program memory, which means that they can be erased by the programmer before burning a new application into them. All of the PIC MCUs except the PIC16F54 have the mid-range processor architecture. I included the PIC16F54 because many early introductory books and web sites reference the PIC16C54, and the PIC16F54 can be used in its place for these applications. Another traditional beginner's part is the PIC16F84A, but I would recommend that you consider more advanced and feature-rich chips such as the PIC16F627A instead. These more advanced chips have additional peripheral features as well as built-in oscillators, eliminating the need for you to add oscillator circuitry to your application.

I chose the PIC16F684 for this book because it is a mid-range architecture part (14 bit), can be programmed by the PICkit 1 starter kit, and has ICD 2 and PICC Lite compiler support. This chip also has a built-in precision oscillator and is tolerant of a wide range of operating voltages. It also has a wide range of peripherals that make it appropriate for use with applications like robotics. To summarize, I feel the PIC16F684 gives the best cost-to-performance ratio (when the PICkit 1 starter kit and PICC Lite compiler is included in the decision) for a mid-range part that could be used in a wide variety of different applications and would be easy for somebody learning to work with the PIC microcontroller.

Table 1-5
Recommended First PIC MCUs

Part Number	Number of Pins	Program Memory	Variable Memory	Peripherals	Programmer and Debugger	PICC Lite Support
PIC16F675	6	1,024 instructions	64 bytes	ADC, Timers	PICkit 1 starter kit, PICSTART Plus, ICD 2	Yes
PIC16F630	12	1,024 instructions	64 bytes	Comparator	PICkit 1 starter kit, PICSTART Plus, ICD 2	No
PIC16F684	12	2,048 instructions	128 bytes	ADC, Comparator, Timers	PICkit 1 starter kit, PICSTART Plus, ICD 2 Using PIC16F688	Yes (Limited Size)
PIC16F54	13	512 instructions	25 bytes	None	PICSTART Plus	No
PIC16F84A	13	1,024 instructions	68 bytes	None	PICSTART Plus	Yes
PIC16F627A	16	1,024 instructions	224 bytes	Comparator, Timers, Serial I/O	PICSTART Plus, ICD 2	Yes (Limited)
PIC16F87x	22 to 33	4,096 to 8,192 instructions	192 to 368 bytes	ADC, Timers, Serial I/O	PICSTART Plus, ICD 2	Just PIC16F877A and Limited

For Consideration

Introductory C Programming

PC/ Simulator

PICkit™ 1

starter kit

Required Parts

1 PIC16F684

I often see the C programming language described as the *Universal Assembly Language* by people trying to put it down. This isn't a very fair characterization because C is an extremely flexible programming language that was the development tool of choice for virtually every operating system in use today, as well as for a long list of successful business applications and games. Along with common computer science applications, C is used to program more artificial intelligence, computer-aided design systems, aerospace control systems, and supercomputer applications than any other programming language. Despite acknowledging C's widespread use and popularity, detractors tend to focus on a few points.

C is an incredibly rich language that makes it suitable for a wide variety of different applications. This richness can be a two–edged sword; instead of forcing every developer into following a common program layout, C allows a wide of programming styles. As I will discuss at the end of this section, multiple C statements can be combined, which can make code unreadable to even the most expert programmer. An acknowledgment of C's capability to be written into incomprehensible code is the single point of the yearly "International Obfuscated C Code Contest" (www.us.ioccc.org/) in which the most confusing code possible is created for prizes.

The richness of the code is often given as a reason for why C is hard to learn. I regularly contest this with proponents of BASIC, JAVA, and other programming languages. All these programming languages have a similar number of statement types, and I do not believe that any one has advantages over another to make

programming with it more efficient. In terms of readability, poor code can be written in any language. Good programs are not the result of the language designer; they are the result of the programmer thinking about how to approach a problem and clearly expressing in their code what the programs are doing.

Additionally, C is heavily dependent on pointers, a programming concept that many people find difficult to work with, debug, and understand, especially when reading other people's code. I admit that I like working with pointers, but this is due to spending many years understanding what pointers can do when programmed in C and how they can be used most efficiently for most applications. In this book, I will introduce you to the basics of C pointers and try to emphasize only the things you have to know to work with the language. For most C applications, pointers are minimally required and are often quite transparent to the operation of the program.

My major complaint about the C programming language is one that few people comment on; in its ANSI standard form, C cannot access data smaller than bytes. The version of C used in this book can read from and write to bits, but you will still require some very convoluted code to carry out some hardware interfacing.

Despite these concerns, C has a number of characteristics that make it well suited for use in this book and for individuals learning about programming. Firstly, because the language is over thirty years old, a plethora of books has been written about learning and coding it. Secondly, it is actually quite easy to write efficient and readable code in C, and a good portion of the text in this book is devoted to teaching you to do just that. Lastly, I believe HT-Soft's PICC Lite™ compiler is the best high-level development tool available for the PIC16F684 microcontroller, and I'm pleased that it can be used in this book to teach you about the PIC® microcontroller.

PICC Lite compiler has three things going for it that I feel are critical to its use in the book and for use by new users. First, it integrates extremely well with MPLAB® IDE. As shown previously, PICC Lite compiler works not only with MPLAB IDE's basic operations, but it produces the necessary information needed by the MPLAB IDE simulator, MPLAB IDE debuggers, and *in-cCircuit emulators* (ICEs) to allow source-code-level debugging. I feel this is critical for producing, testing and analyzing applications.

Second, PICC Lite compiler produces very efficient code. In Section 12, I demonstrate how I would solve a number of math problems using assembly language programming. At the start of each program, I have written out the operation of the program in C; you may

want to cut and paste this example code into separate C source-code files and see how efficient my assembler is to the code produced by PCC Lite. I would be surprised if my code had more than 10 percent fewer instructions than any application, and I wouldn't be surprised if there are cases where PICC Lite version created code with fewer instructions than I did.

Finally, the price of PICC Lite compiler cannot be beat. PICC Lite compiler is not a crippled version of the full product; it includes all the capabilities, including code optimization and simulator/debugger/emulator support, of the full product. This is an important point because with PICC Lite compiler integrated into MPLAB IDE, you have a development environment with capabilities that would normally cost well over $1,000 dollars. Taken together, the C programming language, with applications written with an eye toward readability and efficiency and implemented in the PICC Lite compiler, is the best method of learning about the PIC microcontroller, microcontroller programming, and application debugging.

Experiment 5—Variable Declaration Statements

Declaring variables in PICC Lite compiler is generally as simple as:

```
int VariableName;
```

before the main statement of the application. This will create a 16-bit variable with the label VariableName (or whatever variable name you want) that can be used anywhere in your program. In this experiment, I would like to discuss a few issues regarding variable and hardware register declarations to help you make sure that you can successfully create applications even though you have worked through only a few experiments in the book.

VariableName is a label and can start with any upper- or lowercase letter or the underscore character. After the starting character, the rest of the label can be any letter, number, or underscore character. Blank characters cannot be used in a label; if blanks are encountered, the compiler will try to divide the character strings to determine which of them are programming statements, directives, or defines. These character restrictions are also used for subroutine and function names.

For standard variables, two options exist that you should be aware of. The first is the ability to initialize the variable when it is declared. By adding an equals sign and a constant value, you can set your variable to

a specific value for the program without having to add another line later in the code. To initialize the variable *x* to 47, you would simply key in:

```
int x = 47;
```

Another option that is available to the declaration statement is the const keyword, which converts the declared value from a variable to a constant:

```
const int xConstant = 47;
```

In the declaration of xConstant, anytime the label xConstant is encountered, the compiler replaces it with the value 47. By declaring xConstant as a constant, you can no longer write to it. For example the statement:

```
xConstant = 48;
```

will return an error.

If you were to look at the pic16f684.h file that was put on your hard file when PICC Lite compiler was installed in the \PICCLITE\include folder, you will see a number of statements that look like:

```
static volatile unsigned char    TMR1L    @ 0x0E;
```

These are hardware register declarations and involve additional options that you do not need to be concerned with, because HT-Soft has taken care of declaring all the hardware registers in the PIC16F684 microcontroller for you.

I want to say a few words about variable names. Please try to make an effort to make them representative of what they are being used for. I see many students create programs with variable names like Reg3 when the name Remainder or PWMValue is much more representative of what the variable is used for in the application. Although I recognize that the variable memory is limited in the PIC MCU, please do not feel you have to use one variable for multiple functions; try to use each variable for one purpose. Using the same variable for multiple purposes can make writing your program more difficult and cause problems when different functions change a variable in ways that screw up the operation of other areas.

One area that always gets new programmers confused is in the area of counters. I recommend that you use the conventional variable names of i, j, k, and n. These values came into use with Fortran, the first high-level programming language, and by using them for this function consistently, they are immediately understood when reading the code. By naming variables appropriately, you will find the tasks of tracking how data is being processed by the application and of debugging the code much easier.

As I explain more about the C programming language to you, I will expand the variable definition to include arrays, pointers, and local and global variables. For the time being, if you keep with the simple format of the variable declaration that I have given you here, you will not have any problems.

Experiment 6—C Data Types

PC/ Simulator

Depending on your programming experience, the need to specify variable data types might seem new and somewhat ominous to you. The restricted variable memory available to you in the PIC microcontroller under the PICC Lite compiler can make the decision on what values to specify seemingly more ominous. There really is no need for this apprehension; most variables can be declared using the *int data type* without problem.

The data types available to you are listed in Table 2-1. For the most part, the data types follow *American National Standards Institute* (ANSI) standards, or what I call Standard C values, and are available in C compilers for other processors. This allows you to import programs or parts of programs (usually referred to as snippets) that have been created for other applications, but you would like to use on the PIC microcontroller. Similarly, if you were to come up with some good programs or algorithms, this could be exported to other systems quite easily.

Three deviations exist in the PICC Lite compiler data types to ANSI C that I have marked in Table 2-1.

The bit value is not supported by the C language standard, and I would recommend that you do not use it for variables. I realize that a single bit is useful for *flag variables* in a program, but because the bit is not available in other C implementations, you will not be able

Table 2-1
PICC Lite Compiler Data Types

Type	Bit Size	Comments
bit	1	Boolean value—Note: Not a Standard C data type
char	8	ASCII character/signed integer (-128 to 127)
unsigned char	8	Unsigned integer (0 to 255)
short	16	Signed integer ($-32{,}768$ to $32{,}767$)
unsigned short	16	Unsigned integer (0 to 65,536)
int	16	Signed integer ($-32{,}768$ to $32{,}767$); same as short
unsigned int	16	Unsigned integer (0 to 65,535); same as unsigned short
long	32	Signed integer ($-2{,}147{,}483{,}648$ to $2{,}147{,}483{,}647$)
unsigned long	32	Unsigned integer (0 to 4,294,967,295)
float	24	Real (0 to $1/-6.81(10^{38})$; Assume 3 digits of accuracy/default floating point mode; Note: Not a Standard C data type
double	32	Real (0 to $1/-6.81(10^{38})$; Assume 6 digits of accuracy/specified using PICL -D32 compilation option—Note: In Standard C, this is float, not double

to import the code directly to other implementations of the C programming language, and you may end up getting in a bad habit of working with bit variables. Note that I do not make the same comment for individual bits in a special purpose register. The register bits are unique to the PIC MCU and, as such, would not be involved in a code export to another device.

The *float* and *double (floating point) variable types* are substantially smaller in their PICC Lite compiler forms than in their high-end processor forms. For the most practical applications, you will not see a difference between how they execute, but you may have issues with very large numbers or numbers having many digits of precision.

As I have indicated, the C data types are strongly typed, and when you are equating values between types, you may get a warning or an error from the compiler indicating that a type conversion is required. If the data is stored in a format compatible with the destination, the simplest way of resolving this is to use the type cast, which consists of placing the desired data type in parenthesis before the source variable. For example, if the variables i and j were different types but data could be passed between them, a simple type cast, as shown here, could be used:

```
i = (iType)j;
```

The experiment for this application is quite simple; just a few lines of code demonstrate what happens with type conversion:

```
#include <pic.h>
/* cType.c - Investigate Data Type Operation in
PICC Lite compiler

This Program shows the need for Type Casting of
variables
 when data is transferred.

This program is designed to run under the MPLAB
IDE Simulator
 Only.

myke predko
04.09.15

*/
```

```
char i;
float j;

main()
{

  i = 47;

  j = i; // Can you Place char into float?

  i = j; // Can you Place float into char?

 while (1 == 1); // Loop Forever
} // End cType
```

When you compile this application, you are going to get the message:

```
Warning[000] C:\ Evil Genius\cType\cType.c 27 :
implicit conversion of float to integer
```

This message is telling you that the statement "i = j;" involves a data conversion. When this message comes up, you have to decide whether or not to resolve it by using a type cast or by leaving the warning as is. Remember that the two data types store data in different ways (which can be inferred from Table 2-1); casting the float type to a char is a bad idea because with the type cast, the first byte of the float type could be passed to the char without being modified. In this case, you would leave it as is.

For virtually all the applications in this book and for virtually all the PIC MCU applications that you will do on your own, you will only require the int data type. This 16-bit bit-variable data type will handle a reasonably large range of values as well as work with ASCII characters. This data type is handled well natively in the PICC Lite compiler, but for long and floating-point data types, additional libraries will have to be added to the final application. These libraries take up quite a bit of space and can slow down the execution of the application significantly. To avoid this space issue and reduced execution performance, I recommend that these data types not be used unless absolutely necessary and the impact of their operation is well understood. Later in the book, I will give you a couple of thoughts on how to simulate the operation of these data types using 8- and 16-bit variables.

Experiment 7—Constant Formatting

When I introduced declaring variables, I suggested strongly that variable names should be representative of what the data is used for, and if the variables are used for common functions (such as counters), they should be given conventional names. This plea is common to many programming books, but what isn't all that common is a pointer to specifying constants in the most appropriate data type for the situation. The readability of a program can be enhanced or damaged by program formatting, by variable names, or by comments.

Table 2-2
Constant Formatting Options and Suggested Best Uses

Constant Definition	Format	Best Use
Decimal	##	Default Value
Hexadecimal	0x##	Register Counter Values
Binary	0b########	Noncounter Register Values
ASCII	'#'	Human Interface Values

There are four ways to specify comments, as is listed in Table 2-2 along with the situations where they are best used. In cConfuse.c, I have created a simple application that uses data in different formats; before burning it into a PIC16F684 and running it in a PICkit™ 1 starter kit, try to figure out what it does.

```
#include <pic.h>
/* cConfuse.c - Write to PICkit Interface
Hardware in a Confusing Way

This program performs a simple task, but its
function is obfuscated by poorly chosen data
types.

myke predko
04.11.15

*/
```

```
__CONFIG(INTIO & WDTDIS & PWRTEN & MCLRDIS &
UNPROTECT \ & UNPROTECT & BORDIS & IESODIS &
FCMDIS);


main()
{

 PORTA = 0;
 CMCON0 = 7; // Turn off Comparators
 ANSEL = 0; // Turn off ADC
 TRISA = 0x49; // Enable PORTA LED Outputs

 while(1 == 1) // Loop Forever
 {
 if (0 == (PORTA & (1 << 3)))
  PORTA = '3'; // Turn on Four LEDs
 else
  PORTA = 4; // Turn on Remaining Four LEDs

 } // elihw
} // End cConfuse
```

Chances are, you could not understand what cConfuse does by just looking at it. You will probably have to look at a schematic for the PICkit 1 starter kit LEDs and application input hardware. By making four changes to this program (and turning it into cClear.c), I think you'll agree that the function and what is happening is a lot easier to understand.

```
#include <pic.h>
/* cClear.c - "cConfuse.c" after passing through
a Deobfuscator

This is the same program as "cConfuse.c", but with
more appropriately chosen constant data values.

myke predko
04.11.15

*/
```

```
__CONFIG(INTIO & WDTDIS & PWRTEN & MCLRDIS &
UNPROTECT \ & UNPROTECT & BORDIS & IESODIS &
FCMDIS);


main()
{

 PORTA = 0;
 CMCON0 = 7; // Turn off Comparators
 ANSEL = 0; // Turn off ADC
 TRISA = 0b001001; // Enable PORTA LED Outputs

 while(1 == 1) // Loop Forever
 {
 if (0 == RA3)
  PORTA = 0b010010; // Button Pressed, D0, D2,
                    //                D4, D7 On
 else
  PORTA = 0b000100; // Button Released, D1, D3,
                    //                 D5, D6 On
 } // elihw
} // End cClear
```

Three of the changes were to convert the TRISA and PORTA register assignment values to binary, from decimal, hex, and ASCII. I think you will agree that they were effective and helped you see what was happening in the application and the output values. The fourth change was to eliminate the complex test of the RA3 (button input pin) and simply do a bit compare. With the bit compare, you could easily look at the PICkit 1 starter kit schematic and see that RA3 is connected to a pulled-up push button.

When deciding which constant format to use in your program, you might want to follow these rules:

- Use decimal by default. For basic variable and variable functions, decimal is probably most appropriate and easy to read.

- Use binary when you are working with a register that does not contain counter data. An

example of this is loading the OPTION register with the value 0b10011111.

- Use hexadecimal for register counting or result data. Comparing a counter's current value with a hexadecimal constant would be most appropriate.

- Use ASCII data when human interfaces are involved. This includes text messaging as well as user input.

- Data sizes should be appropriate. Use six bits for the two sets of six-bit PORT and TRIS registers. For other registers, use eight bits.

I think what I am trying to say in this experiment is best summed up by the maxim: "There are 10 types of people in this world: those that understand binary and those that don't."

Experiment 8—Assignment Statements

PC/ Simulator

PICkit™ 1

starter kit

If you are familiar with other high-level programming languages, you are aware of the statement type called the *assignment statement*, which is used to store data of a variable of a specified type. In the previous experiments, I have detailed different data types as well as how variables are declared. In this experiment I will present how assignment statements are written in C.

The basic form for a C assignment statement is:

```
VariableName = Expression;
```

The label VariableName is a variable, declared as shown in the previous experiment. The single equals sign indicates that the variable is going to have a new variable value stored in it. I point out the single equals sign because the appearance of two equals signs is a logic operator, as I will explain in a later experiment.

The *expression* is a data type value that is the same as VariableName. An expression can be a constant, the contents of a variable or a series of mathematical oper-

ations. Although C is somewhat tolerant of assigning different data types, you should always try to make sure that the expression's type is the same as the variable's, and if it isn't, make sure you understand any potential issues.

An example of a potential issue between dissimilar types is saving a 16-bit value in an eight-bit variable. In this case, the upper-eight bits will be discarded, which could be a good thing (you are provided with a simple way of stripping out the most significant bits) or it could be bad (with data lost). To avoid this problem, I tend to declare all variables as int (16 bit).

Finally, the assignment statement (and virtually all other statements) is ended with a semicolon character (;). The semicolon is used by the compiler to indicate the end of the statement. The only statements where the semicolon is not required are the ones that end in a right brace (}) character. When in doubt, put a semicolon at the end statement; even if it is unnecessary, the compiler will treat it as a *null statement* and ignore it.

To demonstrate the different forms of the assignment statement, I have created cAssign.c:

```
#include <pic.h>
/* cAssign.c - A brief Look at Assignment
Statements

This Program Demonstrates how Assignment
Statements work in C.
```

```
myke predko
04.09.28

*/

int i;        // Unitialized Variable Declaration
int j = 23;   // Variable Declared with Initial
              // value assignment (Initialization)
main()
{

 i = 47;  // The variable "i" assigned (or loaded
          // with) the constant value 47
 i = j;   // The variable "i" assigned contents
          // of "j".
 i = j = 1; // "i" and "j" assigned the same value.

 while(1 == 1);

} // End cAssign
```

The first assignment statement is the initialization of the variable j. There is no difference between initializing a variable at its declaration and assigning a value to it in the first line of a program. You may find that initializing a variable at declaration is easier to read, making its function easier to understand in the program. The next two statements are basic assignments of a constant value and a variable value, respectively. These statements are common to other programming languages and their operation should be obvious.

The last assignment statement is probably something that you have never seen before in BASIC or other beginner programming languages; the value of i is given the value of j, which itself is being loaded with a new value. This is one of the many capabilities built into C that can be used to both simplify application programming and obfuscate the function of the application code. This *multiple assignment statement*, which has multiple destinations for the expression, is reasonably clear, but as I will show in the next experiment, multiple assignment statements can become quite complex, with their function being confused with the built-in assignment statements.

Access (read and write) of PIC16F684's registers and I/O Pins is accomplished using standard assignment statements such as the ones demonstrated in cAssign.c. The registers, bits, and pins are all declared in pic16f684.inc, which is loaded in by pic.h. cLight.c is a simple application that simply turns on D0 on the PICkit 1 starter kit and illustrates how registers can be written to 8 bits at a time, just as 8-bit variables (of type *char*). Individual bits are read from and written to in exactly the same way.

```
#include <pic.h>
/* cLight.c - Simple C Program to Turn on an LED
on a PIC16F684 in the PICkit 1

RA4 - LED Positive Connection
RA5 - LED Negative Connection

myke predko
04.11.10

*/

__CONFIG(INTIO & WDTDIS & PWRTEN & MCLRDIS &
UNPROTECT \ & UNPROTECT & BORDIS & IESODIS &
FCMDIS);


main()
{

 CMCON0 = 7;    // Turn off Comparators
 ANSEL = 0;     // Turn off ADC
 TRISA4 = 0;    // Make RA4/RA5 Outputs
 TRISA5 = 0;

 RA4 = 1;       // Turn on LED
 RA5 = 0;

 while(1 == 1); // Loop Forever

} // End cLight
```

Experiment 9—Expressions

PC/ Simulator

In the previous experiment, I really skipped over what an expression is. This is unfortunate because a lot of potential exists in the capabilities of the C *expression statement* that can make your programming easier and more efficient. Unfortunately, this potential comes at a cost: it's easy to create expressions that do not execute as you would expect, and are very hard to debug. In this experiment, along with looking at different types of expressions that can be used in your applications, I will present some of the potential pitfalls you may have to navigate around.

To demonstrate the operation of C expressions, I came up with cExpression.c, which examines different formats for expressions and data. As you read through the description of this program, I suggest that you load and single step through cExpression.c so you can better see the points that I am making.

```c
#include <pic.h>
/* cExpression.c - Look at Expressions

 This Program demonstrates a Variety of
 Different Expressions that can be created in C.

myke predko
04.09.23

 Hardware Notes:
 This Program has been written to run in
 the MPLAB IDE Simulator ONLY.

*/


int i, j;

main()
{

 i = 3;          // "3" is the Expression
 j = i;          // "i" is the Expression

 i = j * 11;     // Simple Arithmetic Expression

 i = j * 0x0B;   // Multiply by a Hex Value

 i = j * 0b00001011; // Multiply by a Binary
                     //                Value

 i = '0' + 8;    // Load i with ASCII "8"
                 // Change "Watch" Window display
                 // Format to ASCII to verify
                 // to verify.
 j = 48 / 4;     // Basic Division

 i = (j - 5) % 7; // Complex Arithmetic Expression
                  // Involving Two Operations and
                  // Forced Order of Operations
// i = (j - 5) % 7
// = (12 - 5) % 7
// = 7 % 7
// = 0
 i = j - 5 % 7;   // Same As previous but no
                  // Forced Order of Operations
// i = j - 5 % 7
// = 12 - 5 % 7
// = 12 - 5          (5 % 7 = 5)
// = 7

 i = (j = i / 6) * 2; // Embedded assignment:
// j = i / 6
// = 7 / 6
// = 1
// i = (i / 6) * 2
// = j * 2
// = 2

 while (1 == 1); // Loop Forever

} // End cExpression
```

The first two statements should be pretty straightforward. In the first statement, the expression is a constant value (3) that is stored in the variable i. Next, the

contents of i are the expression and they are saved in j. These are assignment statements, just like the ones in the previous experiment.

The next three statements have the same expressions (j * 11). The difference between them is the numbering system (radix) used to express the constant 4. This is a review of the C Data Types experiment in which I discussed how using bases other than 10 makes it harder to immediately see what is happening with the code.

In the assignment statements following the three j * 11 statements, I am applying the *basic arithmetic operations* (addition [+], subtraction [−], multiplication [*], and division [/], along with the modulus operator [%]. The modulus of two numbers is the remainder of the division operation. In the next experiments, I will look at some of the other operators available in expressions. The next statement has an expression that you have probably never seen before; I am adding the contents of the variable j to the ASCII character zero (which is specified in single quotes as '0'). The value of this expression (and the value stored in i) is 51 decimal, or if you change the Watch window display format for i to ASCII, you will discover that it is '3'. This may be a surprising result, but if you were to review a table of the ASCII codes, you will see that the numeric and alphabetic characters are defined together; so if you have one character's value, you can jump to another character simply by adding or subtracting the difference between them. This property of ASCII codes is used in a number of places in the book to algorithmically change values instead of relying on decision structures to change them.

The "i = (j − 5) % 7;" statement has what is known as a *complex expression*: that is, multiple operations are performed within the statement. Looking at the first statement, it should seem obvious that five is added to the value of j and the sum is found using modulus 7. Right after this statement, I have repeated it with parenthesis around the "j − 5" expression removed. When the full expression is evaluated, you will see the result (stored in i) is different than in the previous expression.

The reason for this difference is due to the *order of operations* of the operators used in these statements. The multiplication and division operators have a higher order of operations, or what you could refer to as execution priority, than addition and subtraction instructions have. Even though the addition operator is encountered before the modulus operator, the modulus operator executes first because it has a higher order of operations. To avoid these problems, you should enclose higher priority operations in parenthesis (indicating that their contents must be evaluated before they are used for other operations) to ensure they execute before other operators execute. As you look through the code in this book, you will see that I am in the habit of always using parenthesis to ensure expressions are evaluated in the order *I* want them to.

The final statement before the "while (1 == 1);" statement may look like a syntax error or some kind of comparison, but it is an example of what I called a multiple assignment statement in the previous experiment. Neither is true; the expression saves the intermediate calculation of i/6 in the variable j before completing the evaluation and storing the result in the variable i. This ability to embed assignments within expressions can help you simplify your programs and/or make them completely unreadable. When you first start creating your own C programs, I recommend that you do not try to combine statements as I have done here. But when you gain some confidence then you might want to look for opportunities to take advantage of this capability. Especially look in situations where the intermediate value of a calculation is needed elsewhere in your application.

With the five basic arithmetic operators and the ability to create calculations with multiple operators, you have the ability to develop expressions for the vast majority of applications that you are going to be working with. Before going on, try to write your own application like cExpression. Test out different expressions with the five operators listed above and with constant data in decimal, binary, hexadecimal, and ASCII formats.

Experiment 10—Bitwise Operators

PC/
Simulator

As well as being able to perform mathematical operations on the contents of variables, the C programming language has a number of operators that allow you to perform Boolean arithmetic on the bit contents of variables. These operators allow you to easily process bit information, but they must not be confused with the logical operators described in the next experiment.

The four basic bitwise operators are as follows:

- & —bitwise AND. When two values are ANDed together, each bit in the result is loaded with the AND value of the corresponding bits in the two values (set if both the bits are set).

- | —bitwise (inclusive) OR. When two values are ORed together, each bit in the result is loaded with the OR value of the corresponding bits in the two values (set if either bit is set).

- ^ —bitwise XOR. When two values are exclusively ORed together, each bit in the result is loaded with the XOR value of the corresponding bits in the two values (set if only one of the two parameter bits is set).

- ~ —bitwise negation. This operator will return the negated or complementary value for each bit of the single input parameter (invert each bit). This operator must never be confused with the ! logical operator, which will be shown to invert the logical value, not the bitwise value.

Using these four operators, standard bitwise Boolean arithmetic operations can be performed on different values, as I show in cBitwise.c. After compiling cBitwise.c, I suggest that you work through the code (including modifying values) in the MPLAB IDE simulator as much as possible until you fully understand the operations work. When you are doing this, you should display the variables (i, j, and k) as binary values in the Watch window.

```
#include <pic.h>
/* cBitwise.c - Bitwise C operators

This Program Demonstrates how bitwise Boolean
arithmetic operations work in C.

Note: Display the variable values as "Binary".

myke predko
04.10.02

*/

char i, j, k; // Use 8 Bit Variables

main()
{
```

```
i = 47;        // Initialize Values
j = 137;

k = i & j;     // AND Values together

k = i | j;     // OR Values together

k = i ^ j;     // XOR Values together

k = ~j;        // Invert the Bits in "j"

k = (i * 2) - j + (1 << 7); // Mix Binary
                               Operators with
// Arithmetic
while(1 == 1);

} // End cBitwise
```

The operation of bitwise operators is very straight-forward. Going through the example cBitwise experiment in the MPLAB IDE simulator, you should see the Boolean arithmetic operations clearly when you display the variables as binary in the order i, j, k. The Watch window will display the two parameter values directly over the result, allowing you to compare the inputs and outputs to the logic gates directly. As shown in the last statement, bitwise operators can be combined with arithmetic operators without any special considerations.

Although the bitwise operators are straightforward, you can run into some difficult-to-debug situations when you write to registers in a PIC MCU (or any other hardware device). To illustrate what I mean, consider the last statement of cBitwise and have it load a port directly. For this example, assume that the hardware device attached to this port saves the least significant four bits of the PORTA, and bit 7 of PORTA is connected to a device clock. When the clock is pulsed high (or strobed), the hardware device saves the four data bits. The statement could be:

```
PORTA = (i * 2) - j + (1 << 7); // Write Data
                                    to Hardware
                                    Device
```

and knowing that it ends up as 0x055 from the simulation of the program, you are comfortable knowing that bit 7 of PORTA is never loaded with a high value (i.e.,

from j, which has bit 7 set). This is a dangerous assumption as you do not know how the expression is evaluated by the compiler and whether or not the destination is used for storing temporary values. Because the operators outside the parenthesis execute on the same order of operations, you can assume they execute from left to right. In this case, PORTA is loaded with the product of i * 2, and then has the contents of j subtracted from it with the value of 0x0D5 being temporarily stored in PORTA. Finally, 128 ($1 << 7$) is added to the value in PORTA to clear bit 7. In this case, you will have inadvertently strobed the four bits of data into the device connected to PORTA (because when j was subtracted from i * 2, bit 7 was set and the final add cleared it). You have probably strobed in an unwanted value as well. This type of problem is extremely hard to find and debug; simulating the application will not reveal the problem, and you may need an oscilloscope to see bit 7 changing during the instruction's operation.

The fix to this potential problem is to use an intermediate value for all complex expressions that are going to be loaded into a hardware port. This changes the single statement into the two that you can see below:

```
k = (i * 2) - j + (1 << 7); // Write Data to
                               Hardware Device
PORTA = k;
```

As a general rule, *never* write the result of a complex expression directly to a hardware register without first storing it in a file-register-based variable. If you follow this rule, you will guarantee that the value being stored in the register is exactly what you want with no potentially problematic intermediate values.

It will probably be surprising to you, but the operators normally reserved for conditional logic are part of the numeric expressions, just as the arithmetic and bitwise operators discussed in the previous two experiments are. This gives you some unique opportunities for clever programming as well as an opportunity for errors that are very difficult to find and debug.

Experiment 11—Logical Expressions

PC/ Simulator

The logic operators available to you in C are listed in Table 2-3 and return numeric true or false values. It is important to remember that logic operators are different from bitwise operators as they return essentially binary values (True or False), where the bitwise operator returns the logic operation for each bit.

Table 2-3
Logic Operators

Logic Operator Operation

A == B	Return True if the two values are equal.
A != B	Return True if the two values are different.
A > B	Return True if the first value is greater than the second.
A >= B	Return True if the first value is equal to or greater than the second.
A < B	Return True if the first value is less than the second.
A <= B	Return True if the first value is equal to or less than the second.
A && B	Return True if both values are true (not equal to zero).
A !! B	Return True if either value is true.
! A	Return the inverted logic level of the value.

The experiment's program (cLogic.c) tests these arithmetic operators for two different values. This program should be run in single steps through the MPLAB IDE simulator with the Watch window set up and ExpValue displayed. As you step through each C statement, you will see the result of the logic expression saved in ExpValue.

```
#include <pic.h>
/* cLogic.c - Quantify Logic Expression Values

This Program looks at the different logic
expression values and the values that are
produced by them.

This program is to be used with just the
simulator.

myke predko
```

```
04.09.27

*/

int i = 5;
int j = 10;
int ExpValue; // Expression Value

main()
{

  ExpValue = i == j; // What is the value for
                     //      False?
  ExpValue = i != j; // Value for True?

  ExpValue = i > j;  // Value for False?
  ExpValue = i < j;  // Value for True?

  ExpValue = (i != j) && (i > j); // Value for
                     //      True AND
                     //      False?
  ExpValue = (i != j) || (i > j); // Value for
                     //      True OR
                     //      False?

  ExpValue = i && j; // True if Both Values != 0?
  ExpValue = i && 0; // True if One Value == 0?

  ExpValue = i || 0; // What about ORing?

  ExpValue = !i;     // Invert "i" Logic Value
  ExpValue = !(!i);  // Double Invert "i" Logic
                     //      Value

  while(1 == 1);     // Loop Forever

} // End cLogic
```

By stepping through each statement of this program, you will make a few conclusions: the first being that True is given a value of 1, and False a value of 0. You will also see that the first six assignments work exactly as you might expect based on your experience with other programming languages. (Although the idea that the result of a comparison can be used as a numeric may be somewhat novel.)

The last five statements probably are surprising, but they do emphasize the point that logic operators are similar to standard operators: Nonzero values are treated as True, and the zero values are treated as False. From these five statements, you should see that the same rules that are applied to logical values (zero and one) are also applied to standard numeric values. This should be kept in the back of your mind, because it allows for some clever programming tricks such as:

```
i = 0x1234 * (j > 4); // if (j > 4) is True,
                      //      then 1 returned
```

which is equivalent to:

```
if (j > 4)
  i = 0x1234;
else
  i = 0;
```

The advantage of this type of programming trick is its apparent elegance. Although the single line looks like it is much simpler than the full if/else solution, you might find that it uses a greater number of programming statements and file registers and that it takes longer to execute due to the need to include the multiply routine. Additionally, the lack of gotos or diversions to different statements could result in a statement that has a constant execution time, which can be a significant advantage in some applications. And although it may not necessarily be more efficient, you should note that its operation is not intuitively obvious, and you and others will probably not understand what is happening in the statement by simply looking at it. Statements like this are interesting curiosities that should be avoided unless there is a tangible reason requiring their use.

The pitfall that you can run into with C logic operators is with the *equals to* (==) statement. The double equals sign is a good solution to the problem of differentiating between the assignment equals and the comparison equals, but it does not take into account the conditioning of people who work with other languages like BASIC, in which a single equals sign in an *if* or a *while statement* indicates a comparison. It is very easy to forget yourself and put in a single equals sign when two are required for a comparison.

If a single equals sign is used in an expression like

```
i = 47
```

instead of the required double equals sign:

```
i == 47
```

the program will behave the same as if i was always equal to 47. To make matters worse, if you put a breakpoint to check the value of i after the comparison, it will seem like the value is always 47. What makes the error so hard to debug is that it *looks* right. To avoid this problem, generally two approaches can be used. The first is to make sure constant values are placed to the left of the comparison. If a constant is to the left of a single equals sign, the compiler will return an error stating that it cannot write to a constant value.

The second solution will seem more drastic, but it is effective 100 percent of the time. Simply do not perform an equals comparison. Instead, the result of a not equals comparison should be NOTted as shown below in the expression that is equivalent to i == 47:

```
!(i != 47)
```

I am not being facetious by suggesting that you never use the equals logical operator. If you are comfortable programming in BASIC or other programming languages that have a single equals sign as a comparison operator, you will have a great deal of difficulty with the double equals sign in your logical expressions.

Throughout the book, I discuss how you can make your programs more efficient by looking at different ways of solving the application requirements. It is a bit of a fine line to walk because some features of the C programming language allow you to use fewer keystrokes to solve a problem, but often at the cost of program readability. (The example at the end of this section illustrates this very well.) What you should be looking for are optimizations that do improve the readability of the code, improve its efficiency, and reduce the total application size.

A great example of how this is done can be illustrated by looking at the first C program you were given (cFlash.c) and looking at what its basic requirement is (i.e., to toggle the LED at RA4/RA5 on and off). In the original code I gave you, I explicitly turned on and off the LED, but by thinking of the problem from another perspective, for example, that you want the LED to be toggled, you might think in terms of basic Boolean algebra and consider changing cFlash.c to something like the following:

```
#include <pic.h>
/* cFlash 2.c - Simple C Program to Flash an LED
on a PIC16F684

This Program is an optimized version of "cFlash
2.c".

RA4 - LED Positive Connection
RA5 - LED Negative Connection

myke predko
04.11.12

*/

__CONFIG(INTIO & WDTDIS & PWRTEN & MCLRDIS &
UNPROTECT \ & UNPROTECT & BORDIS & IESODIS &
FCMDIS);


int i, j;

main()
{

  PORTA = 0;
  CMCON0 = 7;    // Turn off Comparators
  ANSEL = 0;     // Turn off ADC
  TRISA4 = 0;    // Make RA4/RA5 Outputs
  TRISA5 = 0;
```

```
while(1 == 1) // Loop Forever
{
for (i = 0; i < 255; i++) // Simple 500ms Delay
for (j = 0; j < 129; j++);

RA4 = RA4 ^ 1; // Toggle LED State
} // elihw
} // End cFlash 2
```

The change in the code was to change the two RA4 assignment statements to one in which RA4 is toggled (XORing a digital value with 1 will always invert, or toggle, its state). This allowed me to eliminate one of the 500 ms delay loops, thus eliminating an opportunity for error (i.e., keying in the second delay loop properly), which resulted in a better than 36 percent decrease in final application size. Along with these tangible improvements, the readability of the program has been improved because the comment after the XORing of RA4 states clearly what the statement is doing and it can be directly related to the basic operation of the program.

Experiment 12—Conditional Execution Using the If Statement

The basic form of the *if statement* in C is:

```
if (Expression) // Test to see if "Expression"
                    is Not Zero
Statement // Statement executed if "Expression"
              !5 0
else // Optional "else" Statement which
Statement // Executes if Expression is Zero
```

and is similar to the operation of the if statement in other structured languages, although a few points should be noted. The first is that the test *expression* does not have to be only a simple comparison; it can consist of complex terms, which may or may not have comparison operators in it. The two expressions in the if statements below are equivalent.

```
if (0 != (j / 3)) // Execute following statement
                  // if "j / 3" is not zero

If (j / 3)        // Execute following statement
                  // if "j / 3" is not zero.
```

In both cases, the next statement is executed if the result of j divided by three is not zero. The second case is not that much more difficult to follow, and it avoids any issues with comparisons with constants.

As the if statement is written above, you might think that only one statement can be executed conditionally. This is not true because of the use of braces ({and}) to collect multiple statements into one. To demonstrate the operation of the braces, I modified cAssign.c to create cStatement.c as follows:

```
#include <pic.h>
/* cStatement.c - A quick Experiment regarding
Statements in C

This Program further examines how Statements
work in C.

myke predko
04.10.05

*/

int i;      // Unitialized Variable Declaration
int j = 23; // Variable Declared with Initial
            // value assignment (Initialization)
main()
{

  i = 47;   // The variable "i" assigned (or
            // loaded with) the constant value 47
  {    // Can Braces Be Put in before a Statement?
  i = j;    // The variable "i" assigned contents
            // of "j".
  i = j = 1; // "i" and "j" assigned the same
            // value.
  } // End of Statement

  while(1 == 1);

} // End cStatement
```

Using braces changes the basic form factor of the if statement to:

```
if (Expression)  // Test if "Expression" is
                 // Not Zero
{ // Opening Brace to Collect Statements
Statement;       // Multiple Statements that
                 // Execute if Expression
Statement;       // is Not Zero
Statement;
:
} // Closing Brace to End true "statement"
```

```
else           // Execute if Expression is Zero
{ // Opening Brace to Collect Statements
Statement; // Multiple Statements that Execute if
Statement; // Expression is Zero
Statement;
  :
} // fi
```

This is the recommended format for people just starting out programming in C. The braces can be eliminated if there is only one statement following the if, for example:

```
if (a == b)        C = a * 3;
```

But to avoid problems remembering whether or not braces can be used, you should always put them in and then stop using them when you are more comfortable programming in C.

In the basic form of the if statement, you might have noticed that I indented the statements that execute conditionally (i.e., on the value of the expression). This is a common programming technique to visually indicate which statement's execution is dependant on the higher-level decision structure. I indent by four spaces for each level simply because this is what the Microsoft Visual Studio editor uses, and I am simply following its example.

When you look at other people's code, you may see that the opening braces are not at the start of the next line, they could be placed at the end of the previous line as shown in the following:

```
if (Expression) { // Test if "Expression" is
                  // Not Zero
Statement; // Multiple Statements that Execute
Statement; // if Expression is Not Zero
Statement;
  :
} // Closing Brace to End true "statement"
else {         // Execute if Expression is Zero
Statement; // Multiple Statements that Execute
Statement; // if Expression is Zero
Statement;
  :
} // fi
```

The compiler really doesn't care about the position of the braces. They can appear anywhere after the if statement. (The is also true of the closing braces.) I recommend the placement shown here because it is very obvious visually whether or not the braces are present or missing.

You have probably noticed the "fi" comment placed after the last closing brace of each if statement. The reason for this comment is to remind me of the purpose of the closing brace; when you have a very complex and long program, the reason for the closing brace can be forgotten or confused. I mark all program statements that produce a brace with their letters reversed

to help keep track of what the program is doing. For simple programs, it is hard to see the importance of this trick, but as you work with more complex applications, the need will become obvious. You'll find it necessary when the compiler comes back with the message that there is either a missing or an extra closing brace. This is a good habit to get into and will later save you time and grief debugging syntax errors in your program.

To demonstrate the operation of the if statements, I created the cIf.c application as follows:

```
#include <pic.h>
/* cIf.c - Demonstrate Operation of "if"

This program demonstrates the operation of the
if statement.

myke predko
04.10.18

*/

int i = 44;
int j = 0;
int k = 32;
int n = 21;

main()
{
  if (44 == i) // "i" equals a constant
    {
      n = n + 1; // Increment "n" if "i" == 44
    }
  else
    {
      n = n - 1; // Decrement if Not Equals
    } // fi

  if ((j = (i / 3)) == 7)
    {
      j = j + 1;
    } // fi

  if (k = 22)
    n = n + 1; // Increment "n" if "k" equals 22
  else
    n = n - 1; // Note that there is a single
               // statement, so no braces required.
  while(1 == 1);

} // End cIf
```

When you simulate this application, you should notice that I didn't put in the braces for the last if statement. As I indicated previously, they aren't absolutely necessary, but they are a good idea when you are starting out.

There is a mistake in this program that should have become evident when you simulated it. The final if statement, if (k = 22), always executes as if k is equal to 22. This seems strange until you can see exactly what is going on. Remember, the single equals sign (=) *always* behaves as an assignment, and a double equals sign is required for a comparison. If you change this single equals sign to a double one, you will find that the program now works properly.

The problem of incorrectly keying in one equals sign instead of two is very common with new C programmers. To avoid the problem you can reorder your program so that you never require an equals comparison. For example, the last four lines of the application could be changed to the following:

```
if (k != 22)
 n = n - 1;
else
 n = n + 1; // Increment "n" if "k" equals 22
```

I don't highly recommend this method because it involves negative logic. That is, you must first figure out how your code is supposed to work and then do the reverse.

Another strategy is to always put constants first as in:

```
if (22 == k)
 n = n + 1;   // Increment "n" if "k" equals 22
else
 n = n - 1;   // Decrement if Not Equals
```

This is marginally better but doesn't protect you in cases like "if (a = b)," where the comparisons are both variables. In such a case, forgetting to put in the second equals sign causes the value of one variable to be assigned the value of another.

Experiment 13—Nested Conditional Statements

Some people are born troublemakers. When I was 16, I had to go to a two-hour defensive driving course because I had too many points for speeding. Everyone in this course was given a *Driver's Education Manual* with the basic rules of the road. Several minutes into the course, somebody put up his hand and noted that if a police officer was directing traffic in a manner contrary to a set of traffic lights, then you must follow the policeman's directions, and if a farmer herding animals was directing traffic in a manner contrary to traffic lights, then you should follow the farmer's directions. He then asked the question: "So, what do you do if there is *both* a police officer *and* a farmer that are giving contrary directions to the streetlight?" I'm bringing up this little story because in the previous experiment, I presented the idea that the next statement after the if statement (or the following else statement) would be executed. And when I presented this concept to a set of high-school students, I was quickly asked the question, what happens if you have an if statement following another if or else statement?

A logical suggestion would be to place the second if statement (along with the statements that execute conditionally with it) within a set of braces after the first if. This would look like:

```
if (i > j)
{
 if (k < n)
 {
// Statement(s) Executed if "i > j" and "k < n"
 }
 else
 {
// Statement(s) Executed if "i > j" and "k >= n"
 } // fi
} // fi
```

Another solution is to notice that there are only two areas in the code above that execute conditional statements, and they could be accommodated by two if statements:

```
if ((i > j) && (k < n))
{
// Statement(s) Executed if "i > j" and "k < n"
}
else if (i > j)
{
// Statement(s) Executed if "i > j" and "k >= n"
} // fi
```

This method isn't bad, but could become very long if statements are written for both the if and the else. And, depending on how the compiler generated the code, this method could be very inefficient in terms of code size and execution time.

The generally accepted method of combining conditional execution statements like this is to recognize that the if statement and the conditionally executing statement(s) following it are all one statement and can be nested underneath the original if statement as in the following:

```
if (i > j)
 if (k < n)
 {
// Statement(s) Executed if "i > j" and "k < n"
 }
 else
 {
// Statement(s) Executed if "i > j" and "k >= n"
 } // fi
```

I should point out that this method of nesting does not apply only to the if statement. It applies also to all the conditional execution statements in the C programming language. As you work through the book, you will see many examples of nested programming statements of different types. Actually you've seen one already, the 500 ms delay code that was used in the cFlash.c program consists of a for statement nested as part of another for statement.

To demonstrate how nesting can simplify a program, I created cNoNest.c. This program flashes different LEDs on the PICkit 1 starter kit in a somewhat random order, but it is hard to follow and see immediately what is happening in the application:

```
#include <pic.h>
/* cNoNest.c - Jump Between LEDs with ifs

This Program will Jump between different LEDs
due to different conditions.

The LED values are:

LED Anode Cathode
D0 RA4 RA5
D1 RA5 RA4
D2 RA4 RA2
D3 RA2 RA4
D4 RA5 RA2
D5 RA2 RA5
D6 RA2 RA1
D7 RA1 RA2

myke predko
04.11.15

*/

__CONFIG(INTIO & WDTDIS & PWRTEN & MCLRDIS &
UNPROTECT \ & UNPROTECT & BORDIS & IESODIS &
FCMDIS);

int i, j, k, n;

main()
{

 PORTA = 0;
 CMCON0 = 7; // Turn off Comparators
 ANSEL = 0; // Turn off ADC
```

```
 k = 0; // k & n are Special Test Values
 n = 1;

 while(1 == 1) // Loop Forever
 {
  for (i = 0; i < 255; i++) // Simple Delay Loop
   for (j = 0; j < 129; j++);

  if (0 == k)
  {
   if (0 == n)
   {
    PORTA = 0b000010000; // Path 1
    TRISA = 0b011101011;
    n = -2;
   }
   else
   {
    PORTA = 0b000100000; // Path 2
    TRISA = 0b011001111;
   } // fi
  }
  else // K != 0
  {
   if ((4 == k) && (0 == n))
   {
    PORTA = 0b000000100; // Path 3
    TRISA = 0b011101011;
    n = 2;
    k = -2;
   }
   else if (4 == k)
   {
    PORTA = 0b000000100; // Path 4
    TRISA = 0b011111001;
    n = 1;
    k = -2;
   }
   else
   {
    if (0 == n)
    {
     PORTA = 0b000000100; // Path 5
     TRISA = 0b011011011;
    }
    else
    {
     PORTA = 0b000010000; // Path 6
     TRISA = 0b011001111;
     n = 1;
    } // fi
   } // fi
  } // fi
  k = k + 2; // Next Time, Go to "else"
  n = n - 1;

 } // elihw
} // End cNoNest
```

Once I had the program working, I removed the redundant braces and looked for places where the multiple if statements could be combined into something simpler. I came up with the following cNest.c:

```
#include <pic.h>
/* cNest.c - Jump Between LEDs with ifs

This Program will Nest the various "if"
statements of "cNoNest.c" to try and get a
program that is easier to follow.
```

```
The LED values are:

LED Anode Cathode
D0  RA4   RA5
D1  RA5   RA4
D2  RA4   RA2
D3  RA2   RA4
D4  RA5   RA2
D5  RA2   RA5
D6  RA2   RA1
D7  RA1   RA2

myke predko
04.11.15

*/

__CONFIG(INTIO & WDTDIS & PWRTEN & MCLRDIS &
UNPROTECT \ & UNPROTECT & BORDIS & IESODIS &
FCMDIS);


int i, j, k, n;

main()
{

  PORTA = 0;
  CMCON0 = 7; // Turn off Comparators
  ANSEL = 0; // Turn off ADC

  k = 0; // k & n are Special Test Values
  n = 1;

  while(1 == 1) // Loop Forever
  {
   for (i = 0; i < 255; i11) // Simple Delay Loop
   for (j = 0; j < 129; j11);

   if (0 == k)
   if (0 == n)
   {
    PORTA = 0b000010000; // Path 1
    TRISA = 0b011101011;
    n = -2;
   }
   else
   {
    PORTA = 0b000100000; // Path 2
    TRISA = 0b011001111;
   } // fi
```

```
   else if ((4 == k) && (0 == n)) // Combine the
                                  // Test Conditions
   {
    PORTA = 0b000000100; // Path 3
    TRISA = 0b011101011;
    n = 2;
    k = -2;
   }
   else if (4 == k)
   {
    PORTA = 0b000000100; // Path 4
    TRISA = 0b011111001;
    n = 1;
    k = -2;
   } // fi
   else if (0 == n)
   {
    PORTA = 0b000000100; // Path 5
    TRISA = 0b011011011;
   }
   else
   {
    PORTA = 0b000010000; // Path 6
    TRISA = 0b011001111;
    n = 1;
   } // fi

   k = k 1 2; // Next Time, Go to "else"
   n = n - 1;

  } // elihw
} // End cNest
```

I admit that cNest.c isn't a huge improvement to cNoNest.c in terms of readability, but by reducing the number of unneeded braces, I did manage to reduce the amount of space the main loop of the application takes and made it easier to look through the entire application. This is an example of what you will see if your task is to support an application that has been in use for a long time and has been modified to reflect new requirements and fixes to various problems that have been encountered. In these types of applications, it is often impossible to understand exactly how the code is working, and you end up making small changes that perpetuate its increasing complexity.

Experiment 14—The Switch Decision Statement

PC/ Simulator

Cases exist where multiple statements are required because of multiple constant values against which a single variable must be tested. Multiple if and else statements can be combined to meet the requirements quite simply. The code below demonstrates this:

```
if (4 == i) // Go South if Index at 4
{
 Direction = 180;
}
else if (5 == i) // Go North if Index at 5
{
 Direction = 0;
}
else if (7 == i) // Go East if Index at 7
{
```

```
 Direction = 90;
}
else // Go West for everything else
{
 Direction = 270;
} // fi
```

If you were writing your application in BASIC, you would probably use the *select/case statements* like the *switch/case statements* used in cSwitch.c listed below:

```
#include <pic.h>
/* cSwitch.c - Demonstrate Operation of "Switch"

This program demonstrates the operation of the
switch statement.

myke predko
04.10.18

*/

int i = 4;
int Direction = -1;

main()
{

 switch (i)
 {
 case 4: // Go South if Index == 4
  Direction = 180;
  break; // Leave Switch Statement
 case 5: // Go North if Index at 5
  Direction = 0;
  break;
 case 7: // Go East if Index at 7
  Direction = 90;
  break;
 default: // Go West for Everything Else
  Direction = 270;
 } // hctiws

 while(1 == 1);

} // End cIf
```

This application provides the same function of responding to multiple possible conditions that are listed in the multiple if statements at the start of this experiment. By using the switch statement, the code is actually a lot simpler. The case statement's parameter combined with the switch statement's parameter forms the if statement:

```
if (SwitchParameter == CaseParameter)
{
// Statements after the "Case" Statement to the
// "break" or next "case Statement
} // fi
```

The use of the case statement is obviously a lot easier to key and a lot less likely to have a syntax error like you had with the if statement.

The "default:" condition works exactly the same as the else statement in the if statement; the statements after it execute only if all parameters don't match any of the case statements.

This is all there is to the switch/case statements except for one point: the *break statement*. This statement causes execution to jump out of the current switch statement and execute the statement following it. You can do some interesting things if the break statement inside a switch block is not included. For example, if the switch block was being used to record the direction of motion (keeping in the tradition of the wheel direction), it could increment the counter for each 90 degrees rather than placing a hard value in the counter variable. The example code for this implementation is:

```
i = 0; // Clear Direction Counter
switch (Direction)
{
case 180 // Going South Index = 4
 j = j + 1;
case 0 : // Going North Index = 3
 j = j + 1;
case 90: // Going East Index = 2
 j = j + 1;
 break;
default: // Going West Index = 1
 j = j + 1;
} // hctiws
```

Obviously, you have to plan for situations where you can eliminate the break statement in your switch code, but when you do, you really have a feeling of accomplishment—and there's a good chance you've simplified the amount of application code required for the program. When you are starting out, you will probably use the switch/case statements for situations like this one, where multiple if statements exist, and all of them are comparing to a constant value and executing a break statement at the end of the case.

Experiment 15—Conditional Looping

There are a couple of methods of implementing conditional loops. In this experiment I will look at the most common method: the basic *while loop*. The while loop allows you to repeatedly execute a set of instructions while a test expression is true. That is, the while loop can be used for conditionally repeating code. But it can also be used to implement infinite loops in your applications. Some programming philosophies do not use the basic while loop, but they also do not provide you with the simple readability of the basic while loop.

The while loop is a programming construct that tests an expression before allowing execution to take place within the loop. If the expression is not zero, then execution will take place within the while loop, and at the end of the loop, execution will return to the expression test and the process will repeat. If the expression evaluates to zero, then execution will skip past the loop and continue at the statement after it.

To show how this works, the following statements can be used:

```
i = 0;
while (i < 4)
{
 i = i + 1; // While Loop Code
} // elihw
```

In these statements, the variable i is initialized to zero. Next, it is compared to 4, and if it is less than 4 (i.e., the expression is true or returns a nonzero value), the code inside the while loop (incrementing the variable i) is executed. When i is no longer less than 4, the while expression becomes false (and returns a zero value), the code inside the while loop is skipped over, and execution continues at the statement after the closing brace of the while loop.

To demonstrate the operation of the while statement in an application, I have modified cFlash.c into cFlashWhile.c in which the two for statement delays have been replaced with a single while loop that increments the two variables i and j until they are both greater than 255 and 78, respectively:

```
#include <pic.h>
/* cFlash While.c - Simple C Program to Flash an
LED on a PIC16F684

This Program is a modified version of "cFlash.c"
to use "while" loops
 instead of "for" loops

RA4 - LED Positive Connection
RA5 - LED Negative Connection

myke predko
04.06.19

*/

__CONFIG(INTIO & WDTDIS & PWRTEN & MCLRDIS &
UNPROTECT \ & UNPROTECT & BORDIS & IESODIS &
FCMDIS);


int i, j;

main()
{
  PORTA = 0;
  CMCON0 = 7;        // Turn off Comparators
  ANSEL = 0;         // Turn off ADC
  TRISA4 = 0;        // Make RA4/RA5 Outputs
  TRISA5 = 0;

  while(1 == 1)      // Loop Forever
  {
   i = 0;
   j = 0;
   while ((i < 255) || (j < 78))
   {
    i = i + 1;       // Increment Small Counter
    if (i > 255)
    {                // Roll Over to Large Counter
     i = 0;
     j = j + 1;
    } // fi
   } // elihw

  RA4 = RA4 ^ 1;     // Toggle LED
  } // elihw
} // End cFlash While
```

Although cFlash While.c is a direct copy of cFlash.c, I found if I used the same test values for i and j (255 and 129, respectively), the delay would increase to 833 msecs rather than the standard value of 500 ms. By running cFlash While.c in the MPLAB IDE simulator, I was able to empirically determine the value for j that would result in an approximately 500 ms delay.

As you work through the code, you will discover that I use the while statement

```
while (1 == 1)
```

a *lot*. This is my loop-forever code, and I use it either as the overall loop in an application (like this one) to

---

encompass the I/O and processing code, or I place it at the end of the application to stop it from returning to the caller (and end up executing again repeatedly). The statement could be simplified to:

```
while (1)
```

The PICC Lite compiler can detect statements like this where the test expression is always true (or 1) and replace the statement with something like:

```
Loop:
: // Code Executed inside while loop
goto Loop;
```

There is another form of the while loop. It is the do/while, which takes the following format:

```
do
{
// Code Executed inside the do/while loop
}
while (expression);
```

This is a subtle modification of the original, where during the first time through the loop, the expression is not tested to be true; you are guaranteed to execute the code inside the loop at least once. The advantage of using this form of the while loop is that variables or hardware register values that are tested in the while expression do not have to be initialized to force execution to work through the code at least once. Some C implementations have the *do/until statement*, which I do not like because it forces negative logic into your program (i.e., looping until a condition is true is the logical negative of looping while a condition is true).

Two keywords are also used in while loops: break and continue. "Break" will force an exit of the while loop, and "continue" will force execution to return to the while statement where the expression is evaluated. Personally, I do not use these statements because they act as gotos in the program, changing execution without regard to the structured programming statements. And I do not recommend that you use them in your programs, as they can be difficult to debug and can lead an application to behave unpredictably (especially if you are new to programming).

Experiment 16—The For Statement

starter kit

I try to teach programming as I was taught, and that is to emphasize the capabilities of the different functions built into the language and how to use them appropriately in applications. Some people, however, seem to think they can use the for statement in virtually any situation where conditionally looping code is required. I guess the theory behind using the for statement in different situations is to reduce the number of statement types that are in your programming inventory. In this experiment and at the end to this section, I will show that the for statement is a wonderfully flexible statement, but that it can make code a lot more complex to understand and debug.

The design of the for statement is actually quite elegant and results from the question, "what are the requirements of repeating loops?" The statement format is:

```
for (Initialization; Loop Test Expression; Loop
Increment)
 Statement
```

and its operation is similar to the BASIC code for a loop. Initialization is the process of initializing variables that are (ideally) required for the looping operation, but the process can also include other assignment statements. The *loop test expression* is an expression, similar to that used in the while statement to test whether or not the loop should repeat. Finally, the *loop increment statement* is normally used to increment the loop counter after each iteration of the loop.

The for statement is typically used when you need a loop that repeats a set number of times. It might look like the following:

```
for (i = 0; i < MaxNumber; i++)
 Statement; // Statement Executed Repeatedly by
                "for" Loop
```

and could be modeled as:

```
i = 0;
while (i < MaxNumber)
{
 Statement; // Statement Executed Repeatedly by
               "for" Loop
 i = i + 1; // Equivalent to "i++"
} // elihw
```

In this for statement, a counter is initialized to zero and is incremented (using the "i++" statement, which is equivalent to i = i + 1) until it is equal to MaxNumber. This should be quite easy to understand and use in your own applications.

The use of the for statement becomes more complex when you consider that multiple initialization and loop increment assignment statements can be used (with each separate statement separated by a comma). If commas are not used to separate the assignment statements, the compiler will become confused as to how to parse (convert) the statements correctly. The following for statement is completely valid:

```
for (i = 0, j = 47; i < MaxNumber; i++, j = j - 2)
 Statement; // Statement Executed Repeatedly by
               "for" Loop
```

In this for statement, both i and j are initialized, and both variables are changed in the loop increment portion of the for statement.

To demonstrate how versatile the for statement is, I have created an application that cycles each of the PICkit 1 starter kit's eight LEDs. (It will be explained in more detail later in the book.) Rather than explaining how the for statements work in the application, you should work through them on your own (it's really not very hard—especially with the provided comments).

```
#include <pic.h>
/* cPKLED 2.c - Roll Through PICkit 8 LEDs using
only "for"

This Program will roll through each of the 8
LEDs built into the PICkit PCB.

The LED values are:

LED Anode Cathode
D0 RA4 RA5
D1 RA5 RA4
D2 RA4 RA2
D3 RA2 RA4
D4 RA5 RA2
D5 RA2 RA5
D6 RA2 RA1
D7 RA1 RA2

Using only "for" statements.

The original name was going to be "cFor", but
that seemed too potentially explosive.

myke predko
04.11.09
*/

__CONFIG(INTIO & WDTDIS & PWRTEN & MCLRDIS &
UNPROTECT \
 & UNPROTECT & BORDIS & IESODIS & FCMDIS);


int i, j, k, n;

main()
{

PORTA = 0;
CMCON0 = 7;       // Turn off Comparators
ANSEL = 0;        // Turn off ADC

k = 0;            // Start at LED 0

for(;;)           // Loop Forever
{
 for (i = 0; i < 255; i++) // Simple Delay Loop
  for (j = 0; j < 129; j++);

 for (n = 0; (0 == k) && (0 == n); n++)
 { // Simulate "if (0 == k)"
  PORTA = 0b010000;
  TRISA = 0b001111;
 } // rof
 for (n = 0; (1 == k) && (0 == n); n++)
 { // Simulate "if (1 == k)"
  PORTA = 0b100000;
  TRISA = 0b001111;
 } // rof
 for (n = 0; (2 == k) && (0 == n); n++)
 { // Simulate "if (2 == k)"
  PORTA = 0b010000;
  TRISA = 0b101011;
 } // rof
 for (n = 0; (3 == k) && (0 == n); n++)
 { // Simulate "if (3 == k)"
  PORTA = 0b000100;
  TRISA = 0b101011;
 } // rof
 for (n = 0; (4 == k) && (0 == n); n++)
 { // Simulate "if (4 == k)"
  PORTA = 0b100000;
  TRISA = 0b011011;
 } // rof
 for (n = 0; (5 == k) && (0 == n); n++)
 { // Simulate "if (5 == k)"
  PORTA = 0b000100;
  TRISA = 0b011011;
 } // rof
 for (n = 0; (6 == k) && (0 == n); n++)
 { // Simulate "if (6 == k)"
  PORTA = 0b000100;
  TRISA = 0b111001;
 } // rof
 for (n = 0; (7 == k) && (0 == n); n++)
 { // Simulate "if (7 == k)"
  PORTA = 0b000010;
  TRISA = 0b111001;
 } // rof

 k = (k + 1) % 8; // Increment k within range
                  // of 0-7

} // rof
} // End cPKLED 2
```

For Consideration

At the start of this section, I noted that C is notorious for its ability to allow programmers to create very efficient but very difficult-to-understand program statements. In the last experiment, I showed how the for statement is very versatile and how it can be used to replace all the traditionally used conditional execution statements. The motivation for writing complex statements is usually to minimize the amount of keying required for an application, although sometimes it can seem like the author of the code is simply trying to demonstrate his or her mental superiority.

For example, in looking at an example application, you might run across a statement like the following:

```
for (i = (j = Start) * 7, Match = 0; (Match =
(PORTC != (PORTA ^= Sequence[j++]))) && (i++ <
25););
// Sequence Match Confirmation?
```

At first glance, it is probably impossible to understand what this statement is intended to accomplish, and, to make matters worse, the comment is no help at all as it does not seem to relate to anything in the statement. You may feel like giving up and looking for another example, but you can do this; you can decode statements like this surprisingly easily.

When I presented the for statement, I noted that it was in the following format:

```
for (Initialization; Loop Test Expression; Loop
Increment)
 Statement
```

and each part (which consists of a C assignment statement or expression) of the for statement can be broken out into pieces and rewritten into pieces that make more sense. For example, the initialization assignment statement of the for statement is:

```
i = (j = Start) * 7;
```

and takes advantage of the ability of C to save an intermediate value in a complex expression. Because j is equal to Start and it is a factor in the initialization of i, the author has compressed the two following lines into one:

```
i = Start * 7;
j = Start;
```

Similarly, the comparison expression of the for statement can be broken out and understood by recognizing that comparison values are arithmetic values (zero for false, and not zero for true). To take advantage of this point, the *Match* variable (the first part of the comparison) is loaded with the result of the comparison of PORTA (which has been XORed with a value from the *Sequence array*) to PORTC. When an arithmetic or binary operator is placed before the equals sign in an assignment statement, the line is translated as the destination value operated on by the other parameter. I might write out the comparison part of the for statement as follows:

```
PORTA = PORTA ^ Sequence[j]; // Same as "PORTA
                             ^=
                             Sequence[j];
j = j + 1; // Increment "j", See below
Match = 0;
if (PORTC == PORTA)
Match = 1;
```

Both parts of the comparison use the unary increment (++) operator to increment the variables i and j *after* the expression has finished executing. When the unary increment or decrement operator is put to the left of the variable, as in the following example

```
++j;
```

the variable is incremented *before* the statement executes. Similarly, if the unary operator is on the right side of the variable, the variable is incremented after the statement has executed. I recommend the use of the increment and decrement unary operators in your coding, as they are a lot easier to key than the complete statement

```
j = j + 1;
```

and they are generally accepted as the shorthand version of these statements.

A null statement is used for the increment part of this statement. This is a bit unusual, but the unary increment operators in the comparison expression provide this function.

Just as the null statement is used as the for statement's increment statement, a null statement is used as the looping statement or statements that follow the for statement. If you look at how I have broken out the comparison expression, you will see that there is an assignment statement to PORTA, which could be moved to the looping statement area of the for statement.

If I were to write equivalent code to the for statement given at the start of this discussion, it would look something like this:

```
i = Start * 7;
j = Start;
Match = 0;
while ((0 == Match) && (i < 25))
```

```
{
PORTA = PORTA ^ Sequence[j];
if (PORTC == PORTA)
Match = 1; // Sequence Match Confirmation?
j = j + 1;
i = i + 1;
} // elihw
i = i + 1;
j = j + 1;
```

You should be able to relate this code to the original, and I'm sure the comment makes more sense now. What might not make sense is the incrementing of i and j after the while loop; these statements were put in to make sure the values at the end of the equivalent match the values at the end of the original for statement.

In terms of readability and decodability, I am sure the series of statements I have come up with are vastly superior to the single for statement. In terms of efficiency, the number of instructions created for either solution is not substantially different; nor is the execution speed of the two solutions dramatically different. The major difference between the two statements is the amount of keying required for them; the sequence of statements requires many times the number of the keystrokes of the short for statement.

It should be no surprise that I recommend that when you program, you avoid heavily compressing statements unless a strong reason exists to do so. Although you may save a substantially greater number of keystrokes, you should ask yourself how much time you might later lose debugging or decoding one compressed complex statement.

Simple PIC® MCU Applications

PC/ Simulator

PICkit™ 1

starter kit

Required Parts

```
1  PIC16F684
1  14-pin ZIF socket
   (3M/Textool 214-3339-
   00-0602J recommended)
1  0.01 µF capacitor
1  3-foot length of 28- or
   30-gauge solid core
   wire
```

Before going on, I would like to walk through a modification to your PICkit™ 1 starter kit. This minimizes the chance for damaging either the PIC microcontroller you are programming to put into another circuit or the PICkit 1 starter kit you are using to program the PIC MCU. Although the PICkit 1 starter kit is an excellent tool, the machined receptacle socket that is built into the PCB is not designed for many repeated plug/unplug cycles. Looking at manufacturer's datasheets, military-grade dual in-line chip package (DIP) sockets are qualified for 48 cycles. The specified number of plug/unplug cycles for industrial-grade sockets is 50 times (although they are not tested to see if they meet this specification). As you work through this book and your own experiments, you will easily exceed the maximum number of plug/unplug cycles for a military-grade socket, and chances are at least one or more pin receptacles in the socket will wear out and stop making reliable contact. You will find also that plugging and unplugging parts in the machined receptacle is difficult and that it's easy to bend the pins, have them fall off, or get stuck in the PICkit 1 starter kit's sockets. You can avoid these problems by adding a *zero insertion force* (ZIF) socket to the PICkit 1 starter kit.

Follow the steps outlined here. To add the ZIF socket you will need the following:

    1 PICkit 1 starter kit with snap-off PCB still attached

    1 14-pin ZIF socket (3M/Textool 214-3339-00-0602J recommended)

    1 0.01 µF capacitor

    1 3-foot length 28- or 30-gauge solid core wire

    Weldbond glue

    Solder

The tools you require are as follows:

    Soldering iron

    DMM with audible continuity tester

    Needle-nose pliers

    Clippers

    Wire strippers

A ZIF socket is similar to the machined receptacle socket already on the PICkit 1 starter kit. The difference is that the pin receptacles can be opened or closed by moving the lever on the socket. The open position is shown in Figure 3-1, and the pin receptacles are closed when the lever is pushed down. This socket will be added to the open 14-pin DIP socket area on the prototyping snap-off PCB on the right side of the PICkit 1 starter kit (see Figure 3-2), and each pin will be wired to the corresponding pin of the machined receptacle socket already on the PICkit 1 starter kit. Expect that this task will take an hour.

Figure 3-1 *14-pin 3M/Textool ZIF socket*

Figure 3-3 *Point-to-point wiring used to connect ZIF pins to PICkit 1 starter kit programming socket pins*

Figure 3-2 *14-pin ZIF socket added to the prototyping area of the PICkit 1 starter kit*

The steps for adding the ZIF socket are as follows:

1. Solder in the ZIF socket with its lever *up* (i.e., pin receptacles open). This will ensure proper operation. If you solder the ZIF socket in with the lever down, you will find that the receptacles will not open properly. You may find that you have to prop up the PCB with ZIF socket to make sure the lever stays up during soldering. If, after soldering, you find that some pins stick or don't open easily, move the ZIF socket's lever up and remelt the pin's solder to see if that relieves the stress.

2. Using point-to-point wiring, add the 14 connections between the machined receptacle socket and the ZIF socket (see Figure 3-3). Pin 1 of the PCB socket should go to Pin 1 of the ZIF

socket, pin 2 of the PCB socket should go to pin 2 of the ZIF socket, and so on. When I have done this, I try to keep my stripped pin lengths to $^1/_{32}$ inch (1 mm). There are two rows of holes beside the 14-pin socket holes in the prototyping snap-off PCB; to these holes you can attach one side of the wires rather than soldering them to the pins of the ZIF socket. When you are adding the wires, it is a good idea to leave the ZIF lever up to make sure that, if the ZIF socket pins remelt, there won't be problems later with any of the receptacles.

3. Test your wiring using the multimeter continuity tester function. Each pin of the machined receptacle socket should be tested against the corresponding pin on the ZIF socket, as well as against its adjacent pins to make sure no shorting exists.

4. When you are comfortable that your wiring is correct, solder the 0.01 µF in the two holes above the 14 holes used by the ZIF socket. I soldered the 0.01 µF capacitor on the backside of the PCB because the ZIF socket covered the holes on the topside. When soldering in the capacitor, make sure the leads are as short as possible, that it lies against the PCB, and that it does not extend beyond the rubber feet on the bottom of the PICkit 1 starter kit.

5. The final step is to glue down the wiring using the Weldbond glue. If you put on a reasonably thin bead, the glue should set to a hard, clear consistency in 6 to 12 hours. To make sure the wires don't extend beyond the rubber feet on the bottom of the PICkit 1 starter kit, you may

want to hold down the wires with a weight or tie them down while the glue hardens. I use Weldbond because it can be pulled off later without damaging the PCB.

Once you've completed the six steps and the glue has hardened, you have a ZIF-socket-equipped PICkit 1 starter kit that will stand an indefinite number of plug/unplug cycles. And, it is still connected to the other functions of the PICkit 1 starter kit, which allows you to experiment with the LEDs, buttons, and potentiometer interfaces built into the PICkit 1 starter kit.

Experiment 17—Basic Delays

The cFlash.c program presented in the introduction to this book included a simple two "for" statement delay. For the application, I wanted a delay of a half-second (500 ms) so the LED would flash on and off with a period of one second. Finding the end values of the for statements was done empirically; I used the simulator, as I will show in this program to time the delay and then adjusted the values until the delay was approximately 500 ms. In this experiment, I wanted to go back to the cFlash.c application and see if there was some way in which I could quantify the delay so I could use it in other applications.

To test the application, I modified cFlash.c slightly as you can see in the source code below:

```
#include <pic.h>
/* cDlay.c - Try to Quantify Delay Values

This Program is a modification of "cFlash.c" and
used to quantify the value of the end of the
delay variables and the time delay on the
Flashing D0 LED.

This program is to be used with both the
simulator and the PICkit 1 PCB with PIC16F684
installed.

RA4 - LED Positive Connection
RA5 - LED Negative Connection

myke predko
04.06.19

*/

__CONFIG(INTIO & WDTDIS & PWRTEN & MCLRDIS &
UNPROTECT \ & UNPROTECT & BORDIS & IESODIS &
FCMDIS);
```

```
int i, j;
int iEnd = 235;    // Outside Loop Value
int jEnd = 235;    // Inside Loop Value

main()
{

    PORTA = 0;
    CMCON0 = 7;    // Turn off Comparators
    ANSEL = 0;     // Turn off ADC
    TRISA4 = 0;    // Make RA4/RA5 Outputs
    TRISA5 = 0;

    while(1 == 1)    // Loop Forever
    {

      NOP();        // Breakpoint Here

      for (i = 0; i < iEnd; i++) // Delay Loop
      for (j = 0; j < jEnd; j++);

      NOP();        // Breakpoint Here

      RA4 = RA4 ^ 1; // Toggle LED
    } // elihw
} // End cDlay
```

The first change to cFlash.c for this experiment was to add two variables, iEnd and jEnd, that I could change easily to test the operation of the application. The second change was to place two statements you have never seen before (NOP();) before and after the delay code. The reason for the first change should be apparent: The variables allow the loop values to be changed easily without affecting the program statements. The second modification adds two instructions that don't do anything, and I could use them for breakpoints without affecting the operation of the application or breaking an instruction that is used multiple times in the application.

The points made regarding the NOP(); statements are probably confusing and might not make a lot of sense at this time. First, the NOP(); statements are replaced with "nop" or no-operation assembly language instructions, which I will discuss in more detail later in the book. And stated previously, the PICC Lite™ compiler NOP(); statement can be used as a breakpoint without affecting the operation of the C program. Second, you will find that the PICC Lite compiler has a very efficient built-in optimizer, which looks for opportunities to create executable code that is as small and efficient as possible. This optimizer will try to reuse code that performs the same function in different parts of the program. What the optimizer considers to be the same function is not necessarily the same thing you or I would consider the same function. Therefore, you will find situations where execution will jump around to different locations in the applications without apparent reason. By adding the NOP(); statements, you are putting an instruction explicitly before the start and after the end of the two delay for statements.

To measure the time of the delay, I enabled the MPLAB® IDE simulator and then added the Stopwatch function (see Figure 3-4) to the project. The Stopwatch function will count the number of instructions that execute after the start of the application or after being reset.

To measure the delay for different values of iEnd and jEnd, I put a breakpoint at each of the two NOP(); statements. I did this by moving the cursor to the line where the NOP(); statement was found, right-clicking, and selecting "Set Breakpoint." I then reset and ran the program, and when it stopped at the first NOP(); statement, I clicked on the stopwatch's zero button to reset the stopwatch, then clicked on the run button again, and waited for the next breakpoint to stop execution.

I was expecting that the iEnd and jEnd values could be reversed. This is to say that the delay of iEnd equal to 50 and jEnd equal to 100 would be the same as iEnd equal to 100 and jEnd equal to 50. This turned out not to be the case; the delay varied by several percent when the iEnd and jEnd values were reversed. To try

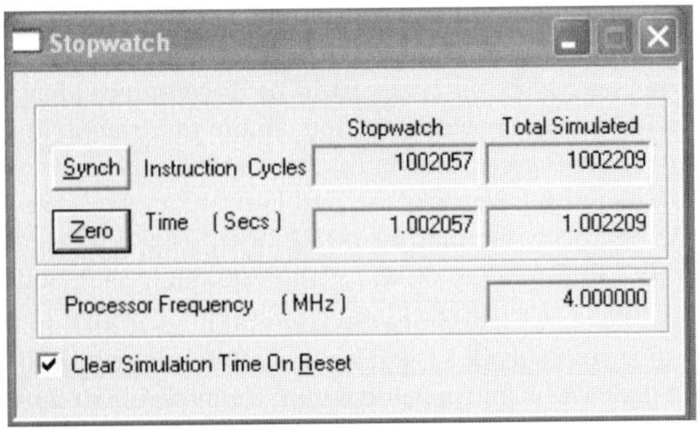

and come up with a simple, repeatable formula that could be used for the application, I tried making both values the same and came up with the following rough formula:

$$\text{Delay(Seconds)} = 1.8(10^{-5}) \times \text{iEnd}^2$$

This formula is reasonably accurate for the range of 50 ms to 2 seconds.

Thinking about the optimizer and thinking about the code, I realized that the optimizer didn't do an obvious optimization and that is why I replaced the two for statements with the following statements:

```
i = iEnd;
j = jEnd;
```

Without anything happening in the inside for loop, the two loops are not doing anything other than exiting with i and j being changed. In this case, the for loop code is still included in the application, but you will find cases where the ultimate optimization that I listed previously will be produced by the compiler and its optimizer.

Delays are critical operations in microcontroller (or any real-time) programming, and I will be showing you a number of methods to implement delays in your applications. Although you might want to get your delays exact (in terms of time or instruction cycles), remember that this is often close to impossible and usually not required. As I discuss different applications, I will point out whether or not the delay's absolute accuracy is critical or if an approximation (usually within 10 percent) is acceptable for the application.

	Stopwatch	Total Simulated
Synch — Instruction Cycles	1002057	1002209
Zero — Time (Secs)	1.002057	1.002209
Processor Frequency (MHz)		4.000000

☑ Clear Simulation Time On Reset

Figure 3-4   *MPLAB IDE Stopwatch function*

PC/
Simulator

PICkit™ 1

starter kit

As you have gained insight into PICC Lite compiler PIC microcontroller programming and the PICkit 1 starter kit, you have probably started asking yourself how did I know to turn on the LED marked "D0" to make the PIC16F684's RA4 and RA5 pins outputs, and then output a 1 (or high voltage) and a 0, respectively. The process that I went through is quite simple and only required looking at the schematics for the PICkit 1 starter kit. It did not require any probing or trial and error.

When you look at the LED circuitry in the PICkit 1 starter kit's schematic (found in the *PICkit 1 Flash Starter Kit User's Guide*), you see that each of the eight LEDs is wired as part of a pair, like I show in Figure 3-5. To turn on one LED, current must flow in one direction, and to turn on the other, current must flow in the opposite direction. When I started this experiment, I had hoped that I could make active more than just the two I/O pins connected to the LED output. But when I followed the various connections, I discovered that if more than two I/O pins were active at any time, there was a good chance that a second LED would be inadvertently lit.

With this knowledge, I came up with Table 3-1, which lists the TRISA and PORTA register values needed to turn on each LED on the PICkit 1 starter

Table 3-1
PICkit 1 Starter Kit LED Display TRISA and PORTA Values

LED	TRISA	PORTA
D0	B'11001111'	B'00010000'
D1	B'11001111'	B'00100000'
D2	B'11101011'	B'00010000'
D3	B'11101011'	B'00000100'
D4	B'11011011'	B'00100000'
D5	B'11011011'	B'00000100'
D6	B'11111001'	B'00000100'
D7	B'11111001'	B'00000010'

kit. To test this knowledge, I came up with the following program, which turns on each LED in sequence. The program uses the same delay that we used for the original cFlash (flashing D0 LED) program to show clearly if each LED lit and if they lit in the correct order. Note that in the program, the information from Table 3-1 is part of the documentation.

```
#include <pic.h>
/* cPKLED.c - Roll Through PICkit 8 LEDs

This Program will roll through each of the 8
LEDs built into the PICkit PCB.

The LED values are:

LED Anode Cathode
D0 RA4 RA5
D1 RA5 RA4
D2 RA4 RA2
D3 RA2 RA4
D4 RA5 RA2
D5 RA2 RA5
D6 RA2 RA1
D7 RA1 RA2

myke predko
04.09.10

*/

__CONFIG(INTIO & WDTDIS & PWRTEN & MCLRDIS &
UNPROTECT \ & UNPROTECT & BORDIS & IESODIS &
FCMDIS);


int i, j, k;

main()
{

 PORTA = 0;
 CMCON0 = 7; // Turn off Comparators
 ANSEL = 0; // Turn off ADC

 k = 0; // Start at LED 0
```

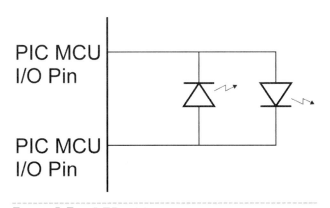

PIC MCU
I/O Pin

PIC MCU
I/O Pin

Figure 3-5   *LED wiring*

```
while(1 == 1) // Loop Forever
{
  for (i = 0; i < 255; i++) // Simple Delay Loop
  for (j = 0; j < 129; j++);

  switch (k) { // Select Which LED to Display
  case 0:
   PORTA = 0b010000;
   TRISA = 0b001111;
   break;
  case 1:
   PORTA = 0b100000;
   TRISA = 0b001111;
   break;
  case 2:
   PORTA = 0b010000;
   TRISA = 0b101011;
   break;
  case 3:
   PORTA = 0b000100;
   TRISA = 0b101011;
   break;
  case 4:
   PORTA = 0b100000;
   TRISA = 0b011011;
   break;
  case 5:
   PORTA = 0b000100;
   TRISA = 0b011011;
   break;
  case 6:
   PORTA = 0b000100;
   TRISA = 0b111001;
   break;
  case 7:
   PORTA = 0b000010;
   TRISA = 0b111001;
   break;
  } // hctiws

  k = (k + 1) % 8; // Increment k within range of
                   0-7

} // elihw
} // End cPKLED
```

Looking at the program, its function should be fairly easy to understand. The large switch block of code passes execution to the series of TRISA and PORTA register writes that are specific to the LED to be written. I could have written to individual bits rather than the entire port, but I felt that the entire port write was easier to follow and understand.

I would consider this application to be quite large, clumsy, and not particularly well written. The use of four register writes per LED took a long time to key in (even with cutting and pasting), and there was a very good chance of making a data entry mistake, or typo. As you work through the experiments in this book, (particularly the ones in Section 4), think about how you could improve the way a program is written. For example, in this previous example I believe I could reduce the total number of lines to less than a quarter and eliminate the repeated data values used to turn on and then turn off each LED.

This program could be considered an initial version of a Cylon Eye from *Battlestar Galactica*, which is also called a "Knight Rider Eye" from the TV show. After trying out the program presented in this experiment, you might want to see if you can make the LEDs reverse direction after reaching one extreme. To do this you will require an additional variable, this one storing the direction of the LEDs and changing its value each time the on LED is either D0 or D7. Along with the direction variable, you will have to make sure that the TRISA and PORTA bits are returned to their original state (i.e., loaded with zeros for output and in input mode) before turning on the next LED in the sequence.

Experiment 19—Binary Number Output Using PICkit 1 Starter Kit LEDs

In the previous experiment, I demonstrated how each LED on the PICkit 1 starter kit could be turned on in sequence. I noted that only one LED could be turned

on at any time, meaning that the PICkit 1 starter kit cannot simultaneously display more than one bit of data at a time. In this experiment, I will show how you can display an incrementing eight-bit counter on the eight bits of the PICkit 1 starter kit.

The method used to display data on all eight bits simultaneously is the same method that I will demonstrate for multidigit LEDs and dot array LEDs. To give the appearance that all the LEDs are active at the same time, we will rotate through each LED in sequence and, if the bit the LED is representing is set, then the LED is turned on for a set period of time. If the bit is not set, then the LED is left off for the same amount of time. This ensures that the brightness of each LED will be constant regardless of the number turned on, and that constant timing for the application is provided. This method is also the basis for controlling multiple motors and servos (as in a robot) and should not be considered applicable only to the devices shown in this book.

The rule of thumb that I use when sequencing LEDs is that each one should be turned on 50 times per second. For this application, I wanted each LED to be on 100 times a second, which means that each LED is turned on over a 0.01 second (10 ms) period. For each LED to be active in the 10 ms time period, it should be turned on for 1.25 ms (10 ms divided by eight). To create this delay, I used a single for loop with the delay value found empirically.

The program I came up with increments a counter once every half a second, and the current counter value is displayed on the eight LEDs of the PICkit 1 starter kit:

```c
#include <pic.h>
/* cLEDDisp.c - Use D0-D7 as an incrementing
Counter

Using "cPKLED.c" as a base, cycle through each
LED at 100x per second (1250 us between LEDs).

myke predko
04.09.12

*/

__CONFIG(INTIO & WDTDIS & PWRTEN & MCLRDIS &
UNPROTECT \ & UNPROTECT & BORDIS & IESODIS &
FCMDIS);

int i, j;
int Value = 0;
int Dlay = 67; // LED Time on Delay Variable

main()
{

 PORTA = 0;
 CMCON0 = 7; // Turn off Comparators
 ANSEL = 0; // Turn off ADC

 j = 0; // Reset the Display Counter

 while(1 == 1) // Loop Forever
 {
  NOP();
  for (i = 0; i < Dlay; i++); // Simple Delay
                                         Loop
   if ((Value & (1 << 0)) == 0)
    PORTA = 0;
   else // Display the Value
    PORTA = 0b010000;
  TRISA = 0b001111;

  NOP();
  for (i = 0; i < Dlay; i++);
   if ((Value & (1 << 1)) == 0)
```

```c
    PORTA = 0;
   else // Display the Value
    PORTA = 0b100000;
  TRISA = 0b001111;

  for (i = 0; i < Dlay; i++);
   if ((Value & (1 << 2)) == 0)
    PORTA = 0;
   else // Display the Value
    PORTA = 0b010000;
  TRISA = 0b101011;

  for (i = 0; i < Dlay; i++);
   if ((Value & (1 << 3)) == 0)
    PORTA = 0;
   else // Display the Value
    PORTA = 0b000100;
  TRISA = 0b101011;

  for (i = 0; i < Dlay; i++);
   if ((Value & (1 << 4)) == 0)
    PORTA = 0;
   else // Display the Value
    PORTA = 0b100000;
  TRISA = 0b011011;

  for (i = 0; i < Dlay; i++);
   if ((Value & (1 << 5)) == 0)
    PORTA = 0;
   else // Display the Value
    PORTA = 0b000100;
  TRISA = 0b011011;

  for (i = 0; i < Dlay; i++);
   if ((Value & (1 << 6)) == 0)
    PORTA = 0;
   else // Display the Value
    PORTA = 0b000100;
  TRISA = 0b111001;

  for (i = 0; i < Dlay; i++);
   if ((Value & (1 << 7)) == 0)
    PORTA = 0;
   else // Display the Value
    PORTA = 0b000010;
  TRISA = 0b111001;

  j = j + 1; // Increment the Counter every 1/2 s
  if (j >= 50)
  {
   Value = Value + 1; // Increment Display Counter
   j = 0; // Reset the Counter
  } // fi
 } // elihw
} // End cPKLED
```

I chose the 10 ms display time because it allows simple calculations for counting a set number of loops for larger delays. I take advantage of this for incrementing the *Value variable*, which is displayed on the LEDs.

Two points should be noted regarding the code I used to determine whether or not a specific LED is turned on in this experiment. The first is the use of shifting the bit ANDed with "Value" to see if the bit in "Value" is set. Rather than putting in the decimal, binary, or hexadecimal equivalent of the bit, I chose to shift one up by the bit number. By doing this, we can see exactly what bit is being ANDed with "Value" rather than having to do a mental calculation. You may feel more comfortable using an equivalent value. Secondly, I chose to OR the value in PORTA with zero if the bit is not set to make sure that either path takes up the same amount of time. This can be extremely important in timed applications, and although it makes the code a bit more complex, it ensures that your timing is as consistent as possible, regardless of the path taken.

This program will be the basis for displaying binary data in future experiments. I must point out that it is actually programmed in a very inefficient manner. In Section 4, I will provide you with features of the C programming language and programming techniques that will make this application more efficient both in terms of space used and execution time.

Experiment 20—Basic Button Inputs

starter kit

The simplest form of user input you can put into a PIC MCU application is a button. This is usually accomplished by a pulled-up pin with a momentary on button that pulls the pin to ground (see Figure 3-6). The PICkit 1 starter kit has a very similar switch circuit on RA3; it can be used for either controlling the reset of the PIC MCU (which will be presented later), or it can be used as an input with which you can experiment. In this experiment as well as the next one I will demonstrate how this switch can be used as an input device.

The software for this experiment consists of simply reading the switch and then determining whether or not the D0 LED on the PICkit 1 starter kit should be turned on. When the button is pressed, the logic level

on RA3 is low, and then RA5 (D0's cathode) is pulled low. When the logic level input on RA3 is high, RA5 is driven high. The program, cButton.c, is quite simple:

```
#include <pic.h>
/* cButton.c - Simple C Program to Turn on an
LED when Button is Pressed

This Program is a modification of "cFlash"

RA3 - Button Connection
RA4 - LED Positive Connection
RA5 - LED Negative Connection

myke predko
04.06.24

*/

__CONFIG(INTIO & WDTDIS & PWRTEN & MCLRDIS &
UNPROTECT \ & UNPROTECT & BORDIS & IESODIS &
FCMDIS);

int i, j;

main()
{

  PORTA = 0x3F; // All Bits are High
  CMCON0 = 7; // Turn off Comparators
  ANSEL = 0; // Turn off ADC
  TRISA4 = 0; // Make RA4/RA5 Outputs
  TRISA5 = 0;

  while(1 == 1) // Loop Forever
  {

   if (0 == RA3) // Set values using "if"
                 statement
   {
    RA5 = 0;
   }
   else
   {
    RA5 = 1;
   } // fi
  } // elihw
} // End cButton
```

cButton.c is written as if the author was familiar with C, but had only recently started working with the PICC Lite compiler. In traditional C programming,

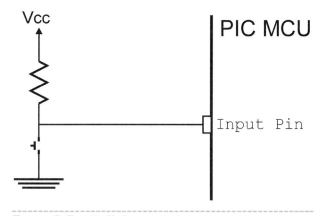

Figure 3-6   *Pulled-up switch*

where no bit inputs are used, the button read and LED output would probably look like this:

```
if (0 == (PORTA & (1 << 3)) // Is RA3 High or
                                         Low?
{
 PORTA = PORTA & (0x0FF ^ (1 << 5)); // Low, Make
                                         RA5 Low
}
else
{
 PORTA = PORTA | (1 << 5) // High, Make RA5 High
} // fi
```

but the author, seeing the defined pins in the include files, probably realized that the pins could be read and written directly rather than having individual bits ANDed and ORed as I do above. What the author probably did not realize is that an input pin value can be passed directly to an output pin as I do in the following cButton 2.c:

```
#include <pic.h>
/* cButton 2.c - Simplified C Program to Turn on
an LED when Button is Pressed

This Program is a modification of "cButton"

RA3 - Button Connection
RA4 - LED Positive Connection
RA5 - LED Negative Connection
```

```
myke predko
04.11.07

*/

__CONFIG(INTIO & WDTDIS & PWRTEN & MCLRDIS &
UNPROTECT \ & UNPROTECT & BORDIS & IESODIS &
FCMDIS);

int i, j;

main()
{

  PORTA = 0x3F; // All Bits are High
  CMCON0 = 7; // Turn off Comparators
  ANSEL = 0; // Turn off ADC
  TRISA4 = 0; // Make RA4/RA5 Outputs
  TRISA5 = 0;

  while(1 == 1) // Loop Forever
  {
   RA5 = RA3; // Simpler: Pass Value through to RA5
  } // elihw
} // End cButton 2
```

This is a fairly simple optimization of cButton, but it illustrates how the pin variable type of the PICC Lite compiler can be used to greatly simplify the source code of an application.

Experiment 21—Debouncing Button Inputs

starter kit

If you have some experience with digital electronic circuits, you'll know that handling button input is a surprisingly challenging task with a bit of science behind it, which you will have to understand before you can successfully process button inputs from the application. The most critical aspect of understanding button operation is to debounce the incoming signals. Knowing the current button state and how you want the application to work is integral. In this experiment, I will demonstrate a simple program to change the state of the D0 LED each time the RA3 button on the PICkit 1 starter

kit is pressed. I will demonstrate also how the MPLAB IDE simulator can be used to verify the operation of an application before a PIC MCU is burned with an application.

Each time a switch (or button) state changes, the contacts literally bounce against each other as shown in the oscilloscope pictured in Figure 3-7. This happens both when the contacts are made and when they are broken. To interpret, or *debounce*, the change in switch state, the line is typically polled until it stays in the same state for 20 ms. The basic code for polling a switch input line and exiting when the line has changed state for 20 ms is as follows:

```
i = 0; // Wait 20 ms for Button Up
while (i < Twentyms)
{
```

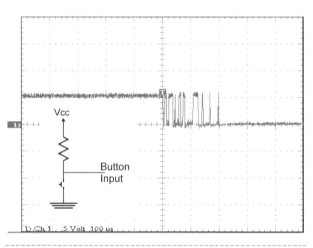

Figure 3-7   *Debounce*

```
if (0 == RA3) // Button Down/Start over
{
  i = 0;
}
else // Button Up/Increment Count
{
  i = i + 1;
} // fi
} // elihw
```

Using this code snippet, the following cDebounce.c application was developed:

```
#include <pic.h>
/* cDebounce.c - Debounce Button Input on RA3

This Program polls the button at RA3 and changes
the state of RA5 after the press has been
debounced. Use a 20 ms debounce period.

RA3 - Button Connection
RA4 - LED Positive Connection
RA5 - LED Negative Connection

myke predko
04.11.07

*/

__CONFIG(INTIO & WDTDIS & PWRTEN & MCLRDIS &
UNPROTECT \ & UNPROTECT & BORDIS & IESODIS &
FCMDIS);

int i;

const int Twentyms = 1150; // Declare a Constant
                           //       for 20 ms Delay

main()
{

  PORTA = 0x3F; // All Bits are High
  CMCON0 = 7; // Turn off Comparators
  ANSEL = 0; // Turn off ADC
  TRISA4 = 0; // Make RA4/RA5 Outputs
  TRISA5 = 0;

  while(1 == 1) // Loop Forever
  {
    i = 0; // Wait 20 ms for Button Up
    while (i < Twentyms)
    {
```

```
      if (0 == RA3) // Button Down/Start over
      {
        i = 0;
      }
      else // Button Up/Increment Count
      {
        i = i + 1;
      } // fi
    } // elihw

    NOP();
    i = 0; // Wait 20 ms for Button Down
    while (i < Twentyms)
      if (1 == RA3) // Button Up/Start over
        i = 0;
      else // Button Down/Increment Count
        i = i + 1;

    RA5 = RA5 ^ 1; // Toggle RA5 to Turn ON/OFF LED

  } // elihw
} // End cDebounce
```

In the cDebounce.c application, I first wait for the RA3 line to be high for 20 ms and then I wait for RA4 to be low for 20 ms before toggling the D0 LED output state. In the application code, for the first debounce loop (debouncing button going high), I added all the braces for conditional code. But for the second debounce loop (debouncing button going low), I eliminated the extra braces because only one statement appears after each conditional execution statement, and these statements take up a relatively large amount of space. For the rest of the book, I will be using braces only when more than one statement executes conditionally.

The NOP(); statement may seem out of place in this application. When I wrote this application, I found that I could not place a breakpoint at the i = 0; statement following the first debounce loop, so I put in the nop instruction, which gave me a place to set the simulator breakpoint. This is a good trick to remember when you are developing your own application.

You might think that the first debounce loop (debouncing the button input going high) is redundant or not needed because we just want to execute when the button is pressed, but you must remember that the PIC MCU is operating at MHz speeds. The debounce and LED toggle operation executes very quickly, and, without the button up debounce loop, you'll discover that the LED toggles at roughly 50 times per second rather than once, each time the button on the PICkit 1 starter kit is pressed.

Before I burned this code into a PIC MCU on the PICkit 1 starter kit MCU, I simulated it using the MPLAB IDE simulator. If you were to enable the simulator and start executing the code, you would discover that it would get stuck in the first debounce loop because the default value of RA3 is 0 (or a low logic level). To change the state of RA3 in the simulator (and add a digital signal input), you can use the

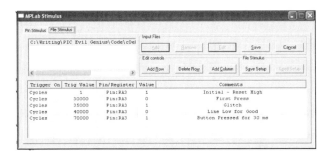

Figure 3-8 *Stimulus window*

MPLAB IDE Stimulus window shown in Figure 3-8. The important part of the stimulus window is the area at the bottom, which looks like a spreadsheet; the changing input pin values are specified along with the time (number of cycles) the change takes place.

To set up the stimulus window, you will go through the following steps:

1. Click on the Debugger pull-down, and click on "Stimulus Controller."

2. In the MPLAB IDE Stimulus window, click on the File Stimulus tab.

3. Next, to start a stimulus file, click on "Add" in the Input File Box, and select a file name.

4. Make sure the stimulus file name is highlighted and click on "Edit" in the Input File box to bring up the spreadsheet-like input area.

5. Click on "Add Row" and then select the desired options (discussed in the following paragraphs) to build an input waveform to the application code. In Figure 3-8, the stimulus list puts the input line high, glitches it down, and then holds the line low.

6. Save the stimulus file and then define and save the project's stimulus file (click "Save Setup" in the File Stimulus box of the Stimulus window).

A number of different options exist for defining stimulus. In this experiment I simply worked with pre-defined pin stimulus, but you can also specify asynchronous pin states using the Pin Stimulus tab. To change or set specific pin states, click on a button on the window. When you specify the predefined pin stimulus for each row, you will have to specify the following:

*   Trigger On—either after executing for so many instruction cycles or when execution hits a specific address. For most pin stimulus applications, you will want to trigger on a set number of instruction cycles. The specific address function is useful for applying a test value to an internal register.

*   Trigger Value—the number of cycles or addresses.

*   Pin/Register—the I/O pin or hardware register to be written to. More than one pin or register can be referenced in a single stimulus file.

*   Value—the 1 or 0 for a pin or a hex value for a register.

*   Comments—gives you a chance to note what is happening.

In previous versions of MPLAB IDE, stimulus files were produced by a text file that was loaded into the simulator. The current Windows version makes it easier to implement simple stimulus inputs like this one, but I feel that for more complex operations, the Windows version is a lot more work and more difficult to implement correctly. If you have a complex stimulus for the MPLAB IDE simulator, I suggest that you write it out first and then enter the data carefully rather than on the fly, as I did for this application.

Experiment 22—_MCLR Operation

Looking back at the first book I wrote about the PIC microcontroller, I was amazed at the amount of space I devoted to the topic of reset and the _MCLR pin, which is used to control whether or not the PIC MCU is to execute. The PIC MCU I wrote about in that book, the PIC16F84, required a separate reset circuit that, ideally, held the PIC MCU reset until the power supply ramped up to the operation voltage, held off clock start until

any initial fluctuations had passed, monitored the power supply, and reset the microcontroller if it fell below a specific value, as well as allowed the user to control the operation of the PIC MCU. Although some shortcuts existed, they could be implemented only if you understood the application and PIC16F84 operation well *and* if you understood what the tradeoffs of the selected method were. One of the big changes since then is the sophistication of the reset circuitry built into PIC microcontrollers; most of these issues are no longer a concern when designing PIC MCU applications.

A *complete reset* solution for the PIC16F84 is shown in Figure 3-9 and takes advantage of a chip like the Panasonic MN18311 reset supervisor chip. This chip will not allow the microcontroller to execute until the input voltage has had a chance to stabilize, and if the power supply voltage drops below a specific point (known as *brownout*), the PIC16F84 will reset. Not only does a modern PIC MCU have the voltage-monitoring hardware built into its reset circuitry, but it is also able to work over a much larger voltage range (from 2.0 volts to 5.5 volts compared to 4.0 volts to 5.5 volts for the PIC15F84), so brownouts are much less of a concern.

The PIC16F684 has the built-in reset features including the brownout detection and built-in start-up delay that were required to be added to PIC16F84 applications. Along with these features, the new chip includes a number of bits in the STATUS and PCON registers (listed in Table 3-2) that can be polled to help determine the cause of a PIC16F684 reset. For most applications, these bits can be ignored, but in some cases it is important to understand why an application has started executing.

For this experiment, I would like to demonstrate how the reset pin (also known as _MCLR) can be used to reset the PIC16F684 while power is active. And by polling the _POD bit of PCON, the program will indicate whether reset was caused by power on or by making the _MCLR (RA3) pin on the PIC MCU active. Normally, I disable the _MCLR function of the RA3 pin and rely on the internal hardware to properly reset the PIC MCU. For this cReset.c experiment, I enabled _MCLR and used the button on the PICkit 1 starter kit to cause a reset of the PIC16F684. The program is based on cPKLED, and when the program first powers

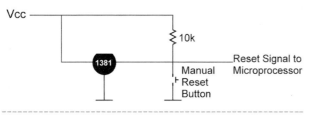

Figure 3-9 *MCU reset*

up and _POR is low, I set the program to run so the LEDs will light in incremental order. On subsequent power ups, if the cPKLED bit is set, then I run the LEDs in decremental order.

```c
#include <pic.h>
/* cReset.c - Roll Through PICkit 8 LEDs

This Program will roll through each of the 8
LEDs built into the PICkit PCB.

The LED values are:

LED Anode Cathode
D0  RA4   RA5
D1  RA5   RA4
D2  RA4   RA2
D3  RA2   RA4
D4  RA5   RA2
D5  RA2   RA5
D6  RA2   RA1
D7  RA1   RA2

myke predko
04.09.10

*/


__CONFIG(INTIO & WDTDIS & PWRTEN & MCLREN &
UNPROTECT \ & UNPROTECT & BORDIS & IESODIS &
FCMDIS);
 // Note "MCLREN" instead of "MCLRDIS"


int i, j, k;
int Polarity;

main()
{

PORTA = 0;
CMCON0 = 7;      // Turn off Comparators
ANSEL = 0;       // Turn off ADC

k = 0;           // Start at LED 0

if (0 == POR)    // Check PCON Register for Start
                 up
{
Polarity = 0;
POR = 1; // Indicate PIC MCU Powered UP Once
}
else     // Reset
Polarity = 1;    // Go Backwards

while(1 == 1)    // Loop Forever
{
  for (i = 0; i < 255; i++) // Simple Delay Loop
   for (j = 0; j < 129; j++);

  switch (k) { // Select Which LED to Display
   case 0:
    PORTA = 0b010000;
    TRISA = 0b001111;
    break;
   case 1:
    PORTA = 0b100000;
    TRISA = 0b001111;
    break;
   case 2:
    PORTA = 0b010000;
    TRISA = 0b101011;
    break;
```

```
 case 3:
  PORTA = 0b000100;
  TRISA = 0b101011;
  break;
 case 4:
  PORTA = 0b100000;
  TRISA = 0b011011;
  break;
 case 5:
  PORTA = 0b000100;
  TRISA = 0b011011;
  break;
 case 6:
  PORTA = 0b000100;
  TRISA = 0b111001;
  break;
 case 7:
  PORTA = 0b000010;
  TRISA = 0b111001;
  break;
 } // hctiws

 if (0 == Polarity)
  k = (k + 1) % 8; // Increment k within range
                   of 0-7
 else
  k = (k - 1) % 8; // Decrement k within range
                   of 0-7

 } // elihw
} // End cReset
```

Table 3-2
Bits Responsible for Indicating Reasons for Reset

Bit Name	Register, Bit	Function
_TO	STATUS, 4	Normally Set, Reset after a Watchdog Timer Reset
_PD	STATUS, 3	Normally Set, Reset after Executing a Sleep Instruction
ULPWUE	PCON, 5	Set to Enable Ultra Low-Power Wake-Up
SBODEN	PCON, 4	Software Brownout Detect Enable
_POR	PCON, 1	Reset on Power Up—after Power On Set Bit
_BOD	PCON, 0	Unknown Value on Power Up—after Power On Set Bit

For virtually all applications that you create, you can safely disable the _MCLR pin and use it as an input-only pin. The only circumstance where I would expect you to enable the _MCLR pin is when you are debugging an application and expecting the application to fail and hang up periodically, requiring a low-reset value.

Experiment 23—Ending Applications

Before going off to more complex applications, I thought it would be useful to discuss how your PICC Lite compiler application executes in the PIC MCU. In particular, the manner in which the hex code is loaded and executed may be confusing and, therefore, worth discussing. You are probably familiar with running programs in your PC: when a program ends, control doesn't return to the operating system. Obviously, this is not true for the PIC MCU, which does not have an operating system; so you might expect that when the program completes, the PIC MCU just stops. This experiment was written to test this expectation.

The cEnd.c application was written for the PICkit 1 starter kit. It will turn on the first three LEDs (D0 through D2) in sequence and then end. I enabled the _MCLR pin so the application could be restarted rather than requiring you to download it again to start the execution process again.

```
#include <pic.h>
/* cEnd.c - Light 3 LEDs and Return

This Program will Light D0 to D2 in Sequence and
then End.

myke predko
04.09.12

*/

__CONFIG(INTIO & WDTDIS & PWRTEN & MCLREN &
UNPROTECT \ & UNPROTECT & BORDIS & IESODIS &
FCMDIS);


int i, j;
int iEnd = 235; // 1 Second Delay

main()
{

 PORTA = 0;
 CMCON0 = 7; // Turn off Comparators
 ANSEL = 0; // Turn off ADC

 PORTA = 0b010000; // D0
```

```
TRISA = 0b001111;
for (i = 0; i < iEnd; i++) // Simple Delay Loop
                                      for D0
for (j = 0; j < iEnd; j++);

PORTA = 0b100000; // D1
TRISA = 0b001111;
for (i = 0; i < iEnd; i++) // Simple Delay Loop
                                      for D1
for (j = 0; j < iEnd; j++);

PORTA = 0b010000; // D2
TRISA = 0b101011;
for (i = 0; i < iEnd; i++) // Simple Delay Loop
                                      for D2
for (j = 0; j < iEnd; j++);

PORTA = 0b000000; // Return PORTA to Initial
                              State
TRISA = 0b111111;

// while(1 == 1); // Normal End
// SLEEP(); // Alternative End

} // End cEnd
```

When you run the program as it is presented here, you will discover that instead of just stopping at the end (D2 turns on and then off), D0 lights again and the program seems to run again. I suggest that you now run it on the MPLAB IDE simulator, put a breakpoint at the first statement (PORTA = 0;) and at the last statement (TRISA = TRISA | 0b011111111;), and start the program executing (with the stopwatch displayed). The first time the program executes, it will stop at "PORTA = 0;," indicating that it has executed in 100 instructions. Click on "Run" again and the program will execute for a few moments, finally stopping at the "TRISA = TRISA | 0b011111111;" statement with 3,006,075 executed instructions on the stopwatch.

Now, click on "Run" a third time. You might expect that the simulated PIC MCU never stops executing or that it immediately stops and resets the simulated PIC MCU. Actually, neither happens; execution returns to the first breakpoint (PORTA = 0;) and the stopwatch displays 3,006,175 instructions. From these results you can conclude that program starts again after 100 instruction cycles. What is happening is most easily illustrated when you are familiar with assembly language, but the short explanation is that when the PIC MCU reaches the end of program memory, instead of stopping, the program counter resets itself and starts executing from the start again.

To prevent this from happening, I normally end my programs with a statement like the following that puts the PIC MCU into a hard loop that execution cannot escape from:

```
while(1 == 1);
```

Although you will see that I have noted this with a comment (or "commented it out") in the previous pro-gram, to test the operation of this statement, delete the two slashes to the left, rebuild, and retry the program. You will find that it will stop after D2 has been turned off and won't start until you press "SW1" (the reset button) on the PICkit 1 starter kit. Along with the while forever loop, you can also use the SLEEP(); PIC MCU instruction, which is set off as a comment after the while statement. This instruction puts the PIC MCU into a low power state, and then allows it to be restarted by pressing the reset button on the PICkit 1 starter kit. In most microcontroller applications, the program is continually looping, so the need for the while statement or sleep instruction is not required. But in applications where a single operation is being carried out (like the examples in this book), you will need the while statement or sleep instruction to prevent the program from executing repeatedly.

For Consideration

So far in this book, I have relied upon the PICkit 1 starter kit exclusively for putting in the circuitry, but you can build these circuits on a breadboard very easily if you have a limited number of PICkit 1 starter kits available (such as in a classroom). Another case in which you might want to build a PIC MCU circuit on a breadboard is when you require different circuitry than is available on the PICkit 1 starter kit. In the following paragraphs, I will quickly introduce you to putting a PIC MCU on a breadboard and some issues that you should be aware of when building your own circuits.

The basic parts required for wiring a PIC MCU on a breadboard include the following:

1 Small breadboard

1 Breadboard wiring kit

2 AA or AAA batteries and clip with breadboard wires

1 0.01 mF capacitor (any type)

1 Power switch (E-Switch's EG1903 is recommended)

LEDs

Capacitors of various values

Optional parts that you will want include the following:

Pushbuttons with breadboard wires

Potentiometers of various values

Small plastic parts box

The tools that you will require include the following:

Small flat screwdriver

Needle-nose pliers

Wire clippers

Wire strippers or knife

This probably seems like a modest list of parts and should not cost you much more than $20.00, and with them you can create a wide variety of different PIC MCU applications outside of the breadboard. What may surprise you is the usefulness of the two "radio batteries" that together provide about 3 volts (for alkaline batteries). The PIC16F684 (and most other new PIC MCUs) will work with voltage inputs ranging from 2.0 to 5.5 volts. Many modern peripheral devices also work at these voltage levels, so rather than using something like a 9-volt battery and a voltage regulator to get 5 volts, you can simplify your circuit by powering with two radio batteries. A complete kit is shown in Figure 3-10 and can be purchased from a variety of sources (including Radio Shack and Digi-Key).

The kit consists of a breadboard, a breadboard wiring kit, some LEDs, resistors or different values (100 Ohms, 220 Ohms, 330 Ohms, 470 Ohms, 1k, and 10k are good values to start with), capacitors (0.01 μF, 10 μF, and 47 μF will be useful), a momentary on push button, an SPST switch, and a two-cell AA or AAA battery pack. As you create more complex applications, you will probably want to expand this kit. It would be a good idea to get a small tool kit as well.

Pin through hole (PTH) chips (like the PIC16F684 microcontroller), resistors, capacitors, and breadboard wiring work without any modifications to the breadboard and only a few to some of the components. Although it will take a bit of work. For a power switch, the E-Switch EG1903 SPDT (available from Digi-Key) will work with a breadboard without modification. Unfortunately, I have not found a momentary on switch or battery pack that will work with a breadboard without modification; I solder wires onto standard parts as shown in Figure 3-11 to get this capability.

Once you have the parts, you can wire together the cFlash.c circuit (Figure 3.12) in just a few minutes, and the resulting circuit should look something like Figure 3-13. Although the circuit seems quite simple, there are three things to be aware of. The first is to wire both sides of the breadboard with power (Vdd/Vcc) and ground (Vss/Gnd) on the common bars on both sides of the breadboard (see Figure 3-13). This will make your wiring simpler when you have multiple devices

Figure 3-11   *AAA battery clip and momentary push button modified with breadboard wires soldered to them*

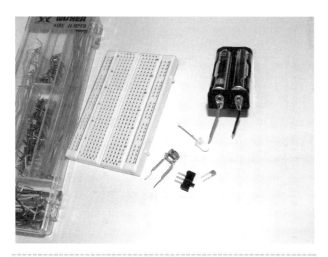

Figure 3-10   *Breadboard with basic parts needed to run the PIC16F84*

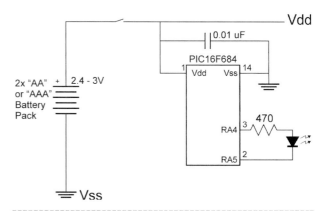

Figure 3-12   *Breadboard circuit*

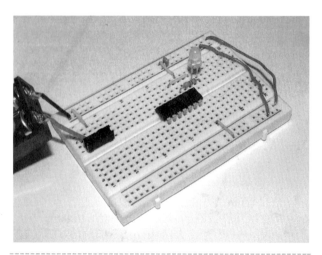

Figure 3-13   *Breadboard wiring to demonstrate the cFlash.c operation*

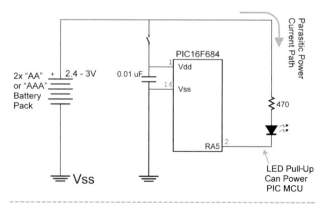

Figure 3-14   *Parasitic power*

on the breadboard and you won't have to cross over chips when you are connecting power and ground.

Secondly, make sure that a switch is located on the positive terminal of the battery pack *before* power is passed to the common bars on the breadboard, and don't just place the switch on the Vdd pin of the PIC MCU chip (Pin 1). If power and ground are available on other pins of the PIC MCU, you will discover that it will run, even if no connection to the Vdd and Vss pins exists. The clamping diodes on the I/O pins of the PIC microcontroller will allow current to pass through them as shown in Figure 3-14, and frequently this will

be enough to run the application. This is often called *parasitic power* and problems with it can be avoided if you control power at the source, rather than at the PIC MCU chip.

It's important to remember to remove gently the PIC MCU chip after programming and insert it before reprogramming. The easiest way I have found to remove chips is to slide a small flat-blade screw driver under the chip and twist; it will pop up from the breadboard. Before inserting the PIC MCU chip, make sure all the pins are aligned before pushing the chip onto the breadboard. You can even roll the chip on a table top to bend the pins inward. Bent pins can be straightened using a pair of pliers, but by taking a bit of care, you'll never have to.

C Language Features

In this section, I will present many of the advanced features of C that will help you more efficiently program your applications. These features allow you to more quickly create applications that are smaller, run faster, and handle data in different formats. Before I go on to the more advanced features of C, I would first like to address the topic of making your applications more readable.

Readability of a computer program is a function of five different parameters:

- Appropriate file name
- Header information
- Intelligent comments
- Adherence to formatting conventions
- Conventional statement usage

This list of parameters should not be surprising, but I am amazed at how few people think to use them when they are under a deadline. Actually, these rules often fall by the wayside in normal programming, causing a lot of grief when it's time to go back and try to fix something or even to use the code in another application.

The filename that I have used for the code presented in this book and the folder in which it is stored follows the convention:

```
"LanguageextFunction[ Iteration].languageext"
```

"Languageext" is "c" or "asm," depending on the language the application is coded in. I put the language type at the start of the file name for two reasons. The first is so I can use the folder name to differentiate between similar programs—those coded in C or assembler. The second reason is that the extension is not displayed in the folder window of Microsoft Windows, and I want to make sure I know at a glance what language the program is written in.

The main part of the file name (Function, in the previous convention line) is simply what the program does. Personally, I would like this to be arbitrarily long so I could create file names like "Single Button Debounce," but the 64-character limit imposed by the MPLAB® IDE simulator makes this difficult. So I stick to a single summarizing word. If I am going to produce multiple versions of the application, I keep track of them by a version code separated from the rest of the file name by a blank.

Remember to keep each application in its own separate folder. Multiple applications in the same folder (or even on your PC's desktop) are difficult to sort through to find specific applications. By keeping every application in its own folder, you will ease the workload required to find a specific application, help you manage the applications, and identify which ones are needed and which aren't.

Every program should have a header block with the following information:

- Summary of program operation
- Detailed description of what the program does
- Hardware information
- Author's name
- Date last updated (ideally with comments explaining what was done)

The purpose of most of this information is to allow you or somebody else to scan through and see if this is the application required. You might be looking through previously written applications for samples that use specific hardware, or for a particular application that requires a fix.

Intelligent comments explain the purpose of a block of code. How the code works should be self explanatory. The following is an example of what not to write:

```
i = i + 1;              // Increment Counter i
```

This comment is not helpful as you would assume that the reader knows C well enough to understand what the code is doing. Instead the comment should explain why the code statement is there. A better comment for the statement might be:

```
i = i + 1;              // Reference next
                        data point to be
                        processed
```

In university, I had a professor that took marks away for every comment in an application. His logic was that the code should be written in such a way that it is self-commenting. I can see his line of thinking, although it is probably the other extreme of voluminous, redundant commenting. I would suggest that you follow the rule that comments should only add information.

When you've been taught to program and when I have presented different statement types, they are formatted in very traditional ways. You may be unaware that C and most other high-level languages are very loose with their formatting, and an if statement like this one:

```
                          if (A > B) i = i + 1;
else

    j =
      j + 1;
```

is just as valid to the compiler as the traditionally formatted statement like this one:

```
    if (A > B)
        i = i + 1;
    else
        j = j + 1;
```

In C, the semicolon (;) is the ending of the statement. Until the semicolon is reached, the statement is parsed as if it were all on the same line, or formatted traditionally. You might think that a different statement format will make the application more efficient, but it will only make it frustrating for others to read and debug.

When I am formatting code, the rules that I use are as follows:

- Each statement is on its own line.
- Nested conditional execution is indented by four spaces.
- If statements are longer than the page (or screen) width, find a natural break and start them on the next line, indented four spaces from the start of the previous line.
- Comments all start on the same columns.

Simple, clear code that is written in a consistent format that is easier to read—even if it is formatted differently than you are used to—than code that does not follow a single format and uses different conventions for indentations and variable names. For example, the number of indentation spaces I suggest is based on what Visual C11 uses and I have become used to. You may find that two or eight spaces for each indentation makes more sense to you.

I do have one warning regarding indentations. Try to resist the urge to rely on tab characters when moving code or comments to a specific column; use space characters instead. Different editors and web browsers interpret tab characters differently, and by using tabs in your code, you will find that your neatly organized program will become virtually unreadable in other tools.

The final comment is similar to the previous one. As shown, C is a remarkably flexible language, and there are a lot of different ways of doing things. My recommendation is this: Do not try to be creative in how you code your applications. Instead be as conventional as you can. The space savings and speed improvements (if any) made by creatively organizing the code of your application will be offset by the increased difficulty reading and understanding how the code works. I put the caveat "if any" in parenthesis because the optimizer in the PICC Lite™ compiler is very sophisticated and very good at identifying situations where repeated code is integrated into a subroutine or even reorganizing the code to avoid redundant blocks of code.

To summarize this introduction, I quote the credo of the poor coder: "If it was difficult to write, it should be difficult to read." And the corollary is also true: If you work at making your code easy to read, you will find that it is much easier to write. The added benefit will be that your teachers, supervisors, and coworkers will appreciate your efforts over those programmers who do not try to make their code easier to read.

Experiment 24—Functions and Subroutines

PC/
Simulator

If you were to Google® the C programming language for a definition, you would discover that it is "procedurally based." This means that the language is designed to implement subroutines easily so that programming tasks can be broken down into smaller pieces for easier programming and reduced opportunities for errors. Subroutines in C will allow you to simplify your program, but I am more excited about C's ability to create *functions*, which offer much greater utility to your applications. Just so the terms are straight, I use the word *invoke* synonymously with *call* when I am referring to subroutines and functions.

Subroutines in C are declared in the following format:

```
SubroutineName(type Var1, ...)
{

//  Subroutine statements here

}  //  End SubroutineName
```

and must either be declared before the subroutine or a prototype has to be placed before the subroutine is used. This prototype consists of the subroutine declaration statement with a semicolon after the input parameters. Prototypes are often placed in include files (ending in .h) that are loaded at the start of the C source code. For the previous example, the prototype is as follows:

```
SubroutineName(type Var1, ...);
```

Personally, I put subroutines and functions above the mainline to avoid the need for prototypes. The only time I use prototypes is if I have previously compiled the subroutine or function and I'm linking them with the mainline. In these cases, I will load them all in a .h include file. One important reason to place subroutines and functions before the mainline is to avoid the bother of maintaining the prototypes if the functions themselves change. If you change the input or output parameters of a function and do not change the prototype, you will get an error message.

Functions are very similar to subroutines except they return a single value. The declaration statement for a function is very similar to the subroutine, but differences between the two relate to the declaration statement's ability to return data to the caller:

```
type FunctionName(type Var1, ...)
{

//  Function statements here

    return ReturnValue;

}  //  End Subroutines
```

To invoke a function, a statement in the following format can be used:

```
i = FunctionName(inputParameter, ...);
```

Or a subroutine invocation can be used:

```
SubroutineName(inputParameter, ...);
```

If a function is invoked as a subroutine, then the return value is ignored upon return.

To demonstrate the operation of a function in a program, I created cFactoral.c, which calculates the factorals of the first eight integers:

```
#include <pic.h>
/*  cFactoral.c - Calculate Factorals of
Integers

This Program Calculates the Factorals from 1 to
10.

myke predko
04.09.16

*/

int i, j;
int FactValue;

int Factoral(int Number)
{

    for (FactValue = 1; Number > 1; Number--)
        FactValue = FactValue * Number;

    return FactValue;

}  //  End Factoral

main()
{
```

```
    for (i = 1; i <= 8; i++)        // Return the
                                    Factoral of
                                    1 to 8
        j = Factoral(i);

    while(1 == 1);
}   //   End cFactoral
```

The program multiplies the integer by each decrementing values until it reaches one. This is not the traditional way of calculating factorals. In C, this function is normally shown as an example of a *recursive* function. I will not be discussing recursive functions in this book for two reasons: (1) It is beyond the scope of this book, and (2) the PIC16F684 microcontroller cannot support it easily.

When you simulate cFactoral.c, I suggest you display the Watch window for the four variables used in the program: i, j, FactValue, and Number. When you run the program, you will notice that each of the four variables, except for Number, is displayed at all times. Unless execution is within the Factoral function, the variable Number has the message, "Out of Scope." The reasons for this message will be explained in the next experiment.

Experiment 25—Global and Local Variables

PC/ Simulator

In C, you have the ability to redefine variable names, so long as they are declared within functions and subroutines and not outside the main statement. This feature is known as *local variables* and allows you to reuse common variable names (such as i, j, k, and n) without fear that the contents of a variable will be changed or corrupted by other functions and subroutines in the application. To demonstrate this capability, I created the program cPrime.c, which has the function PrimeCheck that uses the same variables as the mainline uses. When you run the application and try to monitor the values for the counter and flags (i and j, respectively in each function), you will discover that MPLAB IDE has trouble monitoring the correct value for the variables and may even stop single-stepping correctly!

```
#include <pic.h>
/* cPrime.c - Find the Primes in the first 200
Integers

This Program works through each of the prime
numbers from 1 to 200 and indicates which ones
are prime by attempting to divide every integer
from 2 to ((Number / 2) + 1).

This program is designed to run under the MPLAB
IDE Simulator
  Only.
```

```
myke predko
04.09.16

*/

int PrimeCheck(int Value)
{
int i, j;

    j = 1;                          // Assume the
                                    value is prime
    for (i = 2; (i < ((Value / 2) + 1)) && (0 !=
j); i++)
        if (0 == (Value % i))       // If Divisible
                                    by i, Input
                                    Value is
            j = 0;                  // NOT Prime

    return j;                       // Return the
                                    Prime Flag

}   //   end PrimeCheck

main()
{
int i, j;

    for (i = 1; i <= 200; i++)      // Test Every
                                    Integer
        j = PrimeCheck(i);          // Return != 0
                                    (true) if i
                                    is Prime

    while (1 == 1);                 // Loop Forever
}   //   End cPrime
```

The problems are caused by the confusion the MPLAB IDE simulator has regarding *which* i or j it should be displaying. This doesn't mean the program isn't working properly, but that MPLAB IDE cannot properly display the values of i and j.

In this program the i and j local variables that are specific to each function are unique to one another as well as to any variables that are declared outside all the functions of the program. The variables that are declared outside all the functions are known as *global variables* and they can be accessed within any function. When debugging functions, you will probably want to change the variables that are local to the function into globals so that you can debug them using the full features of the MPLAB IDE simulator. From a programming quality perspective, this is not necessarily a bad idea; by testing and debugging individual functions before integrating them into an application, you to ensure that they are correct before they are used and simplify your application debug.

The big question you are probably asking yourself from this experiment is "When should I use local versus global variables?" When I program in C for the PC, I try to make all my variables local; but I use global variables only for the values that are accessed across the application and/or that would become cumbersome if their parameters need to be passed to other functions. I do not follow this rule when I am programming the PIC® microcontroller, because the PIC16F684 processor architecture does not allow for *parameter strings* of unlimited length to be implemented efficiently. Ideally, you should keep your input and output parameters to 16 bits. Additionally, there could be situations where multiple variables with the same name are declared but only very rarely accessed, and therefore there may never be any danger of them appearing multiple times in *nested subroutines* (subroutines and functions that others call or are called by other subroutines or functions). For these reasons, I tend to declare all the variables as globals, except for counters (i, j, and so on) and temporary values.

Experiment 26—Defines and Macros

In Section 2, when I introduced you to variable declarations, I also noted that constant numeric values could be declared as well. The constant values that were represented by the labels would be substituted in anytime the label was encountered. In this way, commonly used constant values could be stored as an easy-to-remember string that should help the readability of the program. I did not focus on the constant value declaration. I try to avoid using it in my programming, because I believe that its declaration can be easily confused with variable declarations and because it does not encompass all the different types of data that are used in programming.

To declare constant values of all types, I use the #define directive built into the C programming language. The format for the #define directive is as follows:

```
#define Label[(Argument,...)] String
```

When a label that has been declared as a constant is encountered, the *numeric value* associated with the label is returned. When a label declared in a #define is encountered, the *string* that is associated with the label is inserted into the program. A subtle but very important difference exists between the two, as I will show in this experiment.

The string returned by the define label does not have to be numeric only; it can be alphabetic characters or even C program language statements like if or for. These substitutions are allowed because the define substitutions take place in the compiler's *preprocessor*. That is, the strings are substituted into the program source code before compilation starts.

Further enhancing the usability of the #define directive is the ability to specify parameter strings (the arguments in the previous directive description). These parameter strings are substituted in the #define string, allowing you to customize the string for a particular situation. The addition of parameter strings allows the #define to behave as a macro and allows you to easily enhance the operation of your application.

Macros are sections of code that are commonly used, but deemed inappropriate as their own subroutines or functions. In this experiment's demonstration application, cDefine.c, I show how #defines are used to define a constant value as well as, with small pieces of code, to provide simple functions that may be required multiple times in an application. It isn't unusual to refer to constant values, declared by using the #define directive and code strings that are added to the application as *macros*.

```
#include <pic.h>
/* cDefine - Demonstrate the Define Directive
for Equates and Macros

    The C "Define" directive can be used for both
equating values as well as providing basic macro
services.
    These capabilities are demonstrated in this
program.

    Hardware Notes:
    PIC16F684 running at 4 MHz in Simulator
    Reset is tied directly to Vcc via Pullup/
Programming Hardware


    Myke Predko
    04.09.26

*/


// Defines
#define ConstantValue       0x1234

#define HighByte(Value)     Value / 0x0100

#define LowByte(Value)      Value % 0x0100

#define UpperCase(Character)\
    ((Character >= 'a') && (Character <= 'z')) ?\
    Character + 'A' - 'a': Character


main()
{
int Number1, Number2;
char Char1, Char2, Char3;

    Number1 = 0x0123;           // Standard
                                   Initialize
    Number2 = ConstantValue;    // Initialize
                                   with Define

    Char1 = 'A';                // Character
                                   Initialize

    Char2 = 'b';
    Char3 = '7';

    Number1 = HighByte(Number2); // Get High
                                    Byte of
                                    Number2

    Number1 = LowByte(Number2);  // Get Low Byte
                                    of Number2

    Char1 = UpperCase(Char1);    // Test
                                    "UpperCase"
                                    Macro

    Char2 = UpperCase(Char2);    // Only Value to
                                    be Changed

    Char3 = UpperCase(Char3);

    while (1 == 1);              // Finished,
                                    Loop Forever

} // End cDefine
```

The strange format of the UpperCase macro will require some explanation. The two backslashes at the end of the first two lines are line continuance indica-tors; when these characters are encountered, the pre-processor moves the next line to the end of the current one. This allows for multiple-line macros or for situations like this, where you could cram the macro string all onto one line. But by using the line continuance character, the line can be read more easily. Secondly, the macro has a question mark (?) and a colon (:) character inserted into it, neither of which seem to make any sense.

The question mark and colon are part of an alternative conditional execution statement, which has the following form:

```
CondExpression ? TrueExpression :
FalseExpression
```

When "CondExpression" is evaluated if it is true (not zero, which is what "true" means), then "TrueExpression" is evaluated; else "FalseExpression" is evaluated. This statement is useful for situations where you want to set a variable to a specific value, as in the code:

```
if (CondExpression)
        i = TrueExpression;
    else
        i = FalseExpression;
```

Personally, I do not like using this alternative form of the if statement except in macros, as its operation is not always obvious (e.g., the colon can be easily missed or misread as a semicolon), it does not allow multiple conditionally executing statements, it is not logically nested with other statements, and it often ends up running over a single line, which makes its operation even more cryptic. It is, however, an efficient way of adding a conditional operation to a macro, just as I have done in this experiment.

Before ending, I should note that the #define directive can be used for renaming I/O pins in PIC MCU applications. Rather than accessing an I/O pin as the generic label, you can assign a label to it using the #define directive. For example, if you had an LED wired to RC2, you could declare it using the following statement:

```
#define LED RC2
```

Now, when the PICC Lite compiler encounters it, it replaces every instance of "LED" with "RC2." This is a useful trick for making your applications easier to read and write.

Experiment 27—Variable Arrays

Array elements in C start at an index of zero, not one as you might expect. You could create your programs with one extra element and start your index at one, but this would not follow conventional C programming and would make it difficult for others to read your program. I should point out that many other languages (including various flavors of BASIC) all have their first array element as zero, so this is a convention that is transferable to other languages.

To define an array variable, the variable with its same type is specified along with the number of elements of the array. Along with this, the initial values of the array can be optionally specified by placing them within braces as I show in the sample array variable declaration statement shown here:

```
int ArrayVariable[4] = {77, 121, 107, 101};
```

A common program used to demonstrate the operation of arrays is the *bubble sort* (see Figure 4-2). In this simple program, each element of an array is read and compared to the one next to it (this is the bubble). If one element is greater than the next, then the two are swapped using this simple three-statement sequence:

```
Temp = SortArray [i];          // Save the
                               //   Current Value
SortArray[i] = SortArray[i + 1]; // Store Next
                               //   Value in
                               //   Current
SortArray[i + 1] = Temp;       // Put Original
                               //   Current Value
                               //   in Next
```

These three statements, which require an extra Temp variable, are a fast and efficient way of swapping the contents of two variables, not just array variable elements.

Array addressing is a common feature of most programming languages and provides you with a mechanism to allow you arbitrary variable data within your application. Array addressing could also be called *indexed addressing*, and it utilizes an arithmetic value to specify a certain variable element in an array variable. Arrays allow you to handle more data than could be handled normally in an application and to write applications that would be very difficult without this capability.

Using arrays in your program is actually quite easy. In Figure 4-1, I show how I visualize an array—as a series of memory elements that are given a single variable name. Each element is accessed by specifying an index value. The index can be a constant, a variable, or an arithmetic expression. For example, each of the three statements below will load the variable i with the contents of the fourth element of ArrayVariable:

```
i = ArrayVariable[3];     // Index is a
                          //   Constant
i = ArrayVariable[j];     // Index is a
                          //   Variable (j = 3)
i = ArrayVariable[j * 1]; // Index is a
                          //   Expression = 3
```

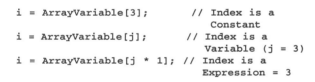

Variable Name

Index

Declared As:
int VariableName[4];

Figure 4-1 *Array*

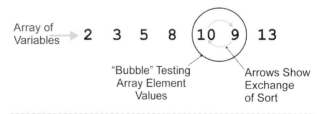

Array of Variables   2   3   5   8   10   9   13

"Bubble" Testing Array Element Values

Arrows Show Exchange of Sort

Figure 4-2 *Bubble sort*

The most basic bubble sort program that I can come up with is as follows:

```
#include <pic.h>
/*  cbSort.c - Simple C Program Bubble Sort

   This Program is a simple "Bubble Sort" of 20
Values.

myke predko
04.08.03

   Hardware Notes:
   This Program has been written to run in the
MPLAB IDE Simulator ONLY.

*/


int i, j, Temp;
char SortValues[20] ={33, 54, 42, 120, 37, 5,
99, 105, 8, 21, 100, 33, 41, 77, 81, 69, 13, 86,
51, 90};

main()
{

 for (i = 0; i < sizeof(SortValues); i++)
  for (j = 0; j < (sizeof(SortValues) - (i + 1));
     j++)
   if (SortValues[j] > SortValues[j + 1]) {
       Temp = SortValues[j];
       SortValues[j] = SortValues[j + 1];
       SortValues[j + 1] = Temp;
   }  // fi

   while(1 == 1);               // FinishedLoop
                                   Forever

}  // End cbSort
```

So far I have discussed single-dimensional arrays (arrays with one index), but you can also implement multidimensional arrays in C by simply adding a second set of square brackets as I do in the following example program. Note that when I came up with the initial values, I kept the values together according to the row in which they appear.

```
#include <pic.h>
/*  cMulArray.c - Multiple Dimensional Array

   This Program demonstrates how a two
dimensional array could be implemented to
simplify Multiplication.

myke predko
04.09.07

   Hardware Notes:
   This Program has been written to run in
   the MPLAB IDE Simulator ONLY.

*/
```

```
int i, j, Product;
int MulArray[5][5] = {0, 0, 0,  0,   0,   // Zero
                                          Row
                      0, 1, 2,  3,   4,   // One
                                          Row
                      0, 2, 4,  6,   8,   // Two
                                          Row
                      0, 3, 6,  9,  12,   // Three
                                          Row
                      0, 4, 8, 12,  16};  // Four
                                          Row

main()
{

    i = 3;   j = 4;
    Product = MulArray[i][j];

    i = 4;   j = 3;
    Product = MulArray[i][j];

    i = 0;   j = 2;
    Product = MulArray[i][j];

    i = 4;   j = 2;
    Product = MulArray[i][j];

    while(1 == 1);

}  // End cMulArray
```

If you are looking at the cMulArray program, you might think that it is a clever way to avoid having to use a complex multiplication routine. Unfortunately, when the compiler is calculating the address, the row index (or the first dimension) is multiplied by five and then added to the column index (or the second dimension) to calculate the index element to be accessed. Another way of putting this is that all arrays are actually single dimensional, and the multidimensional array format is simply provided as a way to avoid requiring the programmer to calculate the address.

Before using a multidimensional array, think about how necessary it is to your application. Two- and three-dimensional arrays can use a lot of space that could be used for other functions. For the cMulArray program, the MulArray variable takes up 50 bytes (five rows times five columns times two bytes for each integer element). If we were to extend it to three dimensions, the total memory requirements would become 250 bytes, which is much more than is available to the application.

From an academic perspective, the bubble sort algorithm is probably the least efficient algorithm for sorting data. Other sorting methods are much more efficient, but they require resources that are generally not available in the PIC MCU processor architecture, and implementing them can be quite difficult. In any case, for most PIC MCU applications, the bubble sort is good enough; in fact for the rest of this chapter, I will look at different ways of implementing it using different C language features.

Experiment 28—Structures and Unions

When processing data or handling multiple data types, simple arrays are often not the answer to the problem. What is needed is some way of combining multiple pieces of disparate data in the same variable. There will also be cases where single pieces of data may be divided in different ways for different data groups. The solution to this problem is actually quite elegant. In the C language you can define your own variable types with specific parameters; you even have the ability to represent data in more than one way.

The *struct language statement* is used to define a variable of arbitrary type that can contain multiple pieces of data. The format of the statement is:

```
StructureName struct {
  ctype Variable1Name;
  ctype Variable2Name;
  ...
}  StructureVariable;
```

When using the struct statement, the Structure-Name and StructureVariable parameters are optional (although both cannot be missing at the same time). StructureName is used as the type for other variables:

```
StructureType Variable;
```

Structures are initialized in declaration statements in the same way arrays are initialized, with values being applied to the appropriate variable within the structure in the same order as the structure is declared:

```
StructureType Variable = {Variable1NameValue,
Variable2NameValue...};
```

Finally, variables within a structure are accessed by connecting them to the structure name with a period as in the following statement:

```
SturctureName.Variable1Name = 47;
```

As I said at the start of the experiment, structures allow multiple pieces of data to be included together. A typical application could be in a database listing names and personal information. The structure declared below is an example of something that could be used by a store to keep track of its customers:

```
CustomerInfo struct {
    char Name[64];              // Customer Name
    char Address[128];          // Address
    int HomePh[10];             // Home Phone
                                   Number
    int BusPh[10];              // Work Phone
                                   Number
    char email[32];
    CDate LastSale;             // Date of Last
                                   Sale
    int LastSameAmt;            // Amount of Last
                                   Sale
};
```

With this information, a store can find out who its best customers are, who the most recent customers are, or the store can query other parameters. Once the store has found the parameter information it wants, it has all the other personal information immediately available in the structure. In CustomerInfo, note that I use the previously defined struct CDate as a type of one of the variables in the struct; you can imbed structs into other structs. This is an effective way to make your application easier to read and work with.

In the previous example, you might be thinking that the structure could easily be built for a single customer, but how would it work for multiple customers? Like numbers can be built into arrays, so can structures. For example, if a store had 2,000 customers, an array of CustomerInfo with 2,000 elements could be defined using the statement:

```
CustomerInfo CustomerDatabase[2000];
```

In the following program, I have defined a 20-element array of a structure in which a value and the initial array position are stored and then sorted using the same bubble sort method that was used in the previous experiment:

```
#include <pic.h>
/*  cstrctSort.c - Structure Based C Program
Bubble Sort

  This Program is a simple "Bubble Sort" of
  20 Values using an array of Structures

myke predko
04.08.03

  Hardware Notes:
  This Program has been written to run in
  the MPLAB IDE Simulator ONLY.
```

```
*/

unsigned i, j;
char InitValues[20] ={33, 54, 42, 120, 37, 5,
99, 105, 8, 21, 100, 33, 41, 77, 81, 69, 13, 86,
51, 90};

struct DataStruct {
    unsigned InitPos;      // Initial Sort Position
    char Value;
} SortArray[20], Temp;    // 20 Element Structure
                             Array

// DataStruct Temp;        // Temporary Value as
                             Structure

main()
{

    for (i = 0; i < sizeof(InitValues); i++) {
     SortArray[i].InitPos = i;
     SortArray[i].Value = InitValues[i];
    }  // rof

    for (i = 0; i < sizeof(InitValues); i++)
     for (j = 0; j < (sizeof(InitValues) - (i + 1));
         j++)
      if (SortArray[j].Value > SortArray[j +
         1].Value) {
       Temp = SortArray[j];
       SortArray[j] = SortArray[j + 1];
       SortArray[j + 1] = Temp;
      }  // fi

    for (i = 0; i < sizeof(InitValues); i++)
     Temp = SortArray[i];

    while(1 == 1);           // FinishedLoop Forever

}  // End cstrctSort
```

In cstrctSort, note that I treat the entire structure like a simple variable and copy it in the same manner as I would with a simple variable. This feature helps to keep the complexity of an application to a minimum.

There will be times that you will want to share data in a structure (or access it as if it were a different variable). This is done using the *union statement*. The union statement has the same syntax as the struct statement and produces a new data type. But instead of each value within the union being unique, each is located at the same memory location. For example, if you had a 16-bit variable (ctype int) and wanted to access each byte individually as separate bytes or together as a 16-bit integer, you could define a structure breaking up and a union as:

```
lowHigh struct {         // 16 Bit value as two 8
                            Bit Values
   char L;
   char H;
};

union {                  // Variable as 16 Bit or 2x
                            8 Bit Values
   lowHigh b;
   int i;
} n;
```

Normally, when you are reading and writing the full 16 bits, you would define the integer n as follows:

```
n.i = 12345;             // Save 16 Bit Value in "n"
```

To access the individual bytes of n, you could use the following statements:

```
ByteVariable = n.b.H;
ByteVariable = n.b.L;
```

which seem very cumbersome, but they are equivalent and often much more efficiently implemented than these statements:

```
ByteVariable = n.i / 256;  // Return the High
                              Byte of "n"
ByteVariable = n.i % 256;  // Return the Low
                              Byte of "n" as
                              Modulo 256
```

Structures and unions are not programming tools that you will use a great deal to program microcontrollers in C; they are usually required for large-system programming. Despite the apparent lack of applicability, they are useful tools and keeping them in the back of your mind could make your programming more efficient or easier to read and follow.

Experiment 29—Pointers and Lists

PC/ Simulator

Two programming concepts seem to strike terror in the hearts of computer science students: working with pointers and working with lists. The concepts themselves are not terribly difficult to understand, but implementing them in a program and debugging the program can be difficult and frustrating; so difficult and frustrating, in fact, that many people question their understanding of the concepts. Philosophically, using pointers in programs is currently out of favor. For example, Java and Visual Basic are designed to hide the operation of data pointers. But in C, understanding the basics of pointers is critical to understanding how to manipulate character strings. Lists are generally considered a "mainframe" computer programming concept for handling large amounts of data, but they do have some useful properties that you may want to consider in microcontroller applications.

If you were to guess that a *pointer* is a variable that points to the location of a data object (variable, struct, or union), you would be correct. Like other variables, pointers are typed, with the asterisk/splat character (*) used in the declaration after the type to indicate that the variable is a pointer to a specific type of data. For example, a pointer to an integer would be declared as:

```
int* IntegerPointer;    // Declare an Integer
                           Pointer
```

The ampersand (&) character is used to indicate the address of a variable. To store the address of a variable in IntegerPointer, the following statement would be used:

```
    IntegerPointer = &i; // IntegerPointer
                            points to =variable
                            <\#34>i<\#34>
```

The value in the variable indicated by the pointer can be accessed by placing an asterisk/splat in front of the variable in a statement. In both of the following statements, the contents of the integer i (which is pointed to by IntegerPointer using the previous statement) are incremented.

```
    i = i + 1;
    *IntegerPointer = *IntegerPointer + 1;
```

When pointers are used with structures, the values within the structures are accessed by using an arrow made up of a dash and a greater than sign (->.) instead of the simple dot, or period, of a struct variable. Creating a struct variable with a pointer to it and assigning a value within the structure using the pointer would be accomplished using the following statements:

```
struct PixelPoint {      // Pixel information
                            structure
    int x;
    int y;
    char color;
} PointValue;

PixelPoint* ptrPixel;    // Pointer to Pixel
                            information structure

ptrPixel = &PointValue; // Point to the Pixel
                           Structure Variable
```

```
ptrPixel->x = xPosition; // Assign a Value to
                            the "X" coordinate
```

Pointers and pointers to structures are often used as parameters to functions. By passing a pointer rather than a value, the original value can be changed and much less data can be transferred, resulting in faster program execution and smaller variable memory requirements.

This is really everything there is to know about pointers—in concept they are quite simple and easy to understand. Debugging programs using pointers is generally difficult because few simulators and emulators provide pointer information in any kind of meaningful context (such as which variable they are pointing to). To effectively debug a program that has pointers in it, two things are required: a good understanding of how variables are stored (and where, most likely, they are in the variable memory) and some imagination in trying to understand how the code works. In the next experiment, I will show how pointers are used with characters and strings, but for now you know all the basics and probably more than you will need for the foreseeable future.

Lists consist of structures with pointers, which point to the structure built into them. They allow each instance of the struct data type to point to another instance and to form a linear chain of structs. The two types of lists most commonly used in programming (see Figure 4-3) are singly linked lists and doubly linked lists. The advantage of the doubly linked list is that it allows movement in either direction easily, while the singly linked list allows movement in only one direction. The end of the lists are usually marked with a null (value 0) or −1. Other ways exist for organizing structures.

For this experiment, I wanted to sort the same selection of numbers as used in the previous two experiments. Here we will use a singly linked list. The code itself, which follows, is not much more complex than the code of the previous experiments, but I experienced a number of challenges debugging and confirming the application's operation that would make this experiment unreasonably difficult for an inexperienced programmer.

```
#include <pic.h>
/*  cpstrctSort.c - Structure Based C Program
Bubble Sort

   This Program is a simple "Bubble Sort" of
    20 Values using pointers to a Linked List of
    Structures

myke predko
04.08.03
   Hardware Notes:
    This Program has been written to run in
```

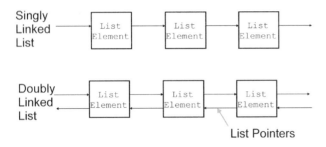

Figure 4-3 *Lists*

```
     the MPLAB IDE Simulator ONLY.

*/


unsigned i;
char InitValues[20] ={33, 54, 42, 120, 37, 5,
99, 105, 8, 21, 100, 33, 41, 77, 81, 69, 13, 86,
51, 90};

struct DataStruct {
 struct DataStruct* NextD;    // pointer to the
                                Next Element
 char InitElement;            // Initial Element
                                Value
 char Value;
} SortArray[20], Start;       // 20 Element
                                Structure
                                Array/Start

struct DataStruct* Previous;  // Previous Pointer
                                Value
struct DataStruct* Current;   // Current Pointer
                                Value
struct DataStruct* Ending;    // Pointer to the
                                Last Element

main()
{

 Start.NextD = &SortArray[0];// Point to the
                                Start of the
                                List
  for (i = 0; i < sizeof(InitValues); i++) {
    SortArray[i].NextD = &SortArray[i+1];
    SortArray[i].InitElement = i;
    SortArray[i].Value = InitValues[i];
  } // rof
SortArray[19].NextD = (struct DataStruct*) -1;
Ending = SortArray[19].NextD;

 while (Start.NextD != Ending) {
  Previous = &Start;  // Point to the Start of
                         the List
  Current = Start.NextD;  // Point to the First
                             Element
  while (Current->NextD != Ending)
   if (Current->Value > Current->NextD->Value) {
    Previous->NextD = Current->NextD;
    Current->NextD = Current->NextD->NextD;
    Previous->NextD->NextD = Current;
    Previous = Previous->NextD;
   } else {         // Nothing Changed, Goto Next
     Previous = Current;
```

```
      Current = Current->NextD;
         } // fi
   Ending = Current;   // Move Back the Pointer
 } // elihw

   while(1 == 1);          // FinishedLoop Forever

} // End cpstrctSort
```

Although building the singly linked list is quite simple and can be found in the first for loop, sorting the list by moving pointers is not exactly obvious. In order to sort the list, I used the Previous pointer as an anchor to the list and then moved the pointers as shown in Figure 4-4. When I developed the code, I made a diagram similar to Figure 4-4 to make sure that I understood how the pointers were moving the two list elements as in the bubble sort swap. Despite making this diagram, it was still a challenge to get the program to work properly. I ended up making a spreadsheet with every element and each pointer and kept track of how they changed and where they ended up pointing. Overall, writing and debugging this application took me about four hours, whereas the other sort applications presented took less than an hour.

The purpose of this experiment is to demonstrate how pointers can be easily created and point to different pieces of data. Lists are generally best suited for large applications with many data elements that might be reordered, inserted, or deleted at random. At first glance lists may not seem to be appropriate for microcontroller applications, but there are cases where they are useful, such as allowing easy editing of a robot program sequence. I suggest you consider the list concept when you have a situation where adding, removing, and moving data elements is required.

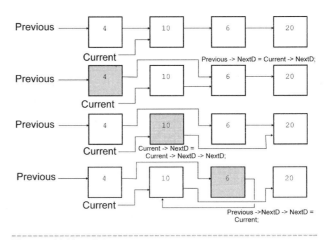

Figure 4-4 *List sort*

Experiment 30—Character Strings

PC/ Simulator

Although working with structs and unions can be difficult, working with pointers and character strings referenced by pointers is quite easy if you follow the basic conventions outlined here. Unlike other languages (such as BASIC), C does not natively handle strings of data. Although it can handle individual ASCII characters, which are treated like eight-bit numbers, each ASCII character is stored in an array element to create a character string. To ease the workload caused by copying each character each time it is used, the starting address of arrays containing ASCII character strings is normally used to reference the entire string. These methodologies combine to make working with character strings in C more complex than in other languages you may have used when learning programming. The same attributes and methodologies make character strings much more efficient when you are comfortable working with them.

When a number of ASCII characters are combined in a string, they normally end with a null (zero) character to form what is known as an "ASCIIZ string." The placement of the null character at the end of the string is automatic when it is enclosed in double quotes like this:

```
char MyName[12] = "myke predko"; // Define a
                                    String with
                                    my name
```

To declare enough space for the string and the null character at the end, you will have to count the number of expected characters of the string and add one to it.

When a string variable is referenced, it is referenced by the variable name and it is a pointer to the first character in the variable array.

```
char* NewName1, NewName2;

    NewName1 = MyName;
    NewName2 = &MyName[0];   // NewName1 ==
                                NewName2
```

Each individual character in a string is accessed as an array element. Individual ASCII characters can either be represented as decimal, binary, or hex numbers or the actual characters in single quotes. In the following code snippet, the third character of MyName is compared against upper case 'K'.

```
    if (NewName1[2] == 'K')
// Statements executed if Equal to Uppercase 'K'
    else
// Statements executed if Not Uppercase 'K'
```

Along with decimal representations of ASCII characters and the actual ASCII characters in single quotes, a number of special characters (preceded by a backslash [\]) are also available in C. These can be placed either in single quotes as a single character or as part of a string listed in Table 4-3 and found in the "For Consideration" part of this section.

Depending on the circumstances, ASCII strings can be located in either program memory of variable memory. To allow the compiler to differentiate which is which, the "const keyword" should be put before the char in the pointer declaration:

```
const char* ConstPointer;
```

The const keyword will make the pointer 16 bits long (instead of the standard eight bits, which can access only variable memory). The extra bits are used to indicate the location of the pointer data as well as the location within the specified memory. My rule of thumb regarding the use of the const keyword is to let the compiler indicate when it is needed; most of the time it is not required.

To demonstrate how strings of characters are implemented and manipulated as arrays of single characters, I have modified the sort program to sort an array of pointers to strings. I created the stringCompare and stringCopy functions, which perform byte-by-byte operations on the bytes in the strings, because C does not have this built-in ability.

```
#include <pic.h>
/* cstrmSort.c - Simple C Program Bubble Sort

   This Program is a simple "Bubble Sort" of
   10 Four Character Strings by moving them
   in the sort array.  The string functions
   were written by Myke Predko.

myke predko
04.08.05

   Hardware Notes:
   This Program has been written to run in
   the MPLAB IDE Simulator ONLY.

*/


unsigned i, j;
char* SortValues[10];
char Temp[5];
char name0[5] = "myke";
char name1[5] = "toni";
char name2[5] = "lori";
char name3[5] = "cara";
char name4[5] = "june";
char name5[5] = "lynn";
char name6[5] = "mary";
char name7[5] = "myra";
char name8[5] = "sara";
char name9[5] = "kira";


int stringCompare(char* String1, char* String2)
{                         // Compare two Strings
int i;
int j = 0;

   for (i = 0; ('\0' != String1[i]) && (0 == j);
       i++)
     j = String1[i] - String2[i];
                      // j < 0 if String1 < String2
                      // j = 0 if String1 = String2
                      // j > 0 if String2 > String2

    return j;         // Return Compare Result

} // End stringCompare

stringCopy(char* String1, char* String2)
{                         // Copy String2 into String1
 int i;

 String1[0] = String2[0]; // Copy the First Byte
 for (i = 1; '\0' != String1[i - 1]; i++)
   String1[i] = String2[i];

} // End stringCopy

main()
{

   SortValues[0] = name0;      // "myke"
   SortValues[1] = name1;      // "toni"
   SortValues[2] = name2;      // "lori"
   SortValues[3] = name3;      // "cara"
   SortValues[4] = name4;      // "june"
   SortValues[5] = name5;      // "lynn"
   SortValues[6] = name6;      // "mary"
   SortValues[7] = name7;      // "myra"
   SortValues[8] = name8;      // "sara"
   SortValues[9] = name9;      // "kira"
```

```
   for (i = 0; i < 10; i++)
    for (j = 0; j < (10 - (i + 1)); j++)
     if (stringCompare((char*) SortValues[j],
        (char*) SortValues[j + 1]) > 0) {
      stringCopy(Temp, SortValues[j]);
      stringCopy(SortValues[j], SortValues[j + 1]);
      stringCopy(SortValues[j + 1], Temp);
     } // fi

   while(1 == 1);          // FinishedLoop Forever

} // End cstrmSort
```

In the program you'll see that I first initialized arrays for each name and then passed the address of each name to the SortValues variable. In standard C, you could initialize a pointer variable to a series of double quoted ASCIIZ strings, but this is not possible in the PICC Lite compiler. I suspect that the reason why it's not possible is due to the memory organization of the PIC MCU; the initial values are stored in program memory, which is completely separate from the variable memory. Before a variable pointer can be assigned to a value, the value must be stored in a variable.

After writing cstrmSort, I noticed that I could improve the performance of the sort routine by simply moving the pointers to the strings in the sort instead of copying the strings themselves. This program is listed here:

```
#include <pic.h>
/* cstrpSort.c - Simple C Program Bubble Sort

   This Program is a simple "Bubble Sort" of
   10 Four Character Strings with moving
   strings.

myke predko
04.08.04

   Hardware Notes:
   This Program has been written to run in
   the MPLAB IDE Simulator ONLY.

*/


char* SortValues[10];
char* Temp;
char name0[5] = "myke";
char name1[5] = "toni";
char name2[5] = "lori";
char name3[5] = "cara";
char name4[5] = "june";
char name5[5] = "lynn";
char name6[5] = "mary";
char name7[5] = "myra";
char name8[5] = "sara";
char name9[5] = "kira"
```

```
int stringCompare(char* String1, char* String2)
{                       // Compare two Strings
 int i;
 int j = 0;

 for (i = 0; ('\0' != String1[i]) && (0 == j); i++)
  j = String1[i] - String2[i];
                       // j < 0 if String1 < String2
                       // j = 0 if String1 = String2
                       // j > 0 if String2 > String2

 return j;        // Return Compare Result

}  // End stringCompare

main()
{
unsigned i, j;

 SortValues[0] = name0;      // "myke"
 SortValues[1] = name1;      // "toni"
 SortValues[2] = name2;      // "lori"
 SortValues[3] = name3;      // "cara"
 SortValues[4] = name4;      // "june"
 SortValues[5] = name5;      // "lynn"
 SortValues[6] = name6;      // "mary"
 SortValues[7] = name7;      // "myra"
 SortValues[8] = name8;      // "sara"
 SortValues[9] = name9;      // "kira"
```

```
   for (i = 0; i < 10; i++)
     for (j = 0; j < (10 - (i + 1)); j++)
        if (stringCompare((char*) SortValues[j],
            (char*) SortValues[j + 1]) > 0) {
         Temp = SortValues[j];
         SortValues[j] = SortValues[j + 1];
         SortValues[j + 1] = Temp;
        }  // fi

   while(1 == 1);        // FinishedLoop Forever

}  // End cstrpSort
```

The updated version of this sort is recommended and demonstrates the important functions of pointers that you should be comfortable with in order to program in C. To recap, these functions include the following:

- Declaring pointers
- Assigning pointers to the start of variable strings
- Accessing (reading and writing) pointer values
- Defining ASCIIZ strings
- Accessing specific characters in ASCIIZ strings as array elements

Experiment 31—Library Functions

PC/ Simulator

In the previous experiment, I showed you how to compare and copy each byte in an ASCIIZ variable array. Knowing how to create these functions in C is important, but it is not always necessary for most applications. As part of the C language specification, you can take advantage of a collection of functions that will take some of the load off you when you are programming. These functions (along with the PICC Lite compiler C language statement definition) are included in the "For Consideration" part of this section and are part of the "C run-time" library that is linked with your application at run time.

For example, I replaced the stringCompare and the stringCopy functions in the previous experiment's code

with strcmp and strcopy, respectively, and added the string.h header to get the following:

```
#include <pic.h>
#include <string.h>
/*  cstrfSort.c - Simple C Program Bubble Sort

   This Program is a simple "Bubble Sort" of
   10 Four Character Strings by moving them
   in the sort array.   Comparison and Movements
   are provided by Library Functions.

myke predko
04.08.05

   Hardware Notes:
   This Program has been written to run in
   the MPLAB IDE Simulator ONLY.

*/


unsigned i, j;
char* SortValues[10];
char Temp[5];
char name0[5] = "myke";
char name1[5] = "toni";
char name2[5] = "lori";
char name3[5] = "cara";
char name4[5] = "june";
char name5[5] = "lynn";
char name6[5] = "mary";
```

```
char name7[5] = "myra";
char name8[5] = "sara";
char name9[5] = "kira";

main()
{
    SortValues[0] = name0;       // "myke"
    SortValues[1] = name1;       // "toni"
    SortValues[2] = name2;       // "lori"
    SortValues[3] = name3;       // "cara"
    SortValues[4] = name4;       // "june"
    SortValues[5] = name5;       // "lynn"
    SortValues[6] = name6;       // "mary"
    SortValues[7] = name7;       // "myra"
    SortValues[8] = name8;       // "sara"
    SortValues[9] = name9;       // "kira"

    for (i = 0; i < 10; i++)
      for (j = 0; j < (10 - (i + 1)); j++)
        if (strcmp((char*) SortValues[j],
            (char*) SortValues[j + 1]) > 0) {
          strcpy(Temp, SortValues[j]);
          strcpy(SortValues[j], SortValues[j + 1]);
          strcpy(SortValues[j + 1], Temp);
        } // fi

    while(1 == 1);       // FinishedLoop Forever

} // End cstrfSort
```

The resulting program takes up just about as much space as the previous one and runs in roughly the same amount of time, which means that my routines are similar in efficiency to the provided library routines. The big difference is that the first program was shorter and required less effort to debug. (I assume that the strcmp and strcpy functions work properly and this assumption is extended to all the built-in functions of the runtime library.) For the most part, I would recommend that the run-time library functions are used in your C programs wherever possible.

One caveat exists to the previous statement and that is the linker brings in only the library functions that are called by the application. If a function is not called, then the function is not linked into the final application hex file. I'm pointing this out because in small devices, like the PIC16F684, you can very quickly use up all of your available space with libraries and application code. It is important to remember to balance the number of built-in and user-written functions so you don't use up the program memory available to your application.

For Consideration

For modern systems, C is the programming language of choice because it is available for a wide range of systems and processors (including the PIC microcontroller). This ubiquity requires anyone who is planning on developing an application for processing systems to have at least a passing knowledge of the language. The following information is specific to the HT-Soft PICC (including the PICC Lite) compiler:

Application mainline/entry point is the main function:

```
main()
{ // Application Code

    :                          // Application Code

} // End Application
```

Functions are defined in a similar format to the main function:

```
Return_Type Function( Type Parameter [, Type
Parameter..])
{ // Function Start

    :                          // Function Code

    return value;

} // End Function
```

The function can be placed before or after the main function (not inside), but if it is placed afterward, there must be a function prototype, like this one, before main:

```
Return_Type Function( Type Parameter [, Type
Parameter..]);
```

PIC MCU configuration register fuse values can be specified using the "__CONFIG(x)" define (found in the pic.h include file). The specified values are ANDed together to form a complete configuration register value. I recommend that a value for each bit be specified to ensure proper operation of the microcontroller. Table 4-1 lists the PIC16F684's configuration fuse bits, the HT-Soft defined constant's option values, the AND values, as well as the Microchip defined assembly language constants. I have noted also recommended values when I felt it was important.

Data declarations are very consistent, starting with the constant declaration:

```
const int Label = Value;
```

Variables declared inside a function (including the *main* variable) are known as *local variables*, are accessible only within that function, and may be different upon entry into the function. Variables declared outside of all functions are termed *global*, as they can be read and updated anywhere in the application.

```
type Label [= Value];
```

Value is an optional initialization constant. The different types are listed in Table 4-2, with the data value ranges are in parenthesis.

Single dimensional arrays are declared using the form:

```
type Label[Size] [= { Initialization Values..}];
```

Strings are defined as single dimensional ASCIIZ arrays:

```
char String[ 17 ] = "This is a String";
```

where the last character is an ASCII null (0x000).

Array dimensions must be specified in the variable declaration statement. Multidimensional arrays are defined with each dimension separately identified in square brackets ([and]):

```
int ThreeDSpace[Dim1][Dim2];
```

Pointers are declared with the asterisk (*) character after the variable type:

```
char* String;
```

Finding the address of the pointer in memory is accomplished using the ampersand (&) character:

```
StringAddr = &String;
```

Table 4-1
PIC16F684 Configuration Fuse Values

Configuration Bit	PICC Lite Compiler Value	Microchip Value	Value—Function
Bit 11—FCMEN, Fail-Safe Clock Monitor Enable	FCMEN (Default), FCDIS (Recommend)	_FCMEN_ON, _FCMEN_OFF	0x03FFF, 0x037FF
Bit 10—IESO, Internal External Switchover	IESOEN (Default), IESODIS (Recommend)	_IESO_ON, _IESO_OFF	0x03FFF, 0x03BFF
Bit 9:8—BODEN, Brownout Detect Enable 0x03DFF, 0x03CFF	BOREN (Default), BOREN_XSLP, SBOREN, BORDIS	_BOD_ON, _BOD_NSLEEP, _BOD_SBODEN, _BOD_OFF	0x03FFF, 0x03EFF,
Bit 7—CPD, Data Memory Protect	UNPROTECT (Default/Rec), CPD	_CPD_OFF, _CPD_ON	0x03FFF, 0x03F7F
Bit 6— _CP, Program Memory Protect	UNPROTECT (Default), PROTECT	_CP_OFF, _CP_ON	0x03FFF, 0x03FBF
Bit 5—MCLRE, External _MCLR Pin Enable	MCLREN (Default), MCLRDIS	_MCLRE_ON, _MCLRE_OFF	0x03FFF, 0x03FDF
Bit 4— _PWRTE, Power Up Timer Enable	PWRTDIS (Default), PWRTEN (Recommend)	_PWRTE_OFF, _PWRTE_ON	0x03FFF, 0x03FEF
Bit 3—WDTE, Watchdog Timer Enable	WDTEN (Default), WDTDIS (Recommend)	_WDT_ON, _WDT_OFF	0x03FFF, 0x03FF7
Bit 2:0—FOSC, Oscillator Selection	RCCLK (Default), RCIO, INTCLK, INTIO (Recommend), EC, HS, XT, LP	_EXTRC, _EXTRCIO, _INTOSC, _INTOSCIO, _EC_OSC, _HS_OSC, _XT_OSC, _LP_OSC	0x03FFF, 0x03FFE, 0x03FFD, 0x03FFC, 0x03FFB, 0x03FFA, 0x03FF9, 0x03FF8

Table 4-2
PICC Lite Compiler Data Types

Type	Bit Size	Comments
bit	1	Boolean Value
char	8	ASCII Character/Signed Integer (-128 to 127)
unsigned char	8	Unsigned Integer (0 to 255)
short	16	Signed Integer (−32,768 to 32,767)
unsigned short	16	Unsigned Integer (0 to 65,536)
int	16	Signed Integer (−32,768 to 32,767); same as short
unsigned int	16	Unsigned Integer (0 to 65,535); same as unsigned short
long	32	Signed Integer (−2,147,483,648 to 2,147,483,647)
unsigned long	32	Unsigned Integer (0 to 4,294,967,295)
float	24	Real (0 to $1/-6.81(10^{38})$; assume 3 digits of accuracy/default floating point mode
double	32	Real (0 to $1/-6.81(10^{38})$; assume 6 digits of accuracy/specified using PICL -D32 compilation option

Table 4-3
C Backslash Characters

Character	ASCII Value	Function
\0	00x00	Null Character
\a	0x007	Alert/Bell
\b	0x008	Backspace
\t	0x009	Horizontal Tab
\n	0x00A	New Line (Line Feed)
\v	0x00B	Vertical Tab
\f	0x00C	Form Feed (New Page)
\r	0x00D	Carriage Return
\\	0x05C	ASCII Backslash Character
\?	0x03F	ASCII Question Mark
\'	0x027	Single Quote
\"	0x022	Double Quote

Calculating the address of a specific element in a string is accomplished using the ampersand (&) character and a string array element:

```
StringStart = &String[n];
```

The variable's type can be overridden, or cast, by placing the new type in front of the variable in brackets:

```
(long) StringAddr = 0x0123450000;
```

The C assignment statement is in the following format:

```
Variable = Expression;
```

Expressions can be constants, variable values, or mathematical statements in the format:

```
[(..] Variable | Constant [Operator [(..]
Variable | Constant ][)..]]
```

Constants can be decimal, binary, hexadecimal, or ASCII characters enclosed in single quotes. Table 4-3 lists the special backslash characters that can be placed in single quotes so they will be evaluated as single characters. Note that a single backslash character is interpreted as a line continuance.

Mathematical operators include the ternary operator and those shown in Table 4-4 and Table 4-5.

The ternary operator is the form:

```
TestExp ? Exp1 : Exp2
```

and executes in the following manner: if TestExp is not zero, then Exp1 is evaluated. If TestExp is equal to zero, then Exp2 is evaluated.

The binary operators listed in Table 4-4 can be simplified to the modified equals statement if the destination variable is one of the parameters of the expression. Table 4-6 lists the resulting compound assignment statements.

C decision structures include the *if* statement,

```
if (Expression)
{
// Execute if Expression is not zero ("true")
} else {
// Execute if Expression is zero ("false")
}   // fi
```

the while statement,

```
while(Expression)
{
// Execute while Expression is not zero ("true")
}   // elihw
```

and the for statement,

```
for (initialization; Expression; Loop Increment)
{
// Execute while the Expression is not zero
   ("true")
}   // rof
```

Table 4-4
C Binary Operators

Operator	Priority	Function
*	Highest	Multiplication
/		Division
%		Modulus
+		Addition
–		Subtraction
<<		Shift Left
>>		Shift Right
<		Less-than comparison (Return not zero if true, zero if false)
<=		Less-than or equals comparison (Return not zero if true, zero if false)
>=		Greater-than or equals comparison (Return not zero if true, zero if false)
>		Greater-than comparison (Return not zero if true, zero if false)
==		Equals-to comparison (Return not zero if true, zero if false)
!=		Not-equals-to comparison (Return not zero if true, zero if false)
&		Bitwise AND
^		Bitwise XOR
|		Bitwise OR
&&		Logical AND
||	Lowest	Logical OR

Table 4-5
C Unary Operators

Operator	Function
-	2's Complement negation
!	Logical negation (Return zero if value is not zero and return not zero if value is equal to zero)
<	Bitwise negation
11	Increment variable *after* expression evaluation if on the variable's right-hand side; increment variable *before* expression evaluation if on the variable's left-hand side)
—	Decrement variable *after* expression evaluation if on the variable's right-hand side; decrement variable *before* expression evaluation if on the variable's left-hand side

Table 4-6
C Compound Assignment Statement Operators

Operator	Function
&=	Bitwise AND
|=	Bitwise OR
^=	Bitwise XOR
+=	Add to Destination
–=	Subtract from destination
*=	Multiply destination
/=	Divide destination
%=	Divide destination and return modulus
<<=	Shift destination left
>>=	Shift destination right

In the for statement, multiple initialization and loop increment statements are separated by commas. Putting in multiple initializations and increment statements can make your applications difficult to follow when you are first programming in C. To avoid this extra complexity, I recommend that you stick to just having a single initialization and loop increment when you first use the for statement.

For looping until a condition is true, the do/while statement is used:

```
  do
  {
// Execute until the Expression is zero ("false")
  }
  while (Expression);
```

To jump out of a currently executing loop, a break statement is used. The continue statement skips over remaining code in a loop and jumps directly to the loop condition (for use with while, for, and do/while loops).

To execute conditionally according to a value, the switch statement is used:

```
switch(Expression) {
   case Value:
// Execute if "Expression" == "Value"
   [break;]
   default:
// If no "case" Statements are True
}
```

Finally, the goto Label statement is used to jump to a specific address:

```
  goto Label;

Label:
```

To return a value from a function, the return statement is used:

```
return Statement;
```

Directives are executed at compilation time and used to modify the source code for the application and are listed in Table 4-7. Each directive is given its own line and, if necessary, continues onto the next line by use of the backslash (\) character.

The PICC Lite compiler pragma options are used by the compiler to change its operation and are listed in Table 4-8. Note that for the code presented in this book, these options are not required.

In the PICC Lite compiler manual, you can find additional information, including error and warning descriptions, information on including assembly language programming in C applications, and file format reference information that is not critical to performing the experiments in this book but may be useful to you. I have not included the list of PICC Lite compiler options in this section because you should be able to successfully run the PICC Lite compiler under MPLAB IDE without having to change any of the operating parameters.

In Table 4-9, I have provided a cursory list of the library functions that are available in the PICC Lite compiler. I have arranged the functions by alphabetical order within the include file rather than by overall alphabetical order, as this should make it easier to find specific and related functions. Unless otherwise specified, assume that the label is for a function. For more information, you should consult with the PICC Lite compiler *User's Guide* (available for download at www.ht-soft.com), the .h files in the PICC Lite compiler INCLUDE folder, and standard texts on the C programming language.

Table 4-7
C Directives

Directive	Function
#	Null directive; do nothing.
#assert Condition	Force an error if condition is false.
#asm	Indicate the start of in-line assembly.
#endasm	Indicate end of in-line assembly.
#define Label (Optional Parameters) Text	Define a label that will be replaced with the specified text when it is encountered. If optional parameters are specified, instances of them in the text will be replaced by the parameters specified when the label is invoked.
#undef Label	Delete the defined label.
#line Number ListFile	Specify the line number and file name for listing.
#include <File>	Load the specified file found in the INCLUDE folder path.
#include "File"	Load the specified file by first looking in the current folder and then in the INCLUDE folder path.
#error Text	Force a compiler error (stop compilation) and output text for error information.
#warning Text	Force a compiler warning and output text for warning information.
#if Condition	If the condition is true, pass following code to next #else, #elif, or #endif.
#ifdef Label	If the label has been defined, then pass following code to next #else, #elif, or #endif.
#ifndef Label	If the label has not been defined, then pass following code to next #else, #elif, or #endif.
#elif Condition	This directive is an else/if, and the condition is tested only if the previous if directive is not true.
#else	Invert pass following code status.
#endif	End a #if, #ifdef, #ifndef, #elif, or #else condition.
#pragma String keywords	Pass the specified string to the compiler (see Table 4-8).

Table 4-8
Available PICC Lite Compiler Pragma Directives

Pragma Directive	Function
interrupt_level 1	Allow interrupt functions to be called by main line code.
Jis	Enable JIS (Japanese) character handling in strings.
Nojis	Disable JIS character handling in strings (default condition).
printf_check(printf) const	Use printf formatting conventions for string checking.
psect text=newpsectname	Change the object file psect name.
regused Registers	Specify registers that are used in interrupt handler.

Table 4-9
PICC Lite Compiler Library Functions and Macros

Library	Function	Description
conio.h	Bit kbhit(void);	Skeleton of keyboard hit function
ctype.h	int isalnum(char c)	Macro testing if c is alphanumeric
	int isalpha(char c)	Macro testing if c is alphabetic
	int isascii(char c)	Macro testing for seven-bit value
	int iscntrl(char c)	Macro testing for ASCII control character
	int isdigit(char c)	Macro testing for ASCII decimal digit
	int islower(char c)	Macro testing for ASCII a through z
	int isprint(char c)	Macro testing for printing character
	int isgraph(char c)	Macro testing for nonblank printing character
	int ispunct(char c)	Macro testing for nonalphanumeric character
	int isspace(char c)	Macro testing for space, tab, or newline
	int isupper(char c)	Macro testing for ASCII A through Z
	int isxdigit(char c)	Macro testing for ASCII hex digit
	int toupper(int c)	Macro to convert ASCII character to upper case
	int tolower(int c)	Macro to convert ASCII character to lowercase
	int toascii(int c)	Macro to ensure parameter is valid ASCII math.h
	double acos(double f);	Return the arc-cosine in radians of the input value
	double asin(double f);	Return the arc-sine in radians of the input value
	double atan(double f);	Return the arc-tangent in radians of the input value
	double atan2(double f, double x);	Return the arc-tangent in radians of the input value
	double ceil(double f);	Return smallest whole number not less than f
	double cos(double a);	Return the cosine of the radian argument
	double cosh(double f);	Return hyperbolic cosine for argument
	double eval_poly(double x, const double* d, int n);	Evaluate the polynomial with coefficients stored in array d at point x
	double exp(double f);	Return the value of e^f
	double fabs(double f);	Return the absolute value of f
	double floor(double f);	Return largest whole number not less than f
	double frexp(double f, int* p);	Break real number into integer and fraction
	double ldexp(double f, int x);	Integer is added to real and result returned
	double log(double f);	Return natural logarithm of f
	double log10(double f);	Return base 10 logarithm of f
	double modf(double f, double* iptr);	Split f into integer and fraction parts with same value as f
	double pow(double f, double p);	Return f^p
	double sin(double f);	Return sine of the radian argument
	double sinh(double f);	Return hyperbolic sine for argument
	double sqrt(double f);	Return the square root of the argument
	double tan(double f);	Return tan of the radian argument
	double tanh(double f);	Return hyperbolic tangent for argument
pic.h	di(void);	Disable interrupts
	eeprom_write(unsigned char addr, unsigned char value);	Store the specified byte at the specified address in the PIC MCU's data memory
	ei(void);	Enable interrupts
	unsigned char eeprom_read(unsigned char addr);	Return the byte at the specified address in the PIC MCU's data memory
stdio.h	unsigned char printf(const char* f, . . .);	Standard C print routine; before using sdtout must be defined for application
	unsigned char sprintf(char* buf, const char* f, . . .);	Convert C printf format parameters to string at buf
	unsigned xtoi(const char* s);	Convert ASCII hexadecimal string to integer

Table 4-9 (continued)

Library	Function	Description
stdlib.h	int abs(int j);	Return the absolute value (positive) of the input
	double atof(const char* s);	Convert ASCII string to double variable type
	int atoi(const char* s);	Convert ASCII string to integer variable type
	long atol(const char* s);	Convert ASCII string to long variable type
	div_t div(int num, int divisor);	Divide and return quotient and remainder
	ldiv_t ldiv (long num, long d);	Divide and return quotient and remainder
	int rand(void);	Return random number from 0 to 32,767
	void srand(unsigned int seed);	Initialize random number generator
string.h	const void* memchr(const void* block, int val, size_t n);	Search string for specific byte
	int memcmp(const void* s1, const void* s2, size_t n);	Compare two blocks of memory
	void* memcpy(void* d, void* s, size_t n);	Copy n bytes of data between pointers
	void* memmove(void* s1, const void* s2, size_t n);	Copy n bytes under all circumstances
	void* memset(void* s, int c, size_t n);	Fill block of memory with specified character
	char* strcat(char* s1, const char* s2);	Concatenate s2 at the end of s1
	char* strchr(const char* s, int c);	Search string for first occurrence of c
	char* strichr(const char* s, int c);	Search string for first occurrence of c in upper- or lowercase
	int strcmp(const char* s1, const char* s2);	Compare two strings
	int stricmp(const char* s1, const char* s2);	Compare two strings without regard to upper- or lowercase
	char* strcpy(char* s1, const char* s2);	Copy s2 into buffer (character array) pointed to by s1
	char* strcspn(const char* s1, const char* s2);	Find number of characters from the start of s1 to the part of the string that matches s2
	size_t strlen(const char* s);	Find the length of the ASCII stringchar* strncat
	(char* s1, const char*s2, size_t n);	Concatenate n characters from s2 onto the end of s1
	int strncmp(const char* s1, const char* s2, size_t n);	Compare two strings for up to n characters
	int strnicmp(const char* s1, const char* s2, size_t n);	Compare two strings for up to n characters without regard to upper- or lowercase
	char* strncpy(char* s1, const char* s2, size_t n);	Copy n characters from s2 to s1
	const char* strbrk(const char* s1, const char* s2);	Return a pointer to the first instance of s2 in s1
	const char* strrchr(char* s, int c);	Search for character starting at the end of the string, rather than from the front
	size_t strspn(const char* s1, const char s2);	Return the length of s1, starting from the start that contains characters from s2
	const char* strstr(const char* s1, const char* s2);	Find the first occurrence of s2 in s1
	const char* stristr(const char* s1, const char* s2);	Find the first occurrence of s2 in s1 without regard to upper- or lowercase
	char* strtok(char* s1, const char* s2);	Break s1 into a series of tokens, separated by s2
time.h	char* asctime(struct tm* t);	Convert the variable of the tm struct type into an ASCIIZ string
	char* ctime(time_t* t);	Convert time in seconds to ASCII string
	struct tm* gmtime(time_t * t);	Break down time into type tm
	struct tm* localtime(time_t* t);	Break down time into type tm
	time_t time(time_t* t);	Skeleton of time read function

PIC16F684 Microcontroller Built-in Functions

PC/ Simulator

PICkit™ 1

starter kit

Required Parts

1 PIC16F684

1 LM317 adjustable volt-
   age regulator in TO-220
   package

1 10-LED bargraph display

1 Red LED

1 Green LED

1 100Ω resistor

1 330Ω resistor

1 470Ω 10-pin resistor
   SIP

1 10k breadboard mount-
   able potentiometer

1 1k breadboard mountable
   potentiometer

2 0.01 µF capacitors

1 Breadboard-mountable
   SPDT switch (E-Switch
   EG1903 recommended)

1 9-volt battery pack

1 9-volt battery

1 Three-cell AA battery
   clip

3 AA batteries

1 Two-cell AA or AAA battery pack

2 AA or AAA batteries

Tool Box

DMM

Needle-nose pliers

Breadboard

Wiring kit

Jeweler's screwdriver

A big question asked by software developers is whether or not to rewrite an application. The reasons for doing a rewrite are usually good and noble; the author may have figured out or learned ways the application could be made more efficient, new hardware may be available that makes the task easier, or there may be too many potential errors in the software that only a complete rewrite of the code will prevent problems in the future. Personally, I am of the opinion that applications should never be rewritten unless a tangible reason exists. The question, "what will happen if the changes are not implemented?" has to be asked, and unless the answer includes something about cost savings (e.g., the application could be used in a cheaper microcontroller), then the rewrite should not be attempted. This may seem like an off-topic way to introduce a section about the built-in features of the PIC16F684, but it is surprisingly relevant because many of the built-in features of the PIC® MCU introduced in this section replace software functions that you may have written to implement different functions.

The built-in features of the PIC MCU can make your application a *lot* smaller, easier to write, and easier to debug. For example, in Section 3, I introduced an application that displayed an incrementing binary value on the eight LEDs of the PICkit™ 1 starter kit. It did so by sequencing through each of the LEDs simultaneously to give the impression that they were all on continuously, when in fact no more than one

LED was on at any given time. I feel that the code was fairly clumsy and unnecessarily long. The length made it very difficult to follow values from top to bottom, and as such errors had a good chance to make their way into the different statements of each group of statements written to optionally turn on an LED.

In Section 4, I showed you arrays, and you might have thought about using them to rewrite the cLED-Disp.c application. I have done this and listed the program here:

```
#include <pic.h>
/* cLEDDisp 2.c - 2nd Version of D0-D7 as an
incrementing Counter

Using "cPKLED.c" as a base, cycle through each
LED at 100x per second (1250 us between LEDs).

This is the second version which is rewritten to
take advantage of Arrays to reduce the amount of
space required by the application.

myke predko
04.09.12

*/


__CONFIG(INTIO & WDTDIS & PWRTEN & MCLRDIS &
UNPROTECT \ & UNPROTECT & BORDIS & IESODIS &
FCMDIS);


int i, j, n;
int Value = 0;
int Dlay = 65;      // LED Time on Delay Variable
const char PORTAValue[8] = {0b010000, 0b100000,
                            0b010000, 0b000100,
                            0b100000, 0b000100,
                            0b000100, 0b000010};
const char TRISAValue[8] = {0b001111, 0b001111,
                            0b101011, 0b101011,
                            0b011011, 0b011011,
                            0b111001, 0b111001};
const char NOTPORTA[8] = {0, 0, 0, 0, 0, 0, 0, 0};

main()
{

    PORTA = 0;
    CMCON0 = 7;      // Turn off Comparators
    ANSEL = 0;       // Turn off ADC

    j = 0;           // Reset the Display Counter

    while(1 == 1)   // Loop Forever
    {
     for (i = 0; i < 8; i++ )
     {    // Loop through Each of the 8 LEDS
        for (n = 0; n < Dlay; n++);
        if ((Value & (1 << i)) == 0)
          PORTA = NOTPORTA[i];
```

```
        else
          PORTA = PORTAValue[i];
        TRISA = TRISAValue[i];
     }  // rof

     j = j + 1;  // Increment the Counter every
                          1/2s
     if (j >= 50)
     {
       Value = Value + 1; // Increment Display
                             Counter
       j = 0;                // Reset the Counter
     }  // fi
    }  // elihw
}  // End cLEDDisp 2
```

This application takes up less than half of the space of the original (both in terms of lines of code and instructions required), although it uses only four more variable bytes. By just about any measurement, this rewrite is substantially better than the original, but still I would consider replacing the original with it only if the application required more program memory than was available. Therefore, you should keep the knowledge regarding using an array to implement the multiple LED display code in the back of your mind until you have to implement a similar function.

The built-in hardware features of the PIC16F684 provides similar opportunities when you are looking over working applications. As you first start programming, you will implement basic input and output functions using the standard digital I/O capabilities of the microcontroller pins. These functions will simplify generating motor control signals, help you process analog data, and help you more easily communicate with other devices. Although these built-in features will seem dauntingly complex in the beginning, they will become easier as you become more familiar with developing applications for them. I can promise you that you will look back and consider rewriting the applications to take advantage of the hardware "behind the pins." But remember my advice: You should rewrite this code only if there is a concrete reason for doing so.

Of course, you can avoid the whole question of whether or not to rewrite to take advantage of hardware features by implementing the features in your application right from the start. In this section, I introduce the different built-in functions of the PIC16F684 along with some sample applications that illustrate how they work and what can be done with them.

Experiment 32—Brownout Reset

PC/ Simulator

PICkit™ 1

starter kit

Required Parts

1 PIC16F684

1 LM317 adjustable volt-
age regulator in TO-220
package

1 1k breadboard mountable
potentiometer

1 Red LED

1 Green LED

1 330Ω 1/4-watt resistor

1 0.01 µF capacitor (any
type)

1 Breadboard-mountable
SPDT switch (E-Switch
EG1903 recommended)

1 9-volt battery pack

1 9-volt battery

Tool Box

DMM

Jeweler's screwdriver (for
1k potentiometer)

Needle-nose pliers

Breadboard

Wiring kit

In a previous experiment I explained that when I first started working with the PIC MCU and writing about it, various specialized circuitry would have to be added to provide reset with brownout protection for the application. The PIC16F684, like other recently released chips, has these features built into it, allowing you to take advantage of these features without adding to the cost or complexity of your circuitry. Unfortunately, you may find that the work required to add the advanced reset functions to your application code is not trivial. In this experiment, I will demonstrate how the brownout detect and reset circuitry built into the PIC16F684 works and provide you with some clues as to how to decipher the datasheets when a hardware function doesn't work.

The PIC16F684 is designed to work in the voltage range of 2.0 volts to 5.5 volts, which makes it ideal for battery power without a voltage regulator. When the voltage goes below 2.0 volts, the PIC MCU can no longer operate reliably, it may stop, it may run different parts of the program memory, or it may execute the instructions incorrectly. To help avoid these problems the *brownout reset* (BOR) can be enabled in the con-figuration fuses to detect low voltage conditions and enable the PIC MCU's reset.

To demonstrate the operation of the PIC16F84's brownout reset, I created the test *brownout detect* (or

BOD) circuit shown in Figure 5-1. This circuit uses a variable-voltage power supply (based on the LM317 chip) to reduce the operating voltage from 5 volts (nominal) down to 1.25 volts. Note that even though I am using the brownout detect reset function of the PIC MCU, I did not enable the MCLR function of RA3.

The LM317 (see Figure 5-2) is a nice little adjustable linear voltage regulator circuit that is ideally suited for a task like powering the PIC16F684 in this circuit to test its brownout detect function. The lower end of the chip's output is 1.25, and the upper end is defined by the V_{in} power source. Its output is defined by the formula:

$$V_{out} = 1.25 \text{ V} \times (1 + (R2/R1)) + i_{ADJ} \times R2$$

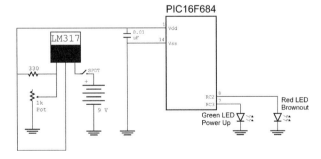

Figure 5-1   *BOD circuit*

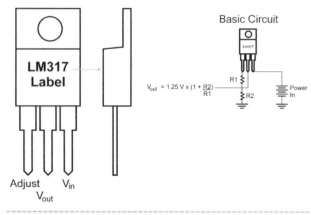

$$V_{out} = 1.25\,V \times \left(1 + \frac{R2}{R1}\right)$$

Figure 5-2   *LM317*

but, as I_{ADJ} is only 100 μA, the product of it times R2 is normally negligible. This is why in Figure 5-2, I have the simplified formula. For this application, I used a 1k potentiometer for R2 and a 330Ω resistor for R1; this gives me an output voltage range of 1.25 volts to 5.00 volts when a 9-volt battery is used for application power. The small size of the breadboard-mountable 1k potentiometer meant that I had to use a jeweler's screwdriver to change the potentiometer (which can be seen in my prototype wiring in Figure 5-3).

The plan was to have the PIC16F684 light the green LED on RC3 after reset, but if the reset was caused by a voltage brownout, the red LED on RC2 would be lit. The original application code was based on cReset.c in which the _BOD bit (known as BOD in PICC Lite™ compiler) of PCON was set after power up. Along with setting and checking this bit, the BOREN label, instead of the typical BORDIS label, was used in the configuration fuse specification.

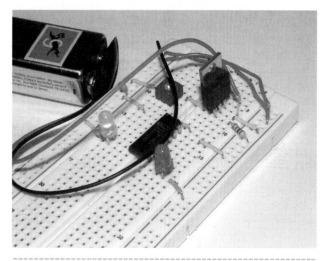

Figure 5-3   *Assembled brownout detect experiment circuitry*

I should point out that a number of different options exist for brownout reset, including ignoring brownout reset during sleep (or low-power) mode and allowing brownout reset to be controlled by application software. Also, a number of bits are set in the factory, which will tune the actual brownout reset voltage. For this application, I avoided the more complex brownout reset options, wanting simply to stay with a hardware reset when the voltage became too low.

With this setup, the red LED always lit on power up, regardless of the voltage input and whether or not I set or reset BOD on power up. Additionally, I tried different configuration register values—all with no success. Unfortunately the use of the brownout reset in the Microchip datasheets, or apnotes, is not shown and a Google search did not reveal any code that uses this feature. As I scoured the datasheets, I noticed a difference in how the PCON register bits were specified for power up. For the BOD bit, in one case it was marked as "unknown" instead of 0 at reset, which meant that for it to have a valid value after a brownout reset, it would have to be set manually. With this information in hand, I then recoded the cBOD.c application so that it would first check for a power on reset, at which time it would set the BOD bit. And on subsequent resets, the state of the BOD would be checked.

```
#include <pic.h>
/*  cBOD.c - Monitor Reset on Brown Out Detect

This Program will light an LED based on how it
  was reset (Power Up or Brown Out)

Power Supplied to the PIC16F684 comes from an
  LM317 wired to provide 1.5 to 5 Volts

LEDs are connected to:
RC3 - Power Up Reset
RC2 - Brown Out Reset

myke predko
04.11.10

*/

__CONFIG(INTIO & WDTDIS & PWRTEN & MCLRDIS &
UNPROTECT \ & UNPROTECT & BOREN & IESODIS &
FCMDIS);                   // <- Note "BOREN"

main()
{
    PORTC = 0;
    CMCON0 = 7;             // Turn off Comparators
    ANSEL = 0;             // Turn off ADC
    TRISC = 0b000110011; // RC3:RC2 as Outputs

    if (0 == POR)         // Power-On Reset Occurred
    {
        POR = 1;           // Indicate Power Active
        BOD = 1;           // Make Sure BOD Set
        RC3 = 1;           // Set Power Up/Good LED
    }
    else if (0 == BOD)     // Check PCON Register
                            // for Start up
```

```
{
    RC2 = 1;      // BOD Happened
    RC3 = 0;      // Not a Good Power Up
    BOD = 1;      // Indicate PIC MCU Powered
                  UP Once
}
else              // No Brown Out Detect Reset
    RC3 = 1;

while(1 == 1);    // Loop Forever

}  // End cBOD
```

cBOD did not simulate well, which added to my problems in trying to figure out what was happening. Now, when cBOD is run in hardware, the hardware powers up with the green LED, and when the potentiometer is adjusted to a lower voltage, the green LED dims a bit and then turns off completely. And, when the LM371's voltage output is raised, the red LED turns on, indicating that the application code (and hardware) has recognized a brownout voltage situation.

In every PIC MCU datasheet, Microchip marks each bit when a register is defined. Each register is defined and then labeled as Register #-x (where # is the datasheet section and x is the number of the register definition in the section), and the ability to read and/or write the bit along with its power up value is specified. Read and write are specified with a R and W, respectively, as would be expected, but as I show in Table 5-1, there are four initial values that you should be aware of.

Microchip documentation is uniformly excellent, and it is very unusual to find a mistake like the one I found. (In case you are wondering, the mistake is in Register 3-1: PCON Register of DS31003A, pages 3–14, and I have notified Microchip of this error. So hopefully it will not appear in later datasheets.) When you are confronted with a problem, as I have been in this experiment, where the code doesn't work as expected, you should do the following:

1. Check over the code to see if you can find the problem. Simulation (although not possible in this experiment) of the program is always a good idea.

2. Reread the datasheet to make sure that you are using the function properly. Often the datasheet will have example code that you can use.

3. Look for apnotes on the Microchip web site that illustrate how the function works. Don't expect there to be an apnote on every feature built into the chip that you are using. You may have to look at how the function works (and is programmed or wired) in another PIC MCU part number and understand any differences.

4. Look for example circuits on the Internet using Google or another search engine.

5. Compare resources to find any discrepancies or clues as to how the feature works by comparing the different descriptions.

It's important to know that these instructions are for getting a hardware feature of the PIC MCU working; they should not be considered the most appropriate way to debug a failing program. Later in the book I discuss some of the techniques for finding a logic problem in a program and making sure that you have fixed it correctly.

Table 5-1
Microchip Datasheet Initial Register Bit Value
Specifications

Bit Power Up Value	Meaning
0	After Reset, the Bit value is 0.
1	After Reset, the Bit value is 1.
u	After Reset, the Bit is the same value as before Reset.
x	After Reset, the value of the bit is unknown.

Experiment 33—ADC Operation

PC/ Simulator

PICkit™ 1

starter kit

If you were to look in the PIC16F684's datasheet, you would discover that the function of the *analog-to-digital* (ADC) function is explained over eight pages. I find this description of the ADC function, although thorough, a lot more complex than it needs to be. Using the ADC itself is surprisingly simple and does not require a lot of code or understanding of how it works. In this experiment I am going to explain the function and characteristics of the ADC and give you an example experiment that samples an incoming analog voltage and displays it on the eight LEDs of the PICkit 1 starter kit.

The ADC function of the PIC16F684 can be seen as a simple block diagram like the one shown in Figure 5-4; the line with voltage to be measured is connected to a capacitor in the ADC. When the ADC operation starts, this capacitor is disconnected from the input and the voltage from the capacitor is passed to a comparator input. The other input to this comparator is connected to a *sweep generator*, which produces a ramping voltage at a known rate. To measure the voltage (which can be a maximum of Vdd), the sweep generator is started and the time when the sweep generator's voltage to be greater than the capacitor's voltage is recorded using a counter. This is really all there is to it, although you should know a few other things, including being familiar with the registers used to control the ADC. The operation of the ADC is quite simple and easy to integrate into your applications.

Because the ADC operates with the sweep generator, it does not execute over a single instruction cycle; it takes at least 20 μs to perform the 10-bit data conversion operation. You should also allow at least 12 ms between samples to ensure the capacitor charge has changed to reflect any change in voltage. The time required for the capacitor to charge is why Microchip recommends that the impedance of the signal should be less than 10k. These delays mean that samples should be made after 32 μs or so, resulting in a maximum sample frequency of about 30 kHz. This is fast enough to implement audio sampling, but not fast for many real-time data-monitoring activities.

Five registers are involved in working with the ADC. The ANSEL register is used to select which of the eight possible ADC inputs (RA0 to RA7) will be connected to analog inputs. On reset, the ANSEL register, which controls which of the eight bits are analog inputs, has all its bits set. If you look back at previous experiments that use the PICkit 1 starter kit, one of the hardware initialization statements used was to clear

the contents of the ANSEL register. For this experiment and any application that requires analog inputs be measured, the bits representing the analog input pins are left set.

The next register is the ADCON0, which controls the operation of the ADC. This register is used to enable and start the ADC hardware, to select which pin will have its analog input sampled, and to set the output format. Table 5-2 lists the function of the different bits used in the ADC.

The sweep generator has an internal counter, which is run from a built-in RC oscillator or the processor's clock and is known as the A/D Conversion Clock. It has the timing signal TAD. Its source and prescaler value are selected by bits 6:4 of the ADCON1 register. In Table 5-3, I have listed the different ADC clock options and the resulting prescaler values. The TAD value should be between 1.6 μs and 6.4 μs for proper operation of the ADC, and this time is calculated by simply multiplying the PIC MCU's clock period by the TAD operation listed in Table 5-3. If you don't want to calculate the TAD operation that will give you the

Table 5-2
ADCON0 Bit Functions

Bit	Name	Function
7	ADFM	Output Format: 0 = Left Justified (ADRESH has 8 Bits); 1 = Right Justified (ADRESH has 2 Bits)
6	VCFG	Voltage Reference bit: 1 = Vref Pin (RA1); 0 = Vdd
5	Unused	
4:2	CHS	Analog Channel Select: 000 = RA0; 001 = RA1; . . . 111 = RA7
1	GO/_DONE	ADC Operation Start/End Indicator
0	ADON	1 = Power ADC; 0 = ADC Off

Table 5-3
ADCON1 Bit Functions

Bit	TAD Operation	Comments
7		Unused
6:4	000—Period * 2	Example: For a PIC16F684
	001—Period * 8	running at 4 MHz.
	010—Period * 32	Clock Period = 250 ns
	x1—Internal 4 μs RC	Desired TAD = 1.6 μs to 6.4 μs
	100—Period * 4	Clock Period * TAD Operation 001
	101—Period * 16	= 250 ns * 8
	110—Period * 64	= 2 μs
3:0		Unused

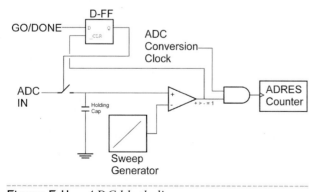

Figure 5-4 *ADC block diagram*

fastest ADC operation, you can simply select the built-in resistor-capacitor (RC) clock—this runs at a nominal 4 μs and will allow the PIC MCU's ADC to operate under all conditions.

The fourth and fifth registers are ADRESH and ADRESL, respectively. These two registers store the 10-bit result of the ADC operation. The 10 bits are displayed as eight and two, with each register having either two or eight bits. I recommend that you left justify the data, or store the eight most significant bits in ADRESH and the two least significant bits in ADRESL. The two least significant bits are generally considered to be in the noise region, and their value is not accurate; by reading only the most significant eight bits, you should be getting a reliable eight-bit analog to digital conversion.

With the registers set up, you will set the GO/_DONE bit (known as GO in PICC Lite compiler) and wait for it to go low. Hopefully, my description of how the ADC works hasn't confused you, because to carry out an analog-to-digital conversion on a PIC16F684's RA0 pin, only the following six C statements are required:

```
ANSEL = 1;     // Just RA0 is an Analog Input

ADCON0 = 0b000000001;  // Turn on the ADC
   // Bit 7 - Left Justified Sample
   // Bit 6 - Use VDD
   // Bit 4:2 - Channel 0
   // Bit 1 - Do not Start
   // Bit 0 - Turn on ADC
ADCON1 = 0b001110000;  // Select the Internal
                       RC Clock
   :
ADCON0 = ADCON0 | (1 << 1); // Could also be
                            "GODONE = 1;"
while(!GODONE);        // Wait for ADC to
                       Complete
 ADCValue = ADRESH;    // Save Sample Value
                       After Operation
```

For this experiment, I have used the scanned eight-bit LED display code to display the value of the ADC. To ensure that it does not negatively affect the operation of the LED display code, I have programmed the ADC operation in cADC.c as a *state machine*, with the sample taking place over several passes of the LEDs. This will avoid spending too much time with any one LED lit longer than the others.

```
#include <pic.h>
/* cADC - Display the PICkit Pot Input Value on
the built in LEDs

This program samples the voltage on RA0 using
the ADC and Displays the value on the 8 LEDs
using "cLEDDisp 2" as a base.

myke predko
04.10.03
*/

__CONFIG(INTIO & WDTDIS & PWRTEN & MCLRDIS &
UNPROTECT \ & UNPROTECT & BORDIS & IESODIS &
FCMDIS);


int i, j;
int ADCState = 0;   // Keep Track of ADC
                       Operation
int ADCValue = 0;
int Dlay = 63;      // LED Time on Delay Variable
const char PORTAValue[8] = {0b010000, 0b100000,
                            0b010000, 0b000100,
                            0b100000, 0b000100,
                            0b000100, 0b000010};
const char TRISAValue[8] = {0b001111, 0b001111,
                            0b101011, 0b101011,
                            0b011011, 0b011011,
                            0b111001, 0b111001};
const char NOTPORTA[8] = {0, 0, 0, 0, 0, 0, 0, 0};

main()
{

 PORTA = 0;
 CMCON0 = 7;        // Turn off Comparators
 ANSEL = 1;         // Just RA0 is an Analog Input

 ADCON0 = 0b00000001;  // Turn on the ADC
   // Bit 7 - Left Justified Sample
   // Bit 6 - Use VDD
   // Bit 4:2 - Channel 0
   // Bit 1 - Do not Start
   // Bit 0 - Turn on ADC
 ADCON1 = 0b00010000;  // Select the Clock as
                          Fosc/8

 while(1 == 1)          // Loop Forever
 {
  for (i = 0; i < 8; i++ )
  {      // Loop through Each of the 8 LEDS
   for (j = 0; j < Dlay; j++);  //Display "On"
                                  Delay Loop
     if ((ADCValue & (1 << i)) == 0)
       PORTA = NOTPORTA[i];
     else
       PORTA = PORTAValue[i];
    TRISA = TRISAValue[i];
  } // rof
  switch (ADCState)    // ADC State Machine
  {
   case 0:             // Finished, Start Next Sample
    GODONE = 1;
    ADCState++;
    break;
   case 1:  // Wait for ADC to complete
    if (!GODONE)
    ADCState++; // Sample Finished
    break;
   case 2:             // Save Sample Value in
                          "ADCValue"
    ADCValue = ADRESH;
    ADCState = 0;
    break;
  } // hctiws
 } // elihw
} // End cADC
```

When you run this code on the PICkit 1 starter kit, you will be able to change the value displayed on the LEDs by changing the potentiometer (marked "RP1" on the PICkit 1 starter kit). You may find that the least significant bit flashes with certain settings: This is due to slightly different readings each time the ADC operates. If the least significant two bits were also shown, you would probably see significantly more flashing at other potentiometer settings. What you should conclude from this experiment is that the most significant seven bits of the ADC output will be accurate and the least significant three bits should be ignored.

For this experiment, I am using the circuit built into the PICkit 1 starter kit shown in Figure 5-5. This circuit consists of a potentiometer wired as a voltage divider with a current-limiting resistor and a capacitor. The voltage output of this potentiometer circuit can be

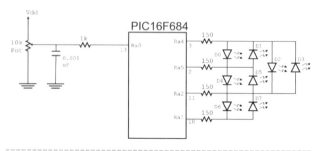

Figure 5-5 *ADC circuit*

read more than one way, as I will show in the next two experiments.

Experiment 34—Comparator Operation

One of the measurements that I use to gauge the difficulty of applications in this book is whether or not the circuit I design and the code I write works the first time. When I started this experiment, I had a perfect record of 72 applications that worked the first time I built them, and burned their code into a PIC16F684. Other factors further bolstered my confidence for first-time success for this experiment: My plan was to base the code on the previous experiment and my familiarity with the PIC MCU's comparator and integrated variable voltage reference circuits. In this experiment's write up, you'll get an explanation of the operation of the comparator, plus you'll get a story of hubris on my part.

The comparator circuit built into the PIC MCU is quite simple. It consists of two comparators with differing input and operating options, which are controlled by the CMCON0 register explained in Table 5-4. Beyond the CMCON0, a COMCON1 register is used to control the operation of one of the built-in timers and synchronize one of the two comparators' output with a timer. I have passed over this CMCON0 register because these options are required only for specific applications.

The CM bits of CMCON0 select the operating mode of the comparator (see Table 5-5). In most

Table 5-4
CMCON0 Register Bits and Comparator Operation

Bit	Name	Function
7	C2OUT	Comparator 2 Output.
		if (0 == C2INV)
		C2OUT = 1 if C2Vin1 > C2Vin-
		else // if (1 == C2INV)
		C2OUT = 1 if C2Vin1 < C2Vin-
6	C1OUT	Comparator 1 Output.
		if (0 == C1INV)
		C1OUT = 1 if C1Vin1 > C1Vin-
		else // if (1 == C1INV)
		C1OUT = 1 if C1Vin1 < C1Vin-
5	C2INV	Set to Invert the Comparator 2 Output Condition (See C2OUT)
4	C1INV	Set to Invert the Comparator 1 Output Condition (See C1OUT)
3	CIS	Comparator Input Switch, Used when CM bits == 010
		if (0 == CIS)
		RA1 is C1Vin-
		RC1 is C2 Vin-
		else // if (1 == CIS)
		RA0 is C1Vin-
		RC0 is C2 Vin-
		When CM bits == 001
		if (0 == CIS)
		RA1 is C1Vin-
		else // if (1 == CIS)
		RA0 is C1Vin-
2:0	CM	Comparator Mode Select Bits (see Table 5-5)

Table 5-5

Comparator Operating Mode Selected by the CMCON0 CM Bits

CM 2:0	Operation of Comparators
111 (7)	Comparators Off. Required setting to use RA0, RA1, RC0, and RC1 as digital I/O.
110 (6)	Comparators On. RC0 is common Vin1 for two comparators. RA1 is C1Vin-, RA2 is C1OUT, RC1 is C2Vin-, and RC4 is C2OUT.
101 (5)	Comparator 1 Off. Comparator 2 On. RC0 is C2Vin1 and RC1 is C2Vin-.
100 (4)	Both comparators operating. RA0 is C1Vin1, RA1 is C1Vin-, RC0 is C2Vin1, and RC1 is C2Vin-.
011 (3)	Both comparators operating with common Vin1 on RC0. RA1 is C1Vin-, and RC1 is C2Vin-.
010 (2)	Both comparators operating with common Vin1 from Vref module. C1Vin- and C2Vin- are selected by CMCON0 CIS (see Table 5-4).
001 (1)	Both comparators operating with common Vin1 on RC0. C1Vin- selected by CMCON0 CIS (see Table 5-4) and RC1 is C2Vin-.
000 (0)	Comparators Off. Analog input on RA0, RA1, RC0, and RC1. Default setting.

applications presented in this book, I have turned the comparator bits off by writing 0b0111 to the CM bits. For this experiment, I want to use RA0 as a comparator input along with the controllable Vref voltage source to measure the voltage coming from the potentiometer built onto the PICkit 1 starter kit.

The Vref source is available only to the comparator. In some other comparator-equipped PIC MCU part numbers, the Vref value can be output to other devices. This circuit consists of a tapped voltage ladder with different voltages passing through a 16-to-1 analog multiplexer. The VREN bit of VRCON controls the power, and VRR selects the operating mode (see Figure 5-6). Figure 5-6 also lists the output voltages based on the state of VRR and the VR bits of VRCON, which are listed in Table 5-6.

To demonstrate the ability of the comparator and Vref circuitry as an analog to digital converter, I used cADC.c as a base for the eighth LED output control loop, and I used the state machine to poll the ADC. I came up with cComp.c in which I found the potentiometer's wiper voltage by using a binary search algorithm.

```
#include <pic.h>
/*  cComp - Read the PICkit Pot Input Voltage
using Comparator/Cref

This program samples the voltage on RA0 using
the Comparator with the Cref Module and Displays
the four bit result on the least significant
LEDs on the PICkit PCB.
```

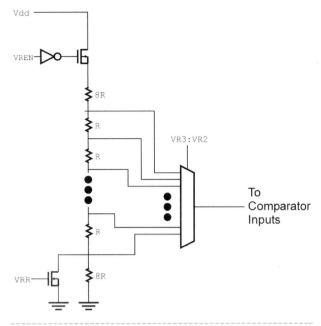

Figure 5-6 *Vref circuit*

Table 5-6

VRCON Register Bit Description

Bit	Name	Function
7	VREN	Vref Module Power Enable
6	N/U	Not Used
5	VRR	Vref Voltage Select (1 = Low Range; 0 = High Range)
4	N/U	Not Used
3:0	VR	Voltage Selection Bits (see Figure 5-6 for formula)

```
The ADC operation consists of a binary search
algorithm

myke predko
04.10.06

*/

__CONFIG(INTIO & WDTDIS & PWRTEN & MCLRDIS &
UNPROTECT \ & UNPROTECT & BORDIS & IESODIS &
FCMDIS);


int i, j;
int CompState = 0;    // Keep Track of Comparator
                             Operation
int NextVrefValue;
int VrefValue;
int DispValue = 0;    // Display Value to avoid
                             Flashing
int Dlay = 63;        // LED Time on Delay
                             Variable
const char PORTAValue[8] = {0b010000, 0b100000,
                    0b010000, 0b000100,
                    0b100000, 0b000100,
                    0b000100, 0b000010};
```

```
const char TRISAValue[8] = {0b001111, 0b001111,
                            0b101011, 0b101011,
                            0b011011, 0b011011,
                            0b111001, 0b111001};
const char NOTPORTA[8] = {0, 0, 0, 0, 0, 0, 0, 0};

main()
{

  PORTA = 0;
  CMCON0 = 0b00011010;  // Initialize Comparators
   // Bit 6 - Comp 1 Output
   // Bit 4 - Invert Comp  (V- > V+ = 1)
   // Bit 3 - RA0 C1 Input (RC0 C2 Input)
   // Bit 2:0 - Comparator with Cref
  VRCON = 0b10100000;    // Vref Module Control
   // Bit 7 - Enable
   // Bit 5 - Low Range (0 - 2/3Vdd)
   // Bit 3:0 - Analog Levels
  ANSEL = 0;             // No ADC Inputs

  while(1 == 1)          // Loop Forever
  {
   for (i = 0; i < 8; i++ )
   {       // Loop through Each of the 8 LEDS
    for (j = 0; j < Dlay; j++);
                          //Display "On" Delay Loop
     if (0 == (DispValue & (1 << i)))
       PORTA = NOTPORTA[i];
     else
       PORTA = PORTAValue[i];
     TRISA = TRISAValue[i];
   } // rof
   switch (CompState)  // Comparator State Machine
   {
    case 0:             // Finished, Start Next
                            Sample
     NextVrefValue = 8; // Start with High Value
     VrefValue = 0;
     VRCON = 0b10101000;
     CompState++;
     break;
    case 1:     // Wait for ADC to complete
    // if (0 != (CMCON0 & (1 << 6)))  // Vin > Vref?
     if (1 == C1OUT)    // Vin > Vref?
       VrefValue = VrefValue + NextVrefValue;
     else          // No - Take Away from Vref
       VRCON = VRCON - NextVrefValue;
                    // Try next bit of binary search
     NextVrefValue = NextVrefValue >> 1;
     VRCON = VRCON + NextVrefValue;
     if (0 == NextVrefValue)
     {              // Finished, Display New Value
       DispValue = VrefValue;
       CompState = 0;
                    // and Restart Analog Measurement
     } // fi
     break;
   } // hctiws
  } // elihw
} // End cComp
```

To measure the potentiometer voltage, you could have used a simple algorithm that started with the Vref voltage at a minimum value (VR = 0b00000) and incremented it until the comparator changed state. I used the binary search algorithm, because it requires the logarithm base 2 comparisons of the number of bits instead of potentially having to run through each bit. In this case, using the binary search algorithm reduced the number of comparisons from a maximum of 16 to four. A binary search algorithm used in comparison operations like this one continually halves the comparison increment until nothing is left. As I show in the following pseudocode, a test value is used, and if the value is greater, it is left as is, else, the value is removed. After the comparison is complete, the test value is divided by two again and added to the comparison value. This is repeated until the comparison value is zero.

```
ComparisonValue = Maximum / 2;
               // Start at 1/2 Maximum
CompareValue = ComparisonValue;
               // Add to the Comparison Value
while (0 != ComparisonValue)
               // Repeat While a Value to Add
{              // Determine if value to be Saved
 if (CompareValue > InputValue)
   CompareValue = CompareValue - ComparisonValue;
 ComparisonValue = ComparisonValue / 2;
               // Repeat Operation
 CompareValue = CompareValue + ComparisonValue;
} // elihw
```

The binary search is ideal for digital systems because it works natively with binary bits. The previous code is not significantly more complex than code that does an incrementing compare, and the operational speed gained is significant and grows in significance for each additional bit to compare.

As I alluded in the introduction to this experiment, when I first turned on the original code nothing happened. I was very sure that the hardware was good (it worked fine for cADC.c), so the problem was clearly a programming problem. This is a very difficult program to debug because the MPLAB® IDE simulator will not simulate the operation of the comparator and Vref module with an analog input. To find the problem I reviewed the operation of all the program's statements (fortunately it's a short program) and found the following one line

```
if (1 == (CMCON0 & (1 << 6)))
```

that never behaved as if the expression was true. This instruction ANDs the contents of CMCON with bit 6 set and returns true if it is equal to 1. My initial response was to reread the comparator section along with previously written applications to figure out if I was doing anything wrong. The code looked great. I then changed the program so every hardware register was renamed to a variable (for example, CMCON0 became _CMCON0) and went on to try different values of _CMCON0 to find out if there was something I wasn't seeing. Using this method I discovered there wasn't a value for _CMCON0 that would cause the expression to be true.

I looked at the expression "1 == (CMCON0 & (1 << 6))" in the if statement, and when I checked its result using the simulator, I discovered the error: the expression can never equal 1—it will equal 0b01000000, but never 1. When I changed the expression to test for not zero the program worked fine (this is commented out statement in cComp.c above). Further improving on the statement, I simply polled the C1OUT bit to be high, rather than manipulating the contents of CMCON0.

This description of what I did to solve the problem wasn't put in to scare you. You probably aren't comfortable yet doing this kind of debug, especially considering the simulator cannot simulate the operation of the comparator and Vref hardware and registers. But I want to point out the steps I followed:

1. Check over the hardware documentation to make sure the hardware interfaces are properly coded.

2. Test the operation of each statement. If necessary, change from a register to a variable; a variable can be declared by using a register name with an underscore (_) character preceding it.

3. Clearly articulate what the problem was. For me the problem was that the register expression "(CMCON0 & (1 << 6))" could never be equal to the comparison value (1).

4. Fix the problem according to what you feel the problem is.

5. Test the fix using the simulator tools created for the task of discovering the problem.

6. When you are satisfied, burn the new program into the PIC16F684 and see if this works better for you.

You should be able to follow these steps in debugging your own applications, and although they may seem slow and tedious, you will find the problem and have an idea how to fix it.

I want to point out that the root cause of my problem was my dogged insistence on programming applications using standard C and not using PICC Lite compiler extensions where they make sense. As I have said, C is not well designed to manipulate individual bits, but PICC Lite compiler gives you the ability to access bits individually, and all the register bits are coded in the include file.

Experiment 35—Watchdog Timer

PC/ Simulator

PICkit™ 1

starter kit

Parts Bin

1 PIC16F684

1 10-LED bargraph display

1 0.01 µF capacitor (any type)

1 Breadboard-mountable SPDT switch (E-Switch EG1903 recommended)

1 Two-cell AA or AAA battery pack

2 AA or AAA batteries

Tool Box

DMM

Needle-nose pliers

Breadboard

Wiring kit

The *Watchdog Timer* (WDT) is built into all PIC microcontrollers to provide a reliable method of resetting the microcontroller if execution ever stops following the expected execution path. This can happen when the PIC MCU runs in a high electrical-noise environment, such as near the flyback transformer of a TV set or near a car's ignition system. When the WDT is enabled, it continually counts down a set delay, and if it is not reset within this period (using the clrwdt instruction), the WDT resets the PIC MCU and execution starts over. Personally, I have never used the WDT in any of my applications, and I can think of only one commercial application that takes advantage of it. Along with this, the advanced reset functions of modern PIC MCUs eliminate much of the practical need for the WDT. Chances are you will never require it in your applications, but you should be aware of it and how it operates as there is one trap you will probably fall into when you work on your own applications.

That trap is to not correctly disable the WDT in the configuration fuse specification. Many new PIC MCU application developers will specify only the labels for the configuration fuse functions that they believe are required for their applications. Unfortunately, the configuration fuse functions are both positively (bit set) and negatively (bit reset) active, and not specifying a value for the bits could result in an unwanted function being active. The most common unwanted function being made active is the Watchdog Timer.

When the Watchdog Timer is enabled inadvertently in a simple application, with no clrwdt repeatedly executing in the application, the PIC MCU will reset itself after two seconds or so. The reason for the reset is indicated as the _TO bit of the STATUS register being reset on power up. The problem is, that the application will have executed for a couple of seconds, giving the appearance that the application started correctly, but a software problem caused it to reset itself and start over.

The reason for the reset can be clearly seen by running the MPLAB IDE simulator, by setting a breakpoint at the first executing statement and adding the Stopwatch function. If the WDT is active, the program will execute the first statement a second time, about 2.1 seconds after the start of the simulation. If you see this behavior at any time during your application development, you can fairly confidently assume that the WDT is becoming active.

The circuit that I used to test the operation of the WDT reset is shown in Figure 5-7 and wired on a breadboard as shown in Figure 5-8. The test code is cWDT.c and increments the number of LEDs that are lit on the 10-LED bargraph display once every 500 ms. To disable this function, simply change the WDTEN argument to WTDIS in the __CONFIG macro. Unless you are going to use the WDT in your application, then the WDTDIS argument must be present in every application that you create.

You should find that after you turn on power to the application, the first four LEDs light but are then turned off when the tenth LED is lit, and the process repeats itself over and over again. The 2.1-second delay between WDT resets is determined by hardware within the PIC MCU. I will discuss it in more detail later in this chapter.

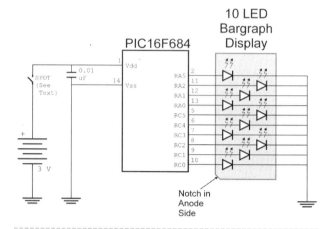

Figure 5-7 *WDT circuit*

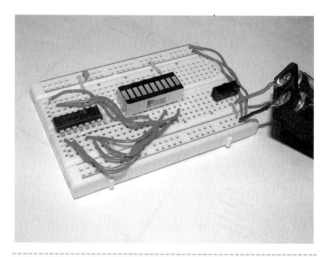

Figure 5-8 *Test application to demonstrate the operation of the WDT reset*

Previously I stated that I know of only one application that takes advantage of the WDT. That is the Parallax BASIC Stamp (and BASIC Stamp 2). In these PIC MCU–based devices, the WDT is used to time the *sleep* statement, which puts the BASIC Stamp to sleep for increments of 2.3 seconds (the maximum WDT reset interval for the PIC MCU chips that are used as the basis of the BS2).

Experiment 36—Short Timer Delays Using TMR0

PC/ Simulator

For most of the experiments presented in this book, I use some kind of incrementing or decrementing loop for producing delays. In the latter half of the book, I show how you can create very precise delays by counting out the execution time of the instructions and loops of instructions. But when a delay is required for programming in C, I tend to create a simple loop and then tune the delay values empirically using the MPLAB IDE simulator's Stopwatch function. The empirical approach works well for most large (i.e., tens of ms or longer) delays, but for shorter delays that require some precision, this method just isn't good enough.

A better way of producing accurate delays is to use a timer. The PIC16F684 has three built-in timers that you can use for producing delays of known lengths as well as for some of the hardware peripheral functions built into the microcontroller. In this experiment, I will introduce you to the operation of TMR0 (or Timer 0) and how it can be used to produce accurate delays up to 256 μs. In the following experiments, I will present the other two timers, their standard operation, and how they work with peripheral devices.

TMR0 is an eight-bit timer that can be driven either from an external source or from and internal instruction clock. Most applications use the instruction clock as a method of providing a set delay. In TMR0's basic configuration while running in a 4 MHz PIC MCU, delays of up to 256 μs (2^8) can be added to the application. An external source can also be used. And, as some people have noticed in the datasheets for different parts, the PIC MCU TMR0 module can take up to 50 MHz of input, leading to some interesting possibilities for measuring very high-speed signals.

Control for the input source for TMR0 is the T0CS bit of the OPTION register (see Figure 5-9). The OPTION register provides a number of execution control bits for the I/O pin internal pull ups, Watchdog Timer/TMR0 operation, and interrupt operation (Table 5-7). When you are working with the different hardware capabilities of the PIC MCU, you will find the OPTION register to be a very useful resource.

When TMR0 overflows, it sets the T0IF bit of the INTCON (interrupt control) register; to do so is an interrupt request. I do not discuss how interrupts are implemented in this book, but this bit can be useful to poll to find out if TMR0 has overflowed. In the low-end PIC MCUs, no interrupt capability exists and no T0IF bit exists so the contents of TMR0 will have to be continually polled.

The T0IF overflow bit is set when TMR0 changes from 0xFF to 0x00 and is the basis for basic delay timing. Assuming that the internal instruction clock is used for TMR0 and this clock is 1 MHz (the instruction clock speed of a 4 MHz PIC MCU), the formula for determining the delay is:

```
Delay (in μs) = (256 - TMR0 Initial Value)
```

So, to specify a 200 μs delay, the formula can be rearranged to find the TMR0 Initial Value:

```
TMR0 Initial Value = 256 - Delay
                   = 256 - 200
                   = 56
```

In cTMR0.c, the operation is demonstrated first by the while loop in which the incrementing of TMR0 can be seen. This is done by single-stepping through each

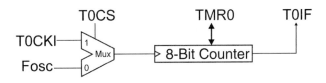

Figure 5-9 *Basic TMR0*

Table 5-7
OPTION Register Bits

Bit Name	Number	Function
_RAPU	7	Enable PORTA pull ups
INTEDG	6	Select the edge for RA2 interrupt request (1 = Rising)
T0CS	5	Select the TMR0 clock source (1 = RA2, 0 = Internal Clock)
T0SE	4	Select the edge for TMR0 updating (1 = Rising)
PSA	3	Select the prescaler device (1 = WDT, 0 = TMR0)
PS2:PS0	2:0	Select prescaler value (Prescaler value = $2^{PS2:PS0}$)

statement. Each C statement is made up of multiple PIC MCU assembly language instructions, so the different values by which TMR0 increments should not be a surprise.

```
#include <pic.h>
/*  cTMR0.c - TMR0 Operation and Short Delays

This program demonstrates the operation of the
TMR0 Hardware.

myke predko
04.11.28

*/

__CONFIG(INTIO & WDTDIS & PWRTEN & MCLRDIS &
UNPROTECT \ & UNPROTECT & BORDIS & IESODIS &
FCMDIS);

int i;

main()
{

    OPTION = 0b11011111;
            // Run TMR0 from the Clock Oscillator

    i = 0;
            // Demonstrate Operation of TMR0
    while (i < 1000)
        i = i + 1;

// Demonstrate a fairly precise 200 us delay

    TMR0 = 56;
            // Initialize TMR0/Breakpoint Here
    T0IF = 0;
            // Turn off Pending Interrupt Requests
    T0IE = 1;
```

```
        // Enable TMR0 Overflow to Request Ints
    while (!T0IF);
        // Wait for TMR0 to Overflow
    NOP();
        // Breakpoint Here - Time to Previous

    while (1 == 1);      // Loop Forever

}   // End cTMR0
```

In the second part of cTMR0.c, I demonstrate how to code a 200 μs delay controlled by TMR0. The first statement is putting in the TMR0 Initial Value, which is followed by a statement resetting the T0IF bit, and then one setting the T0IE bit to enable the interrupt request. In the while loop, I am simply waiting for T0IF to become set when TMR0 overflows.

By clicking on "Debugger" and "Stopwatch," and then placing breakpoints where I have indicated in cTMR0.c, I was able to time the execution from the setting of the TMR0 Initial Value to the final NOP();. I found the delay to be 207 instruction cycles, which is an error of seven instruction cycles. This translates to a delay of 3.5 percent. For most practical applications, this error is acceptable. Although some applications require absolutely precise timing, when providing a basic delay, an error of a few percent will not negatively affect the operation of the application and will keep you from carrying out what is essentially useless debugging.

Experiment 37—Using the TMR0 Prescaler

When you looked at the OPTION register definition (see Table 5-7), you probably noticed the PSA bit, which selected whether the prescaler was passed to TMR0 or the Watchdog Timer (see Figure 5-10). The prescaler is a programmable counter that can be used to change the TMR0 (or Watchdog Timer) timeout

interval. This converts these delay intervals from just a few hundred ms to several μs or even seconds.

The prescaler can be used with either TMR0 or the Watchdog Timer, and its use is selected by the PSA bit of the OPTION register. When the prescaler is selected for the timer, it divides the incoming clock signal according to the OPTION register's PS bits, with the divider being the $2^{PS2:PS0}$. When the prescaler is used with TMR0, the signal is synched with the internal instruction clock, which results in another division by two. So essentially, the prescaler divides the TMR0 input by $2^{PS2:PS0+1}$.

In cTMR0Pre.c, I specify that TMR0 is driving by the internal instruction clock, and I also reset the T0CS bit, which passes the clock signal through the prescaler.

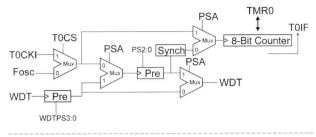

Figure 5-10 *Complete TMR0*

Because the prescaler bit selection (from PS2 to PS0) is still set, the prescaler delay is 1:256. Or, put another way, TMR0 is updated once every 256 instruction cycles.

```c
#include <pic.h>
/*  cTMR0Pre.c - TMR0 Operations with the
Prescaler

This program demonstrates the operation of the
TMR0 and Prescaler.

myke predko
04.11.28

*/

__CONFIG(INTIO & WDTDIS & PWRTEN & MCLRDIS &
UNPROTECT \ & UNPROTECT & BORDIS & IESODIS &
FCMDIS);

int i;

main()
{

 OPTION = 0b11010111;
        // Run TMR0 with 1:256 Prescaler

// Use Same Code as fairly precise 200 us delay
   of "cTMR0.c"

 TMR0 = 56;      // Initialize TMR0/Breakpoint
                        Here
 T0IF = 0;       // Turn off Pending Interrupt
                        Requests
 T0IE = 1;       // Enable TMR0 Overflow to
                        Request Ints
 while (!T0IF);  // Wait for TMR0 to Overflow
 NOP();          // Breakpoint Here - Time to
                        Previous

 OPTION = 0b11010001;  // Run TMR0 with 1:4
                            Prescaler
 TMR0 = 0x6;     // Execute 1000? Cycle
                        Delay/Breakpoint Here
 T0IF = 0;
 T0IE = 1;
 while (!T0IF);
 NOP();          // Breakpoint Here - Time to
                        Previous

 OPTION = 0b11010111;  // Run TMR0 with 1:256
                            Prescaler
 TMR0 = 0xFF;    // Execute 256 Cycle
                        Delay/Breakpoint Here
 T0IF = 0;
 T0IE = 1;
```

```c
 while (!T0IF);
 NOP();          // Breakpoint Here - Time to
                        Previous

 TMR0 = 0xFF;    // Execute 256 Cycle
                        Delay/Breakpoint Here
 T0IF = 0;
 T0IE = 1;
 for (i = 0; i < 10; i++); // Wait a Few Cycles
  TMR0 = TMR0 | 0x80;        // Set Bit 7 Again
 while (!T0IF);
 NOP();          // Breakpoint Here - Time to
                        Previous

 while (1 == 1); // Loop Forever

}   // End cTMR0Pre
```

Like cTMR0.c, cTMR0Pre.c requires the MPLAB IDE Stopwatch to understand what is going on. In the first TMR0 delay, TMR0 is loaded with an initial value of 56, and the time taken to reach the NOP(); is 51,208 cycles. This is 256 times 200 plus 8. The product of 256 and 200 is exactly what was specified by the TMR0 Initial Value, and the eight additional cycles are similar to the seven-cycle overflow of the previous experiment. Using the prescaler, a delay can be specified as:

$$\text{Delay} = 2^{\text{Prescaler} + 1} \times (256 - \text{TMR0 Initial Value})$$

where the prescaler value is defined as the base 2 logarithm of the delay divided by 256 (the maximum TMR0 value). This is a complex way of saying which power of 2 would produce a number large enough to store the value of delay. For example, if you wanted a 1 msec delay, you would use the following process:

$$\begin{aligned}\text{Prescaler} &= \text{Log}_2 \text{ Int}(1,000/256)\\ &= \text{Log}_2 \text{ } 3\\ &= 1\end{aligned}$$

Now, to find the TMR0 Initial Value, the formula is rearranged again to the following:

$$\begin{aligned}\text{TMR0 Initial Value} &= 256 - (\text{Delay} / 2^{\text{Prescaler} + 1})\\ &= 256 - (1000 / 2^{1 + 1})\\ &= 256 - (1000 / 2^2)\\ &= 256 - (1000 / 4)\\ &= 256 - 250\\ &= 6\end{aligned}$$

The second delay in cTMR0Pre.c shows the delay produced by this calculation and executed in 1,009 cycles according to the MPLAB IDE simulator.

The final two timer delays demonstrate what happens if you incorrectly poll TMR0 during its operation. In both cases, I set TMR0 to 0xFF, and with a 1:256

prescaler, you would expect that the delay would be 256 plus 7 or 8 extra cycles. And, this was what I got (264 cycles specifically). In the second case, a few cycles in, I write to TMR0 again: This could be considered a no operation because I am writing TMR0 with the same value as the one already stored in it. When you time the second case, you will discover that the delay is significantly different than the first. In my case, it was 429 cycles instead of the 264 of the first case.

The difference is that I am writing to TMR0. Even though I am not changing the value of TMR0 (and this can be confirmed in the simulator), the write operation always resets the prescaler. This is true for writes to any timer that has a prescaler (e.g., TMR1). For high-level programming, this is not an issue, but it will be when you are programming in assembler and you just want to test the value in the timer. Make it a rule to always save a timer value that has been read somewhere else.

Experiment 38—Long Timer Delays Using TMR1

Once you have worked with TMR0, you will not have any problems using TMR1 to provide the same delay functions. Looking at a block diagram of TMR1 (see Figure 5-11), you should see a lot of similarities to TMR0. The major differences are that TMR1's operation is controlled by a dedicated register (TMR0 has some functions specified in the OPTION register), TMR1 can have a unique oscillator assigned to it, and, most important, TMR1 is 16 bits in size. These differences make TMR1 quite a bit more flexible than TMR0 and one that you should consider using in all your applications.

The T1CON register (see Table 5-8) is used to control the different execution aspects of TMR1, including its source and prescaler value. In many applications, a 32.768 KHz watch crystal is connected to TMR1, giving the PIC MCU a real-time clock capability. The overflow output of the clock can be used as an interrupt request source, similar to TMR0. Or it can be used in

Table 5-8
T1CON Register

Bit Name	Number	Function
T1GINV	7	Timer1 Gate Invert (1 = External TMR1 control bit is active low)
TMR1GE	6	External TMR1 Control Bit Enable (1 = TMR1 controlled by external control it when TMR1ON is set)
T1CKPS1:0	5:4	TMR1 Prescaler Select (0b11 for 1:8. 0b10 for 1:4. 0b01 for 1:2, 0b00 for 1:1)
T1OSCEN	3	Enable External Oscillator when set
_T1SYNCH	2	Reset to Synchronize External Clock to Instruction Clock
TMR1CS	1	TMR1 Clock Source Select (1 = External clock, 0 = Internal instruction clock)
TMR1ON	0	Set to Enable TMR1

two functions of the *Enhanced Capture/Compare/ PWM* (ECCP) module, which will be demonstrated in more detail in the next experiment.

I created cTMR1.c to demonstrate the code required to create a 300 ms delay:

```
#include <pic.h>
/*  cTMR1.c - TMR1 Demonstration

This program demonstrates the operation of the
TMR1 Hardware.

myke predko
04.11.29

*/


__CONFIG(INTIO & WDTDIS & PWRTEN & MCLRDIS &
UNPROTECT \ & UNPROTECT & BORDIS & IESODIS &
FCMDIS);
```

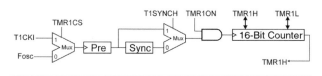

Figure 5-11   *Basic TMR1*

```
main()
{
    // Use TMR1 for a 300 ms delay

    T1CON = 0b00110001; // TMR1 On/Internal Clock,
                             8x Prescaler
    TMR1H = (65536 - (300000 / 8)) >> 8;
                        // Initialize a 300 ms delay
    TMR1L = (65536 - (300000 / 8)) & 0xFF;
    PEIE = 1;           // Enable Peripheral
                             Interrupts
    TMR1IF = 0;         // Turn off Pending Interrupt
                             Requests
    TMR1IE = 1;         // Enable TMR1 Overflow to
                             Request Ints
    while (!TMR1IF); // Wait for TMR1 to Overflow
    NOP();      // Breakpoint Here - Time to Previous

    while (1 == 1);  // Loop Forever

}  // End cTMR1
```

Looking at cTMR1.c, I was impressed with its simplicity and wondered how it compared to the standard for loop delay that I use in most of this book. The test case I created is the following cTMR1Ana.c:

```
#include <pic.h>
/*  cTMR1Ana.c - Code Based 300 ms Demonstration

This program demonstrates the equivalent code
solution to cTMR1.c.

myke predko
04.11.29

*/

__CONFIG(INTIO & WDTDIS & PWRTEN & MCLRDIS &
UNPROTECT \ & UNPROTECT & BORDIS & IESODIS &
FCMDIS);

unsigned int i;

main()
{
    // Use code for a 300 ms delay

    NOP();      // Breakpoint Here
```

Table 5-9
Comparing TMR1 to for Loop Delays

Parameter	cTMR1.c	cTMR1Ana.c
PIC MCU Instructions	22	34
File Register Bytes	0	2
Programming Iterations	1	15

```
    for (i = 0; i < 27272; i++);
    NOP();      // Breakpoint Here - Time to Previous

    while (1 == 1);  // Loop Forever

}  // End cTMR1Ana
```

Table 5-9 compares the two programs over a number of parameters. The Programming Iterations parameter is the number of times I changed the final value of the constant in both programs to get a delay that was within 300,000 cycles. This measurement consists of how many times I had to change the code, rebuild it, and retest it with the simulator's Stopwatch to get the specific delay. I wasn't surprised that using TMR1 for longer delays was better than the for loop; I was surprised at how *much* better TMR1 was. Aside from producing code that is two-thirds the size of the for loop, the TMR1 delay was also much faster and easier to get working correctly.

The obvious question that arises from this analysis is why people use the for loop to create delays. I believe two answers exist to this question. First, most low-end PIC MCUs (and low-end microcontrollers in general) do not have a 16-bit counter available for delays, so they have gotten into the habit of creating a delay using the for loop. Second, simply nobody has thought to do this analysis before. But if they had, the better delay would be obvious, and maybe somebody would have written a macro to produce the delay.

Experiment 39—Comparing Clock Oscillators

PC/ Simulator

PICkit™ 1

starter kit

Tool Box

DMM
Needle-nose pliers
Breadboard
Wiring kit

Required Parts

1 PIC16F684
1 PIC12F675
1 LED
1 2 MHz oscillator
1 4 MHz crystal
1 4 MHz ceramic resonator
   with internal capacitors
1 4.7k resistor
1 470Ω resistor
1 10k breadboard-mountable
   potentiometer
2 0.01 μF capacitors
2 33 pF capacitors
1 22 pF capacitor
1 SPST breadboard-mount-
   able switch
1 Three-cell AA battery
   clip
3 AA batteries

Part of the evolution of the PIC microcontroller has been for Microchip to provide accurate built-in oscillators, freeing the application developer from the chore of selecting and wiring a clock circuit into his or her applications. For people learning how to develop applications for the PIC, this was another potential problem area: If the clock circuitry was not correctly wired, then the application wouldn't work. This is not a concern for you. The oscillators built into the PIC16F684 and PIC12F675 are surprisingly accurate, allowing you to use them instead of external oscillators. In this experiment, I want to look at the accuracy of the different oscillators available to the PIC MCU.

The most efficient method of testing different clocking circuits is to use a frequency counter or an oscilloscope with a measurement option. I realize that most people do not have this equipment available to them, so in this experiment I will use a very simple circuit that will give you a good visual indication of the accuracy of the different clocking methods. I will run two PIC MCUs (a PIC12F675 and PIC16F684) at the same time, using the same program, and then pass a counted-down output to the different pins of an LED (see Figure 5-12). For simplicity, I put both PIC MCUs on the same breadboard as shown in Figure 5-13, using a long breadboard to allow for lots of space for the different clock options.

If the two frequencies are identical, the LED will light and the brightness will not change. If the frequencies are different, the LED will flash on and off. The slower the flashing, the closer the two frequencies are. If you are familiar with tuning a musical instrument, you will understand exactly how this works.

In this circuit, I will use a *canned oscillator* to drive the PIC12F675; this device is a highly accurate clock (usually to within 10 Hz for a 4 MHz device) that is used as a reference clock. The PIC16F684 clock will be changed to one of the four different types available to it. Along with the internal clock, or canned oscillator like the PIC12F675 uses, the PIC16F684 can be clocked using one of the three circuits show in Figure 5-14. These circuits are connected to the oscillator pins of the PIC16F684. For this experiment, I am assuming that each clock is running at 4 MHz (the most common clock speed for PIC MCU applications).

The code I used for the PIC12F675 was called c675Clk.c and uses TMR1 to count down 5 ms before toggling the output of GPIO0. The PIC16F684 uses appropriate register names and similar code, except it divides the constant by 4 rather than the 8 of C675Clk.c.

To test the different oscillators, the only changes made to the PIC16F684 code were in the configuration

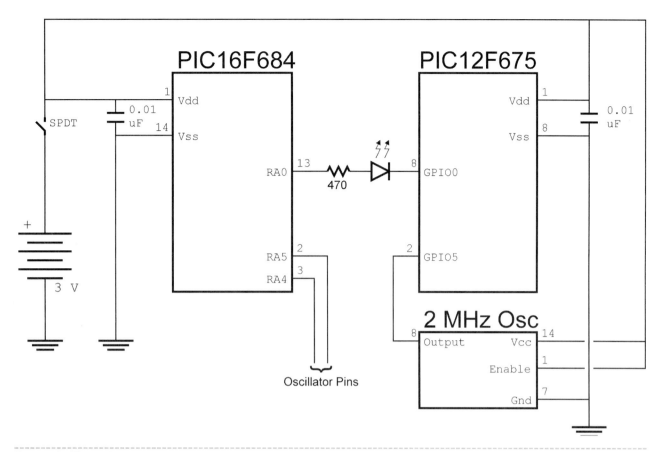

Figure 5-12   *Clock circuit*

fuse clock specification. The clock polling loop does not produce an exact 100 Hz output; it actually outputs 99.78 Hz due to overhead of the loop and the polling of the TMR1 registers. I could have tuned this to an exact frequency, but by keeping the code constant between the two PIC MCUs, I didn't feel this was necessary. The differences in clock speeds would show up as flashing LEDs. In Table 5-10, I have listed the different PIC16F684 oscillators used, their output frequencies measured by my oscilloscope (nominally 100 Hz), their LED flashing rate, and the configuration fuse specification used for each.

To try and improve upon the accuracy of the *resistor/capacitor* (RC) oscillator, I replaced the 4.7k resistor with a 10k potentiometer and tried to tune the signal. This proved to be quite difficult, and I never was able to closely match the PIC16F684's output frequency to that of the PIC12F675.

Next, I drove the PIC16F684 using the canned oscillator to look at the accuracy of the PIC12F675's internal oscillator. The PIC16F684's internal oscillator automatically uses a programmed calibration value,

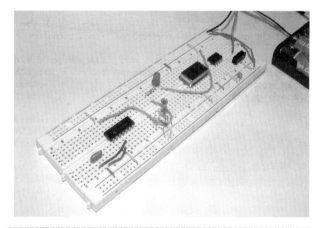

Figure 5-13   *Long breadboard used to test different PIC MCU 4 MHz clocking options*

but the PIC12F675's internal oscillator can optionally run with calibration. I was curious to find out how accurate this clock was. The results are in Table 5-11. To enable the PIC12F675's calibration, the following

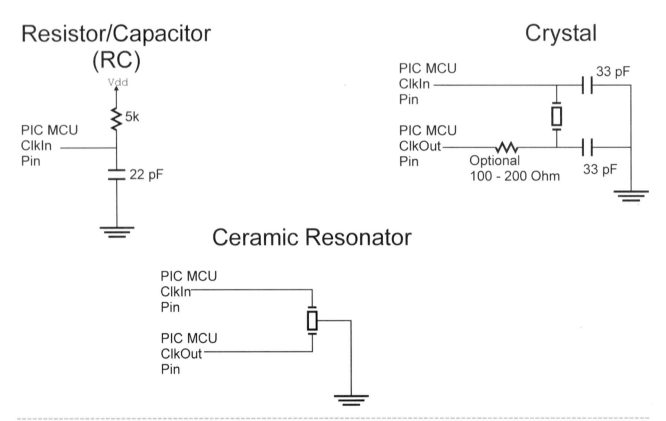

Resistor/Capacitor (RC)

Crystal

Ceramic Resonator

Figure 5-14 *Clock options*

code statements (taken from the PIC12F675 datasheet) were added to the start of the application:

```
            // Load Clock Calibration
#asm
 bsf    3, 5             ;  bsf    STATUS,
RP0
 call   03FFh            ;  call
GetCalibrateValue
 010h             ;  movwf OSCCAL ^ 0x90
 bcf    3, 5             ;  bcf    STATUS,
RP0
#endasm
```

To get the PIC MCU's actual operating frequency, the scaling factor 40,080 is multiplied by the 100 Hz output. This is the ratio between 4 MHz and the 10,022 cycles that were measured in the MPLAB IDE simula-

tor for one complete output cycle. Using this factor, the PIC16F684's internal oscillator runs at 3,987,976 Hz (0.3% error); the resonator runs at 3,995,992 Hz (0.1% error); the crystal runs at 4 MHz; and the RC oscillator runs at 3,330,662 Hz (16.7% error). The PIC12F675's uncalibrated clock is 4,224,449 Hz (5.6% error), and its calibrated clock runs at 4,092,184 Hz (2.3% error).

The obvious conclusion from this experiment is that the PIC16F684's internal clock and the PIC12F675's clock are quite good and are sufficiently accurate for virtually all applications. The applications where these clocks are not adequate would be timing measurement (e.g., in frequency counters and real-time clocks). In these cases, either a canned oscillator or an external crystal would have to be used. In older parts without a

Table 5-10
Comparing Different PIC16F684 Clocking Options to a Reference

Clock Type	Measured Frequency	LED Flash Rate	__CONFIG Clock Parameter
Internal	99.5 Hz	3 flashes/second	INTIO
Resonator	99.7 Hz	2 flashes/second	XT
Crystal	99.8 Hz	1 flash/6 seconds	XT
RC	83.1 Hz	many/second	RC

Table 5-11
PIC12F675 Internal Oscillator Accuracy

Calibration	Measured Frequency	LED Flash Rate
None	105.4 Hz	several/second
Using Calibration Code	102.1 Hz	several/second

built-in oscillator, the RC oscillator can be used. But remember, you can't expect the accuracy of the oscillator to be in the same range as that of the crystals and ceramic resonators.

I should point out that the results of this experiment, although reasonable, are not completely accurate due to the limited sample size of parts. Still the purpose of the experiment was to show the different oscillators available to the PIC MCU, how they are wired to the chip, and how they compared. When you repeat this experiment, you may find that different parts behave differently, and although the overall results will be similar, the measured values will most certainly be different from mine. For this experiment to be completely valid, a large sample size of the different parts would be required, and these parts should be chosen over a wide range of manufacturing dates.

Experiment 40—Timed I/O Pin Resistance Measurements Using the CCP

PC/ Simulator

PICkit™ 1

starter kit

Parts Bin

1	PIC16F684
1	10 LED bargraph display
1	100Ω resistor
1	10k breadboard-mountable potentiometer
1	470Ω 10-pin resistor SIP
2	0.01 μF capacitors
1	SPST breadboard-mountable switch
1	Three-cell AA battery clip
3	AA batteries

Tool Box

Needle-nose pliers

Breadboard

Breadboard wiring kit

Even if you are a teenager, I'm sure you've seen the early video games (like "Pong"), in which the user was given an analog control. The first home and personal computers also had analog controls, the joystick being the most common. But did you also know that 20 years ago, analog-to-digital converters (like the ones built into the PIC MCUs) often cost *more* than basic microprocessors? The ability to read position information was very important in these early games and simple computers, but adding an ADC chip was out of the question because the cost.

Imagine yourself as a designer of one of these systems. How would you provide the ability to read a potentiometer (which was normally used to read the position information)? You could use a comparator and a programmable voltage reference as I have used, but there is a much easier way to do this—by measuring the time it takes a charge to pass from a capacitor through the potentiometer. If you have read my previous *123 Robotics Experiments for the Evil Genius*, you would know that the Parallax BASIC Stamp II has the ability to time how long it takes for a charge to pass through a resistor (see Figure 5-15). In Figure 5-16, I show what the electrical signal looks like. First the capacitor is charged by an I/O pin, and then the I/O pin changes to an input to allow the capacitor's charge to pass through the resistor, with the delay from being fully charged to reaching the I/O pins high-to-low threshold being approximately proportional to the potentiometer's resistance.

As a rule of thumb, the maximum time delay for the circuit in Figure 5-15 is:

$$TimeDelay = 2.2 \times R_{Pot} \times C$$

So, for a 10k potentiometer and a 0.01 μF capacitor, you would expect there to be a maximum time delay of approximately 220 μs. I say "approximately" because potentiometers and small capacitors are not known for

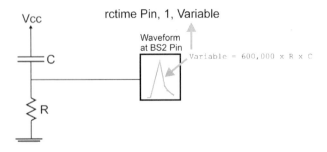

rctime Pin, 1, Variable

Waveform at BS2 Pin

Variable = 600,000 x R x C

Figure 5-15 *rctime function*

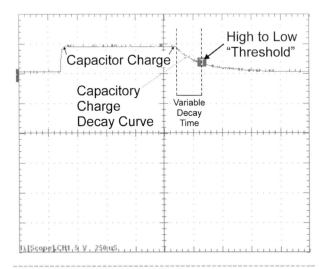

Figure 5-16 *Pot read*

their accuracy and you could have variances of up to 20 percent.

Unfortunately, the rctime function isn't available in the PICC Lite compiler library of functions, and it can not safely be programmed in by simply using a few C statements like the following:

```
RCPinTRIS = 0;      // Make RCPin an Output
RCPin = 1;          // Charge Capacitor
for (i = 0; i < onems; i++);  // Wait for
                                    Charging
RCPinTRIS = 1;      // Make RCPin and Input
for (i = 0; ((i < fivems) && (0 != RCPin)); i++);
// "i" contains a value proportional to the
   Resistance
```

This is true for two simple reasons: the execution time of the second for loop is not known and the returned value for i is proportional to the resistance of the potentiometer. If you have looked at some introductory PIC books, you may think this statement is wrong; many PIC books present code that can be used to find the resistance of a potentiometer, but the example code and projects in these books are written in assembler where each instruction (and the time required to execute) is known, not in a high-level language like C.

The obvious way of timing the code would be to start a timer and save its value when RCPin changed state as in the following code:

```
RCPinTRIS = 0;      // Make RCPin an Output
RCPin = 1;              // Charge Capacitor
for (i = 0; i < onems; i++);
                                  // Wait for Charging
RCPinTRIS = 1;      // Make RCPin and Input
TMR0 = 0;              // Reset the Timer
while (0 != RCPin);
                            // Wait for Pin to Go Low
PotValue = TMR0;  // Save the Pot Value
// "i" contains a value proportional to the
   Resistance
```

But a better way is to use the Capture mode of the ECCP Timer Generator. This function will save the contents of TMR1 at an external event. Using the ECCP module, there's no need to save values or worry about how many instructions take place in a loop or before or after the loop. The circuitry for this experiment (see Figures 5-17 and 5-18) is quite simple as is the code, which I call cPotTime.c.

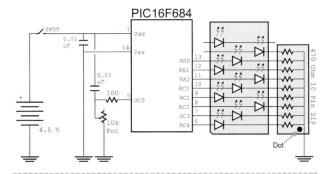

Figure 5.17 *Pot read circuit*

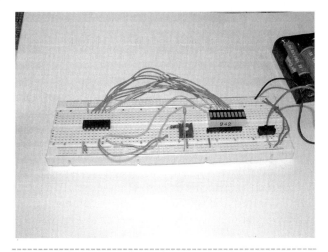

Figure 5.18 *Breadboard wiring for the potentiometer read circuit*

When you look through the cPotTime.c code you'll notice that it was written without using any variables. All the delays are provided by TMR1, which is loaded with constant values instead of the contents of variables. I have not explained the operation of the Capture mode of the ECCP, because when you are first starting out, there aren't that many applications like this one where you can take advantage of it. But you should be able to figure out how it and Compare mode work if you feel they have capabilities that you would want to take advantage of. The lesson I want to impart with this experiment is that if you have a problem, it can often be solved by looking at the datasheet.

Experiment 41—Generating PWM Signals Using the CCP and TMR2

PC/ Simulator

PICkit™ 1

starter kit

Parts Bin

1 LED

The PWM generator function built into the PIC MCU's ECCP is the latest and probably the most sophisticated and useful pulse width modulation (PWM) generator that I have seen on any microcontroller. The PIC16F684's PWM generator departs from the typical design by providing the capability to intelligently interface directly to high-current motor drivers without intervening logic or sophisticated software. As I will show in a later experiment, it is an ideal base for an intelligent DC motor controller. This experiment, however is somewhat more modest. It will simply introduce you to the operation of the PWM generator hardware and demonstrate how an LED's brightness can be changed with a PWM signal.

Before explaining how the PWM generator works, I should probably spend a few lines explaining what a PWM signal *is*. A PWM signal (see Figure 5-19) consists of a repeating digital signal in which the time on is varied to control the amount of power that passes to a device. PWMs are typically used in DC motors to control their speed because they are much more efficient than a variable-voltage output power supply. The *time on* is usually referred to as the *duty cycle* and is measured as a percentage of the total cycle time of the repeating signal.

The heart of the PWM is the TMR2, which is a repeating timer driven by the PIC MCU's instruction clock. When enabled, this timer will count to the value in the PR2 register and then reset itself. TMR2's operation is controlled by the T2CON register (see Table 5-12) and has a prescaler, which divides down the incoming clock, and a postscaler, which divides the reset signal down, all to provide a delayed event indication. The postscaler is not required to use the PWM generator. When TMR2 is run in a PIC MCU that has a 4 MHz signal, a prescaler value of 1:1, and PR2 equal to 0xFF (255), the TMR2 (and PWM) period is 255 ms (or 3,922 kHz),

The PWM generator circuit (see Figure 5-20) uses the reset command from the PR2/TMR2 magnitude comparator to set the output in an RS flip flop. The current value of TMR2 is continually compared

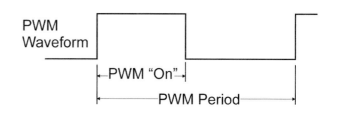

$$\text{Duty Cycle} = \frac{\text{PWM On}}{\text{PWM Period}}$$

Figure 5-19   *PWM*

Table 5-12
T2CON Register Definition

Bits	Label	Function
7		Not Used
6:3	TOUTPS	TMR2 Output Postscaler, 0b0000 is a 1:1 Postscaler, 0b1111 is a 1:16 Postscaler
2	TMR2On	Must be set for TMR2 to be operational
1:0	T2CKPS	TMR2 Prescaler
		0b1x for 1:16 Prescaler
		0b01 for 1:4 Prescaler
		0b00 for 1:1 Prescaler

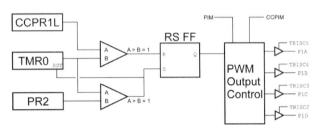

Figure 5-20   *PWM block diagram*

Table 5-13
CCP1CON Register Bits Relating to the PWM Generator

Bits	Label	Function
7:6	P1M	PWM Output Configuration Bits
		0b11 for Full-Bridge reverse output (P1B Modulated, P1C Set, P1A/P1D Reset)
		0b10 for Half-Bridge output (P1A, P1B Modulated, P1C/P1D Port Pins)
		0b01 for Full-Bridge forward output (P1D Modulated, P1A Set, P1B/P1C Reset)
		0b00 for Single (standard) output (P1A Modulated, P1B/P1C/P1D Port Pins)
5:4	DC1B	Least significant two bits of PWM compare value
3:0	CCP1M	ECCP mode select bits
		0b1111 for P1A/P1C/P1B/P1D Active Low
		0b1110 for P1A/P1C Active Low, P1B/P1D Active High
		0b1101 for P1A/P1C Active High, P1B/P1D Active Low
		0b1100 for P1A/P1C/P1B/P1D Active High
		0b0000 for CCP/PWM Off

against the value in the CCPR1L register, and when it is greater than the value in CCPR1L, the RS flip flop is reset. If the value in CCPR1L is greater than the value in PR2, the PWM output will always be high. This is what the basic PIC MCU PWM generator provides. But the PIC16F684 has additional circuitry known as the *output controller*, which can be used to simplify the control of different types of motor controllers. In Table 5-13, I have listed the relevant bits of CCP1CON as they relate to the PWM generator.

When I work with the PIC MCU CCP PWM, I generally work only with the most significant eight bits of the PWM Counters and ignore the two least significant bits. The extra accuracy provided by the least significant two bits is generally not required. In Figure 5-20, I imply that the PWM compare register (CCPR1L) value is compared against the value in PR2. This isn't strictly the case; the value of CCPR1L is buffered and used until PR2 is reset. This is done to ensure a larger value for CCP1RL is loaded in, causing the PWM output to go high unexpectedly. To keep everything simple for your initial PWM applications, I recommend that you either pass a single PWM signal from the module (on P1A, or RC5) or work with full H-bridges controlled as shown in Figure 5-21. When the I/O pins are described in the Microchip PIC16F684 datasheet, the labels P1x (where x is A, B, C, or D) are used exclusively. To try and avoid some confusion, in this experiment, I have included these labels but also included the traditional pin names as well.

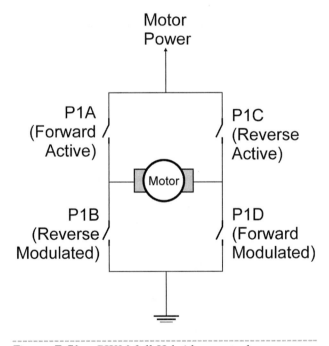

Figure 5-21   *PWM full H-bridge control*

I realize that I have thrown a lot of information at you regarding the PWM generator built into the PIC16F684, but it is actually very easy to set up and work with. I will show just how easy in the following experiment, which varies the brightness of an LED according to the position of the PICkit 1 starter kit's potentiometer. Unfortunately, none of the PWM outputs are connected to any of the LEDs on the PICkit 1 starter kit, and, further complicating manners, P1A (RC5) is not available on the PICkit 1 starter kit's in-line socket. Fortunately, RC2 (P1D) is close to the socket's ground pin, so you put an LED into the PICkit 1 starter kit to demonstrate the operation of the PIC MCU's PWM by varying the perceived brightness of the LED.

In this experiment, I wanted to take advantage of cADC.c, which reads the voltage coming from the PICkit 1 starter kit's potentiometer but also displays it as a binary value on the eight LEDs. A single LED can be wired from RC2 (P1D) to ground on the PICkit 1 starter kit (see Figure 5-22) to demonstrate the operation of the PIC16F684's PWM generator. The code written for the application (cPWM.c) is quite simple, with the only additions to cADC being the TMR2 enable, the PWM enable in *full-bridge output forward mode*, and the ADCValue read by the PIC MCU's ADC passed to the PWM value register (CCP1RL).

```
#include <pic.h>
/*  cPWM - Display the PICkit Pot Input Value as
an LED Brightness Level

This program samples the voltage on RA0 using
the ADC and Displays it using a PWM value on
"P1D" (RC2). This Program is designed for the

myke predko
04.11.16

*/

·__CONFIG(INTIO & WDTDIS & PWRTEN & MCLRDIS &
UNPROTECT \ & UNPROTECT & BORDIS & IESODIS &
FCMDIS);

int i, j, k;
```

Figure 5-22   *LED anode wired into the PICkit 1 starter kit connector at RC2 (third pin from bottom) with its cathode wired to ground (bottom pin)*

```
int ADCState = 0;     // Keep Track of ADC
                         Operation
int ADCValue = 0;
int Dlay = 63;        // LED Time on Delay
                         Variable
const char PORTAValue[8] = {0b010000, 0b100000,
                            0b010000, 0b000100,
                            0b100000, 0b000100,
                            0b000100, 0b000010};
const char TRISAValue[8] = {0b001111, 0b001111,
                            0b101011, 0b101011,
                            0b011011, 0b011011,
                            0b111001, 0b111001};
const char NOTPORTA[8] = {0, 0, 0, 0, 0, 0, 0, 0};

main()
{

  PORTA = 0;
  CMCON0 = 7;   // Turn off Comparators

  ANSEL = 1;    // Just RA0 is an Analog Input
  ADCON0 = 0b00000001;
                // Turn on the ADC
                // Bit 7 - Left Justified Sample
                // Bit 6 - Use VDD
                // Bit 4:2 - Channel 0
                // Bit 1 - Do not Start
                // Bit 0 - Turn on ADC
  ADCON1 = 0b00010000;    // Selemct the Clock as
                             Fosc/8

  PR2 = 0x0FF; // Work the Full Range
  T2CON = 0b00000100;     // TMR2 On with No
                             Prescaler

  CCP1CON = 0b01001100;   // Enable PWM with P1D
                             Active High
  PWM1CON = 0; // No Turn on/Off Delay
  TRISC2 = 0; // Make RC2/P1D Output Active
  CCPR1L = 0x080;
                // Start PWM at Intermediate Value

  k = 0;

  while(1 == 1) // Loop Forever
  {

    for (i = 0; i < 8; i++ )
    {          // Loop through Each of the 8 LEDS
      for (j = 0; j < Dlay; j++);
                // Display "On" Delay Loop
      if ((ADCValue & (1 << i)) == 0)
        PORTA = NOTPORTA[i];
      else
        PORTA = PORTAValue[i];
      TRISA = TRISAValue[i];
    }  // rof

    switch (ADCState)  // ADC State Machine
    {
      case 0:    // Finished, Start Next Sample
        GODONE = 1;
        ADCState++;
        break;
      case 1:    // Wait for ADC to complete
        if (!GODONE)
          ADCState++;    // Sample Finished
                break;
      case 2:    // Save Sample Value in "ADCValue"
        ADCValue = ADRESH;
        ADCState = 0;
        break;
    }  // hctiws
```

```
k = (k + 1) % 10;      // Update PWM Every 100
                          ms
if (0 == k)
  CCPR1L = ADCValue;   // Simply Change the PWM
                          Value to Change the
                          LED Output

} // elihw
} // End cPWM
```

Two points should be noted about this application. First, because I decided to use the full-bridge output mode of the PWM generator, pins RC5 (P1A), RC4 (P1B), and RC3 (P1C) cannot be used in the application. If this were a stand-alone application, I would recommend that you use the *single-output mode* of the PWM generator (CCP1CON loaded with 0b00001100) and the LED connected to RC5 (P1A). The second issue is that when I simulated the application (in MPLAB IDE version 6.60), there wasn't a CCPR1L register available; instead there was a CCPRL register, which performs the same function. This discrepancy will probably be fixed in later versions of MPLAB IDE.

Experiment 42—Storing and Retrieving Data Using EEPROM Memory

PC/ Simulator

PICkit™ 1

starter kit

An often overlooked feature of the PIC MCU is the EEPROM data memory that is built into the flash-based chips. EEPROM is an acronym for *electrically erasable programmable read-only memory*. In the PIC MCU, EEPROM consists of a number of bytes that can be written to with the expectation that the value written will still be there when the chip is powered down and then powered back up some time later. This expectation cannot be made of the static random access memory (SRAM)-based file register's that loose their values when the PIC MCU is powered down. EEPROM data memory (often referred to as simply *EEPROM*) is called *nonvolatile* to indicate that it retains its contents even when powered down. Not surprisingly, the SRAM-based file registers are said to be volatile. The flash program memory used in the PIC MCU is a variation of EEPROM with slightly different operating characteristics.

To access the EEPROM data memory, a series of assembly language instructions are used. Fortunately, the PICC Lite compiler provides two library functions that can be used to read and write the EEPROM. Their prototypes are as follows:

```
unsigned char EEPROM_READ(unsigned char
EEPROMAddress);
```

```
EEPROM_WRITE(unsigned char EEPROMAddress,
unsigned char EEPROMData);
```

The PIC16F684 has 256 bytes of EEPROM while the PIC12F675 has 128 bytes of EEPROM. Therefore, the maximum EEPROMAddress is 0xFF and 0x7F for the PIC16F684 and PIC12F675, respectively. The data value is a full eight bits, which means the value stored at each EEPROM location is in the range of zero to 0xFF (255 decimal).

The typical uses for the EEPROM include saving calibration values or saving constants instead of creating constants. To demonstrate the operation of the EEPROM, I created cEEPROM.c, which will initialize the EEPROM memory of a PIC16F684 with a standard pattern (0x00, 0xFF, 0xAA, 0x55) so that on subsequent power on cycles, the stored value will be recognized as valid. The value at the next address (or zero if this is the first power cycle) will be displayed on the eight LEDs of the PICkit 1 starter kit. When the button is pressed, the ADC will be read and the value will be displayed on the LEDs, and when the button is released, the ADC value will be stored in EEPROM for the initial display when the PIC MCU powers up.

```
#include <pic.h>
/*  cEEPROM - Display the Saved EEPROM Value/ADC
if Button Pressed

This program Displays a Saved EEPROM value
Unless the Button on RA3 is pressed and then the
ADC is Displayed.  When the Button is Released,
the ADC Value is Saved in EEPROM.  This code is
based on "cADC.c".

myke predko
04.12.20

*/

__CONFIG(INTIO & WDTDIS & PWRTEN & MCLRDIS &
```

```
UNPROTECT \ & UNPROTECT & BORDIS & IESODIS &
FCMDIS);


int i, j;
int ADCState = 0; // Keep Track of ADC Operation
int ADCValue = 0;
int ButtonState = 0;
int Display, OldDisplay;
int Dlay = 63;     // LED Time on Delay Variable
const char PORTAValue[8] = {0b010000, 0b100000,
                            0b010000, 0b000100,
                            0b100000, 0b000100,
                            0b000100, 0b000010};
const char TRISAValue[8] = {0b001111, 0b001111,
                            0b101011, 0b101011,
                            0b011011, 0b011011,
                            0b111001, 0b111001};
const char NOTPORTA[8] = {0, 0, 0, 0, 0, 0, 0, 0};

main()
{

  PORTA = 0;
  CMCON0 = 7;   // Turn off Comparators
  ANSEL = 1;    // Just RA0 is an Analog Input

  ADCON0 = 0b00000001;     // Turn on the ADC
                 // Bit 7 - Left Justified Sample
                 // Bit 6 - Use VDD
                 // Bit 4:2 - Channel 0
                 // Bit 1 - Do not Start
                 // Bit 0 - Turn on ADC
  ADCON1 = 0b00010000;     // Selemct the Clock as
                              Fosc/8

  if ((0 == EEPROM_READ(0)) &&
      (0xFF == EEPROM_READ(1)) &&
      (0x55 == EEPROM_READ(2)) &&
      (0xAA == EEPROM_READ(3)))
      Display = EEPROM_READ(4);
  else
  {
    EEPROM_WRITE(0, 0);
    EEPROM_WRITE(1, 0xFF);
    EEPROM_WRITE(2, 0x55);
    EEPROM_WRITE(3, 0xAA);
    EEPROM_WRITE(4, 0);        // No Initial Value
    Display = 0;
  }   // fi

  while(1 == 1)                 // Loop Forever
  {
   for (i = 0; i < 8; i++ )
   {  // Loop through Each of the 8 LEDS
    for (j = 0; j < Dlay; j++);
      // Display "On" Delay Loop
     if ((Display & (1 << i)) == 0)
      PORTA = NOTPORTA[i];
     else
      PORTA = PORTAValue[i];
     TRISA = TRISAValue[i];
   }  // rof
```

```
   if (0 != RA3)     // Button Released
   {
     ADCState = 0;    // Reset ADC Operation
     if (0 != ButtonState)  // Display Has ADC
                               Value
     {
       EEPROM_WRITE(4, Display);
       Display = OldDisplay;
       ButtonState = 0;
     }  // fi
   }
   else                       // Button Pressed
     switch (ADCState)      // ADC State Machine
     {
       case 0:    // Finished, Start Next Sample
         GODONE = 1;
         ADCState++;
         break;
       case 1:    // Wait for ADC to complete
         if (!GODONE)
           ADCState++;          // Sample Finished
         break;
       case 2:    // Save Sample Value in "ADCValue"
         if (0 == ButtonState)
           OldDisplay = Display;
         ButtonState = 1;
         Display = ADRESH;
         ADCState = 0;
         break;
     }  // hctiws
  }  // elihw
}  // End cADC
```

To test the application rather than continually plug the PICkit 1 starter kit or the PIC16F684 in and out, you can control the power to the PIC device in the PICkit 1 starter kit by clicking on "Programmer" and then on "Properties" in the MPLAB IDE. The window that comes up will give you the option of turning on or off the power to the PIC MCU in the PICkit 1 starter kit's socket as well as passing a 2.5 kHz clock into the PIC MCU's RA4 pin.

To test this application, burn the cEEPROM.c code into the PIC16F684 and let the PICkit 1 starter kit execute as it would normally. Next, press the PICkit 1 starter kit's button and turn the PICkit 1 starter kit's potentiometer to some value (which will be displayed on the PICkit 1 starter kit's eight LEDs). When the button is released, the power up value will be displayed, but after cycling the power on, the PICkit 1 starter kit the PIC16F684 will display the value previously set into the EEPROM using the button and the potentiometer.

Interfacing Projects for the PIC® Microcontroller

PC/ Simulator

PICkit™ 1

starter kit

Tool Box

DMM

Needle-nose pliers

Breadboard

Wiring kit

Required Parts

1 PIC16F684

1 74S138 3 to 8 decoder

1 74LS174 hex D flip flop

2 2N3906 PNP transistors

1 Two-line, 16-column LCD

1 1N914 (1N4148) silicon diode

1 Seven-segment common cathode LED display

1 Dual seven-segment common anode LED display

1 Seven-row, five-column LED matrix display

1k resistor

7 470Ω resistors

9 100Ω resistors

2 0.01 µF capacitors (any type)

1 16-button switch matrix keypad with an eight- or nine-pin breadboard interface

1 Breadboard-mountable SPDT switch (E-Switch EG1903 recommended)

1 Breadboard-mountable momentary On push button

1 Three-cell AA battery pack

1 Two-cell AA battery pack

3 AA alkaline batteries

Before I start demonstrating how the PIC MCU interfaces to other digital electronic chips, I feel it would be useful to review the basics of digital logic and note some of the important characteristics to keep in mind when you are creating applications with the PIC MCU. There are literally dozens of different logic technologies, each with different operating character-istics. I've listed the most popular ones in Table 6-1. For the different varieties of TTL, C, AC, and HC/HCT logic families, the part number starts with 74, and the 4000 series of CMOS chips have four-digit part numbers, starting with 4. Table 6.1 lists the characteristics of the different types of logic chips you will want to work with:

Table 6-1
Digital Logic Technologies with Electrical Characteristics

Chip Type	Power Supply	Gate Delay	Input Threshold	0 Output	1 Output	Output Sink
PIC MCU	Vdd = 2 to 5.5 V	N/A	50% Vdd	0.6 V	Vdd − 0.7	25 mA
TTL	Vcc = 4.5 to 5.5 V	8 ns	N/A	0.3 V	3.3 V	12 mA
L TTL	Vcc = 4.5 to 5.5 V	15 ns	N/A	0.3 V	3.4 V	5 mA
LS TTL	Vcc = 4.5 to 5.5 V	10 ns	N/A	0.3 V	3.4 V	8 mA
S TTL	Vcc = 4.75 to 5.25 V	5 ns	N/A	0.5 V	3.4 V	40 mA
AS TTL	Vcc = 4.5 to 5.5 V	2 ns	N/A	0.3 V	Vcc−2 V	20 mA
ALS TTL	Vcc = 4.5 to 5.5 V	4 ns	N/A	0.3 V	Vcc − 2 V	8 mA
F TTL	Vcc = 4.5 to 5.5 V	2 ns	N/A	0.3 V	3.5 V	20 mA
C CMOS	Vdd = 3 to 15 V	50 ns	0.7 Vcc	0.1 Vcc	0.9 Vcc	3.3 mA*
AC CMOS	Vdd = 2 to 6 V	8 ns	0.7 Vcc	0.1 V	Vcc−0.1 V	50 mA
HC/HCT	Vdd = 2 to 6 V	9 ns	0.7 Vcc	0.1 V	Vcc−0.1 V	25 mA
4000	Vdd = 3 to 15 V	30 ns	0.5 Vdd	0.1 V	Vdd−0.1 V	0.8 mA*

The *output sink* currents are specified for a power voltage of 5 volts. If you increase the power supply voltage of the indicated CMOS parts (noted with an asterisk), you will also increase their output current source and sink capabilities considerably.

In this table, I marked TTL input threshold voltage as not applicable because TTL is current driven rather voltage driven. You should assume that the current drawn from the TTL input is 1 mA for a 0 or low input. CMOS logic is voltage driven, so the input voltage threshold specification is an appropriate parameter.

Knowing that each TTL input requires a current sink of just over 1 mA and most TTL outputs can sink up 20 mA, you might expect the maximum number of TTL inputs driven by a single output (which is called *fanout*) to be 18 or 19. The actual maximum fanout is eight to ensure there is a comfortable margin in the output to be able to pull down each output in a timely manner. Practically, I would recommend that you try to keep the number of inputs driven by an output to two but never exceed four. Some different technologies that you will work with do not have the same electrical drive characteristics and may not be designed to pull down eight inputs of another technology; so, to be on the safe side, always be very conservative with the number of inputs you drive with a single output.

The output current source capability is not specified because many early chips were only able to sink current. This was all that was required for TTL and it allowed external devices, such as LEDs to be driven from the logic gate's output without any additional hardware, and it simplified the design of the first MOSFET-based logic chips. The asterisk (*) in Table 6.1 indicates that the sink current specification is for 5 volts of power; changing the power supply voltage will change the maximum current sink capability as well.

There are three basic output types: totem pole, open collector, and the tristate driver, which is presented later in this chapter. In cases where multiple outputs are combined, different output types should *never* be combined due to possible bus contention.

The TTL output (see Figure 6-1) is known as a *totem pole output* because of its resemblance to its namesake. If you were to connect a totem pole output to a TTL input and measure the voltage at the input or output pins, you would see a high voltage, which the gate connected to the input would recognize as a 1. When a low voltage is output, the TTL gate will respond as if a 0 was input. What you are not measuring is the current flow between the two pins.

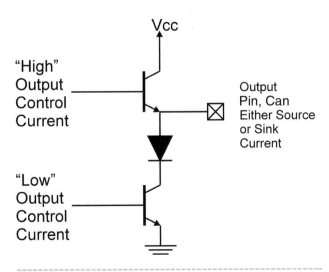

Figure 6-1 *TTL totem pole output*

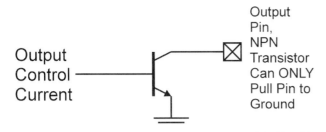

Figure 6-2   *TTL open-collector output*

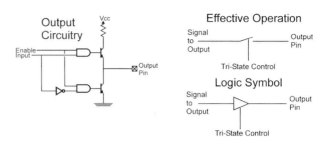

Figure 6-3   *Tri-state driver*

There is another type of output that does not source any current and is known as the *open-collector* (or *open-drain* for CMOS logic) output (see Figure 6-2). This output typically has two uses. The first pulls down voltages that are greater than the positive voltage applied to the chip. Normally these voltages are less than 15 V and can source only 10 to 20 mA. For higher currents and voltages, discrete transistors must be used. The second use is for dotted AND busses in which multiple open-collector/open-drain outputs are tied together with a single pull-up—the output is at a high-voltage level as long as all the output transistors are off.

Totem pole outputs are the recommended default gate output because you can easily check voltage levels between intermediate gates in a logic string. You cannot use a volt meter or logic probe to check the logic levels if a TTL gate is driven by an open-collector output. Additionally, if a CMOS input is connected to an open-collector/open-drain output then there will be no high voltage for the gate to operate. An open-collector/open-drain output can be used with CMOS inputs if a pull-up (typically 10k in value) is added to the output. The only cases where an open-collector/open-drain output should be used is when you are wiring a dotted AND gate or are switching a power supply that is operating at a voltage different from the gate's power.

The third output type, the *tri-state driver*, can not only source or sink current but can be turned off to electrically isolate itself from the circuit that it is connected to (see Figure 6-3). This type of output drivers is used on the PIC MCU I/O pins.

Except in the condition of open-collector/open-drain drivers being used to drive CMOS inputs without a pull-up resistor, these three output types can work with all of the digital logic inputs listed in Table 6-1.

Experiment 43 — Driving a Seven-Segment LED Display Directly from the PIC16F684

PC/ Simulator

PICkit™ 1
starter kit

Parts Bin

Tool Box

DMM
Needle-nose pliers
Breadboard
Wiring kit

1 PIC16F684
1 Seven-segment common
   cathode LED display
1 0.01 μF capacitor (any
   type)
1 Breadboard-mountable
   SPDT switch (E-Switch
   EG1903 recommended)
1 Two-cell AA battery
   pack
2 AA batteries

I found this experiment to be very satisfying. In about a half-hour, I was able to make it display each of the 10 digits on a seven-segment LED, with a half-second delay in between. This included modifying code that I stole from other applications to execute the application finding out the wiring was nonstandard on the seven-segment LED I used for my prototype. I expect that you will be able to create your own version of this application and see how easy it is to create a useful display using a single PIC microcontroller in short order.

The basis for the experiment is the seven-segment LED display (see Figure 6-4), which is normally wired in a *common cathode* or *common anode configuration*. The schematic diagram for the seven-segment LED display in Figure 6-4 shows how a common anode display would be wired. Even if you have a nonstandard seven-segment LED display (as I have), you can plot out the common pins and the pins for each segment in just a few minutes.

The basic 10 digits can be displayed by turning on the LEDs for the different segments as shown in Figure 6-5. Additional characters can be produced using different segments. Although other than the six most significant hexadecimal digits (A, B, C, D, E, and F), you will be hard pressed to come up with all the alphabetic characters that look good on this display (for example, M). The liquid crystal display (LCD) display discussed later in this section is a much better tool for this task.

The circuit for this application, not unexpectedly, consists of a PIC16F684, powered by two alkaline radio batteries and driving a seven-segment LED display (Figure 6-6), and can be wired very simply on a breadboard (Figure 6-7). The actual circuit is very similar to others that you have created although with a different LED display. Similarly, the application code (c7Segment.c) should not hold any surprises for you.

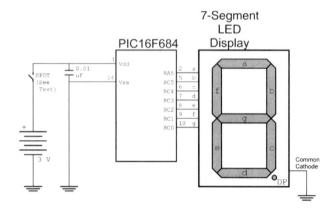

Figure 6-6   *Seven-segment LED circuit*

Figure 6-7   *Single seven-segment LED experiment circuit on a breadboard*

```
#include <pic.h>
/*  c7Segment.c - Roll through 10 Digits on 7
Segment LED Display

This program will display each of the Decimal 10
digits on a 7 Segment Common Cathode LED
Display.

Hardware Notes:

RA5 - Segment a
RC5 - Segment b
RC4 - Segment c
RC3 - Segment d
RC2 - Segment e
RC1 - Segment f
RC0 - Segment g

myke predko
04.11.10

*/

__CONFIG(INTIO & WDTDIS & PWRTEN & MCLRDIS &
UNPROTECT \
   & UNPROTECT & BORDIS & IESODIS & FCMDIS);

int i, j, k;

const char LEDDigit[10] = {
```

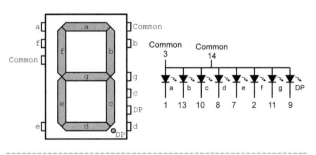

Figure 6-4   *Seven-segment LED display*

Figure 6-5   *Seven-segment numbers*

```
// RRRRRRR - PIC16F684 Pin
// ACCCCCC
// 5543210
// abcdefg - LED Segment
   0b1111110,              // Digit Zero
   0b0110000,              // Digit One
   0b1101101,              // Digit Two
   0b1111001,              // Digit Three
   0b0110011,              // Digit Four
   0b1011011,              // Digit Five
   0b1011111,              // Digit Six
   0b1110000,              // Digit Seven
   0b1111111,              // Digit Eight
   0b1111011};             // Digit Nine

main()
{

  PORTA = 0;
  PORTC = 0;
  CMCON0 = 7;              // Turn off Comparators
  ANSEL = 0;               // Turn off ADC
  TRISA = 0b011111;        // RA5 is an Output
  TRISC = 0b000000;        // All Bits of PORTC are
                           // Outputs

  k = 0;                   // Start at Digit 0

  while(1 == 1)            // Loop Forever
  {
    for (i = 0; i < 255; i++) // Simple Delay Loop
      for (j = 0; j < 129; j++);

      RA5 = LEDDigit[k] >> 6; // Pass Data Bits to
                              //            LED Bits
      PORTC = LEDDigit[k] & 0x03F;

      k = (k + 1) % 10;       // Increment k within
                              // range of 0-9

  } // elihw
} // End c7Segment
```

In the application code, I do want to point out that I stored the digit patterns in a 10-element array and that I tried to make the bit specifications as simple as possible for the programmer (including marking the segment letters and the I/O pin over each bit of the array). With only seven wires running from the PIC MCU to the display, I felt I could work at making the program as simple as possible for the programmer to correctly define the set pin outputs required for each digit.

Experiment 44 — Multiple Seven-Segment LED Displays

PC/ Simulator

PICkit™ 1

starter kit

Parts Bin

Tool Box

Needle-nose pliers

Breadboard

Wiring kit

1 PIC16F684

  2N3906 PNP transistors

1 Dual seven-segment common anode LED display

9 100Ω resistors

1 0.01 µF capacitor (any type)

1 Breadboard-mountable SPDT switch (E-Switch EG1903 recommended)

1 Two-cell AA battery pack

2 AA alkaline batteries

There are a number of ways to add multiple seven-segment LED displays to an application. You could route seven (or eight, if the decimal point is required) PIC MCUs to each display. You could use a commercially available seven-segment LED driver chip (the 7447 is most commonly used) for each display. You could also use a commercially available multidisplay controller chip (such as the MAXIM MAX7219). Or you could program the application's PIC MCU to sequence through the displays. The latter method is surprisingly easy to do and involves the least amount of hardware. There are a couple of points that you should be aware of, however, that I will present in this experiment.

The first is, you cannot drive the common anode or cathode pin using a PIC MCU I/O pin. The PIC MCU I/O pin cannot source or sink enough current to light all seven segments. Instead, I use a common NPN bipolar transistor for common cathode (see Figure 6-8) displays or a PNP bipolar transistor for common anode displays. These transistors can be controlled by either a PIC MCU I/O pin with a proper base-current-limiting transistor or by a data-selector logic chip (as I will show in the next section). You might be surprised to discover that common anode displays are most often used because they can be used with 74xx series logic data-selector chips and PNP bipolar transistors.

To sequence through multiple seven-segment LED displays, the controlling interface device outputs the value for a specific display while enabling only its common pin transistor as shown in Figure 6-8. To give the appearance that all the displays are active at the same time and avoid flickering, you must run through the entire sequence at least 50 times per second. This means that for the four seven-segment display example, each display can be active for a maximum of 5 ms.

To demonstrate the operation of multiple seven-segment LED displays, I created the simple circuit shown in Figure 6-9. I have not been specific on the type of dual common anode LED display to use because quite a bit of variance exists in different parts and you may find some inexpensive ones that are wired differently than the standard displays. All displays should have connections for the segments as well as common anodes for each digit display. The displays that I used wire a single pin to each segment, necessitating two connections for each segment, which resulted in the rather messy wiring of the circuit as seen in Figure 6-10.

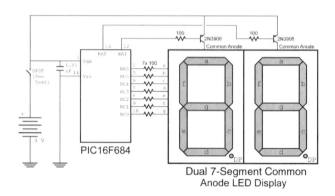

Dual 7-Segment Common Anode LED Display

Figure 6-9 *Dual seven-segment LED circuit*

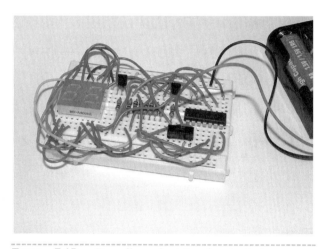

Figure 6-10 *Breadboard wiring for dual seven-segment LED display circuit*

```
#include <pic.h>
/*  c2x7Seg.c - Roll through 16 Digits on a Dual
7 Segment LED Display

This program will display each of the Hex 16
digits on two 7 Segment Common Cathode LED
Displays.

The Code/Wiring is based on "c7Segment.c" Noting
that Negative Active LED Wiring is used.

Hardware Notes:

RA5 - Segment a
RC5 - Segment b
RC4 - Segment c
RC3 - Segment d
RC2 - Segment e
RC1 - Segment f
RC0 - Segment g

RA0 - Right 7 Segment Display
RA1 - Left  7 Segment Display

myke predko
04.12.09

*/
```

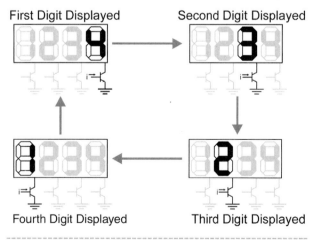

First Digit Displayed   Second Digit Displayed

Fourth Digit Displayed   Third Digit Displayed

Figure 6-8 *Multiple LED operation*

```
__CONFIG(INTIO & WDTDIS & PWRTEN & MCLRDIS &
UNPROTECT \
     & UNPROTECT & BORDIS & IESODIS & FCMDIS);


int i, j;
int DisplayValue, DisplayLED;

const char LEDDigit[] = {
// RRRRRRR - PIC16F684 Pin
// ACCCCCC
// 5543210
// abcdefg - LED Segment
     0b0000001,                    // "0"
     0b1001111,                    // "1"
     0b0010010,                    // "2"
     0b0000110,                    // "3"
     0b1001100,                    // "4"
     0b0100100,                    // "5"
     0b0100000,                    // "6"
     0b0001111,                    // "7"
     0b0000000,                    // "8"
     0b0001100,                    // "9"
     0b0001000,                    // "A"
     0b1100000,                    // "b"
     0b0110001,                    // "C"
     0b1000010,                    // "d"
     0b0110000,                    // "E"
     0b0111000};                   // "F"

main()
{

  PORTA = 0;
  PORTC = 0;
  CMCON0 = 7;         // Turn off Comparators
  ANSEL = 0;          // Turn off ADC
  TRISA = 0b011101;   // RA5/RA1/RA0 are Outputs
  TRISC = 0b000000;   // All Bits of PORTC are
                         Outputs

  DisplayValue = 0;   // Start Displaying at 0x00
  DisplayLED = 0;     // Display the 1s first

  while(1 == 1)       // Loop Forever
  {
   if (0 == DisplayLED)
   {
    RA5 = LEDDigit[DisplayValue & 0x0F] >> 6;
    PORTC = LEDDigit[DisplayValue & 0x0F] & 0x03F;
   }
   else
   {
    RA5 = LEDDigit[(DisplayValue >> 4) & 0x0F] >> 6;
    PORTC = LEDDigit[(DisplayValue >> 4) & 0x0F] &
                0x03F;
   } // fi
   TRISA = TRISA ^ 0b011111;       // Swap Left/Right
   PORTA = PORTA & 0b111100;       // Make Sure Bits
                                      are Low
   DisplayLED = DisplayLED ^ 1; // Other Digit
                                      Next

   NOP();             // Used for 10 ms Timing
   for (i = 0; i < 660; i++);   // 10 ms Delay
                                      Loop
   NOP();             // Used for 10 ms Timing

   j = j + 1;      // Increment the Counter?
   if (25 == j)    // 1/4 Second Passed?
   {
     DisplayValue++;               // Increment the
                                      Counter
     j = 0;         // Reset for another 1/4 Second
   } // fi
  } // elihw
} // End c2x7Seg
```

The basic application, incrementing the display once every 250 ms to count from 0x00 to 0xFF, is quite simple and is called c2x7Seg.c. There really should be no surprises in the code other than the fact that segment values for each bit are reversed from c7Segment.c because the PIC MCU are connected to the display's cathodes instead of anodes. The two NOP(); instructions around the final for loop were placed there to give me somewhere to break and to ensure the application delayed 10 ms, giving a 50 times per second update speed for the two displays.

After creating c2x7Seg.c, I thought it would be fun to come up with a c2x7SegP.c in which the lighted segment seems to run around the two displays. These types of displays are quite easy to write and a lot of fun to watch. After watching this, I can't help but think there must be a simple graphical game that could be created using seven-segment LED displays.

Experiment 45—LED Matrix Displays

PC/Simulator

PICkit™ 1

starter kit

Parts Bin

1 PIC16F684

1 74S138 3 to 8 decoder

1 Seven-row, five-column LED display

470Ω resistors

0.01 μF capacitor (any type)

1 Breadboard-mountable SPDT switch (E-Switch EG1903 recommended)

1 Three-cell AA battery pack

3 AA alkaline batteries

Tool Box

Needle-nose pliers

Breadboard

Wiring kit

Seven-segment LED displays are excellent for displaying numeric data, but when you have character data, their usefulness reaches the end very quickly. A number of different *alphanumeric LED displays* exist. The most basic type is the 14-segment LED display, which is similar to the seven-segment display that you just experimented with. However, due to the specific positions of different segments, it's difficult to make all the different characters recognizable on a 14-segment display. The *dot-matrix LED display* allows arbitrary characters to be displayed while using fewer I/O pins than the 14-segment LED display, but it requires more work to use and plan how to build your characters.

The dot-matrix LED display (which I call just an *LED matrix*) consists of a number of LEDs arranged with their anodes and cathodes as shown in Figure 6-11. By pulling a column connection to positive voltage, individual LEDs can be lit by connecting the appropriate row anodes to ground. For this experiment, I used a 5-column by 7-row LED matrix, but if you look around, you will discover a number of different sizes are available. Regardless of their size, they will be driven using the same methodology as shown here.

Using a 470Ω resistor, the battery clip, and a couple of wires, you can figure out the wiring of your display. I started by connecting pin 1 (the top left corner) to negative voltage and then scanning the pins to try and identify the seven anodes. I was not successful at lighting any LEDs, so I repeated the procedure with pin 1 connected to positive voltage. Once I started getting LEDs to light, it was a simple task of identifying the functions of the various pins and mapping them out on

a sheet of paper. In doing this, I discovered that the rows consisted of the LED's anodes and the columns were the LED's cathodes.

The 5×7-LED matrix display that I used requires 12 pins to operate; although the PIC16F684 is quoted as having 12 I/O pins, it actually has only 11 I/O pins and one input pin. RA3 can only be used as an input, so to take advantage of all 35 LEDs of the matrix that I used, I had to come up with some way to multiply the

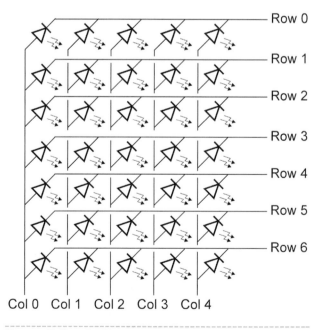

Figure 6-11   *LED matrix*

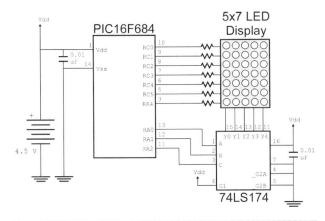

Figure 6-12 *Matrix circuit*

number of operational pins. In this application, I used a 74S138 decoder to provide the column negative power supply select. The selection of the S technology decoder was selected over the more common LS technology to sink the most current (LEDs fully lit) as possible. This allowed me to use three pins to select between five columns, requiring a total of 10 pins to control the LED matrix. The circuit that I came up with for this application is shown in Figure 6-12. Note that I tried to arrange the wiring in such a way that the anodes (rows) are all driven by PORTC (with the seventh bit being passed to RA4), while the cathode's (columns) pull-down decoder is selected by the three least significant bits of PORTA.

To test out my mapping of the LED matrix and how I presumed to wire the application (see Figure 6-13), I created cMatrix 1.c, which turns on each LED in the matrix in sequence.

With the LED matrix and driving circuit behaving as I expected, I then created cMatrix 2.c in which a col-

umn of LEDs is written at the same time and each column is scanned. This is similar to the dual seven-segment LED display experiment and gives the appearance that each column is active at the same time. cMatrix 2.c runs through the first four characters of the alphabet.

```c
#include <pic.h>
/*  cMatrix 2.c - Display Some Letters on 5x7
LED Display

RA0 - Decoder 0 Select
RA1 - Decoder 1 Select
RA2 - Decoder 2 Select
RA4 - Anode 6
RC0 - Anode 0
RC1 - Anode 1
RC2 - Anode 2
RC3 - Anode 3
RC4 - Anode 4
RC5 - Anode 5

myke predko
04.12.04

*/

__CONFIG(INTIO & WDTDIS & PWRTEN & MCLRDIS &
UNPROTECT \
  & UNPROTECT & BORDIS & IESODIS & FCMDIS);

int i, j, k, n, Dlay, CurLetter;

const char Letters[] = {0b1111110, // Start with
                                   //         "A"
                        0b0010001,
                        0b0010001,
                        0b1111110,
                        0b0000000,
                        0b1111111, // Next Letter
                                   //         "B"
                        0b1001001,
                        0b1001001,
                        0b0110110,
                        0b0000000,
                        0b0111110, // "C"
                        0b1000001,
                        0b1000001,
                        0b0100010,
                        0b0000000,
                        0b1111111, // Final
                                   //     Letter "D"
                        0b1000001,
                        0b1000001,
                        0b0111110,
                        0b0000000};

main()
{

    PORTA = 0;              // All Bits are Low
    PORTC = 0b000001;       // Start with Top Left
    CMCON0 = 7;             // Turn off Comparators
    ANSEL = 0;              // Turn off ADC
    TRISA = 0b101000;       // RA5/RA3 Inputs
    TRISC = 0;

    CurLetter = 0;          // Start with "A"
```

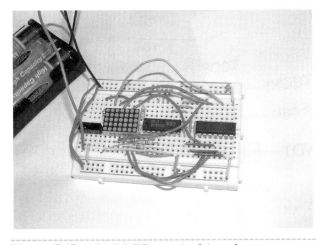

Figure 6-13 *5×7 LED matrix driven by a PIC16F684 and 74S126*

```
while(1 == 1)          // Loop Forever
{

  for (Dlay = 0; Dlay < 50; Dlay++)
  for (i = 0; i < 4; i++)
  {    // 20 ms to Display Character
  j = Letters[(CurLetter * 5) + i];
  k = (j >> 2) & 0b010000;
  PORTC = j & 0b111111;
  PORTA = k + i;
  for (n = 0; n < 259; n++);    // 4 ms Delay
  }  // rof

  CurLetter = (CurLetter + 1) % 4; // Increment
                                   Letter

}  // elihw
}  // End cMatrix 2
```

It is interesting to see that the character display is not substantially longer or more complex than the first set of code that moves the lit LED across the LED matrix. Most of the work was spent tuning the 4 ms delay in the display loop (the concept used was to cycle through each character over 20 ms, requiring 50 repetitions to display the character for 1 second). After mapping the LED matrix pins and before wiring the application, I spent a few moments thinking how to most efficiently wire the application and writing out some code to figure out the wiring arrangement that would keep the code to as few instructions as possible.

Experiment 46—LCD Display

PC/ Simulator

PICkit™ 1

starter kit

Parts Bin

1 PIC16F684

1 Two-line, 16-column LCD

1 10k breadboard-mountable potentiometer

1 0.01 μF capacitor (any type)

1 Breadboard-mountable SPDT switch (E-Switch EG1903 recommended)

1 Breadboard-mountable momentary On push button

1 Three-cell AA battery pack

3 AA alkaline batteries

Tool Box

Needle-nose pliers

Breadboard

Wiring kit

Creating applications with LCDs has a reputation for being difficult. I would disagree with this statement, noting that if you follow a simple set of guidelines you really won't have any troubles. The circuit presented in this application can be built and programmed in as little as 5 minutes. And, if you would like to enhance (or change) the output message or add more functionality to the application, you will find it to be quite easy (as I will demonstrate in the next experiment).

In *123 Robotics Experiments for the Evil Genius*, I included a fairly comprehensive explanation of how to interface to Hitachi 44780 controlled LCDs. I am not going to repeat that information here but will point you to my web site (www.myke.com) where you'll find information regarding interfacing to these LCDs and

programming them. Along with these two references, you'll find additional references on 44780-based LCDs. I should point out that 44780-based LCDs are text based only and have a 14- or 16-pin interface, and you can probably find some surplus or old products for just a dollar or two. Graphic LCDs have a different interface and require a completely different interface.

The circuit that I used for this experiment is shown in Figure 6-14, and there are two things I want to bring to your attention. Most LCDs require 4.5 to 5.5 volts, not the 3 volts I have used up to this point on breadboard circuits. To provide the LCD with the correct voltage level, I used three alkaline radio batteries in a series clip to provide 4.5 volts. If rechargeable batteries are to be used (which produce 1.2 volts per cell, instead

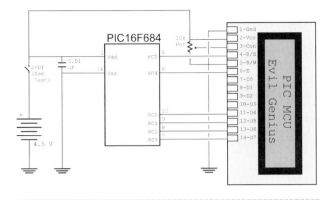

Figure 6-14   *LCD circuit*

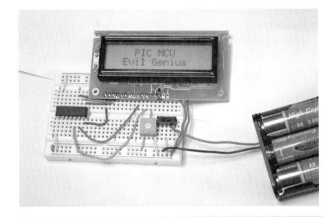

Figure 6-15   *LCD breadboard*

of the 1.5 volts per cell of alkaline batteries), you will have to use four in series.

Normally when I present an LCD application for the first time, I use the full eight-bit interface. In this case, because the PIC16F684 only has 11 pins with output capability, there would be no pins left for interfacing. Therefore, I use a four-bit interface. To reduce the required pin count to six, I also tied the *positive active read, negative active write* (R/W) low and don't poll the LCD. The six-bit interface uses all of PORTC, leaving PORTA free for interfacing to other devices.

When you look at the schematic in Figure 6-14, you might be surprised at some of the pin wiring choices. I wanted to make sure that the circuit could be wired easily and clearly on the breadboard (see Figure 6-15). Secondly, I wanted to arrange the bits so that the program could handle them easily with a minimum of processing.

```
#include <pic.h>
/*  cLCD.c - Write a String to a 4 Bit Hitachi
44780 LCD I/F
```

```
This Program Initializes Hitachi 44780 Based LCD
in 4 Bit Mode and then writes a simple string to
it.  The simulator was used to time delay values.

LCD Write Information can be found at
http://www.myke.com

RC3:RC0 - LCD I/O D7:D4 (Pins 14:11)
RC4     - LCD E Clocking Pin
RC5     - LCD R/S Pin

myke predko
04.11.08

*/

__CONFIG(INTIO & WDTDIS & PWRTEN & MCLRDIS &
UNPROTECT \
   & UNPROTECT & BORDIS & IESODIS & FCMDIS);

int i, j, k, n;  // Use Global Variables for
                      Debug

// 1234567890123456 <- Used to Organize Display
const char TopMessage[] = "    PIC MCU     ";
const char BotMessage[] = "   Evil Genius  ";

#define E   RC4              // Define the LCD
                               Control Pins
#define RS RC5

const int Twentyms = 1250; // Declare a Constant
                             for 20 ms Delay
const int Fivems = 300;
const int TwoHundredus = 10;

LCDWrite(int LCDData, int RSValue)
{

  PORTC = (LCDData >> 4) & 0x0F; // Get High 4
                                  Bits for
                                  Output
  RS = RSValue;
  E = 1;   E = 0;              // Toggle the High 4
                                 Bits Out

  PORTC = LCDData & 0x0F;     // Get Low 4 Bits for
                                Output
  RS = RSValue;
  E = 1;   E = 0;              // Toggle the Low 4
                                 Bits Out

  if ((0 == (LCDData & 0xFC)) && (0 == RSValue))
    n = Fivems;               // Set Delay Interval
  else
    n = TwoHundredus;

  for (k = 0; k < n; k++);   // Delay for Character

}  // End LCDWrite

main()
{

  PORTC = 0;      // Start with Everything Low
  CMCON0 = 7;     // Turn off Comparators
  ANSEL = 0;      // Turn off ADC
  TRISC = 0;      // All of PORTC are Outputs

// Initialize LCD according to the Web Page
  j = Twentyms;
  for (i = 0; i < j; i++);  // Wait for LCD to
                               Power Up
```

```
PORTC = 3;                // Start Initialization
                          //   Process
E = 1;   E = 0;           // Send Reset Command
j = Fivems;
for (i = 0; i < j; i++);

 E = 1;   E = 0;          // Repeat Reset Command
 j = TwoHundredus;
 for (i = 0; i < j; i++);

  E = 1;   E = 0;         // Repeat Reset Command
                          //   Third Time
  j = TwoHundredus;
  for (i = 0; i < j; i++);

   PORTC = 2;             // Initialize LCD 4 Bit
                          //   Mode
E = 1;   E = 0;
j = TwoHundredus;
for (i = 0; i < j; i++);

  LCDWrite(0b00101000, 0);  // LCD is 4 Bit
                            //   I/F, 2 Line

LCDWrite(0b00000001, 0);    // Clear LCD

LCDWrite(0b00000110, 0);    // Move Cursor After
                            //   Each Character
```

```
LCDWrite(0b00001110, 0);    // Turn On LCD and
                            //   Enable Cursor

for (i = 0; TopMessage[i] != 0; i++)
  LCDWrite(TopMessage[i], 1);

LCDWrite(0b11000000, 0);    // Move Cursor to
                            //   the Second Line

for (i = 0; BotMessage[i] != 0; i++)
  LCDWrite(BotMessage[i], 1);

while(1 == 1);              // Finished

}   // End cLCD
```

LCD displays in electronic applications are useful from a number of perspectives; first they allow you to create verbose, useful messages either to yourself (as debugging information) or to the user. Secondly, they add a level of commercialization to an application that gives the appearance that a lot more work has gone into it than actually has. Although requiring a bit more work to add to a microcontroller project than, say a few LEDs, LCDs can add a lot to your application, and I recommend that you consider them whenever it is possible.

Experiment 47—Producing Random Numbers

PC/ Simulator

PICkit™ 1

starter kit

Parts Bin

1 PIC16F684

1 Two-line, 16-column LCD

1 10k breadboard-mountable potentiometer

1 0.01 µF capacitor (any type)

1 Breadboard-mountable SPDT switch (E-Switch EG1903 recommended)

1 Breadboard-mountable momentary On push button

1 Three-cell AA battery pack

3 AA alkaline batteries

Tool Box

Needle-nose pliers

Breadboard

Wiring kit

People are usually surprised to find out that producing random numbers in a computer system is an incredibly hard thing to do. Probably one of the most creative methods for generating random numbers was created by Silicon Graphics, Inc. engineers who pointed a video camera at a "Lava Lamp" and used the digitized output as a random number. Computers are poor random number generators simply because they are designed to work the same way every time, and for most com-

puter applications, random results are not a good thing. Although very few large (on the order of billions of bits) random numbers are required within microcontroller applications, you'll discover that there are a number of applications where a few bits of random data would be useful. Computer games are a common computer application that uses random numbers, as are other applications that interface to humans. The element of randomness can make a game more interesting or help keep an application fresh to the user. In this experiment, I will show you how the users themselves can be used to create the random numbers for an application.

This might seem like a paradox: How can a user generate random numbers for an application that apparently behaves randomly for the user? The answer comes from the unexpected time the user takes to respond to program output. By running the TMR0 eight-bit timer at full speed and adding a debounced button on a pulled-up input pin to the LCD display application, a computer system that generates and displays random numbers can be built very easily (see Figure 6-16).

In cRandom.c, I have enabled the PORTA pull-ups to allow me to wire to ground a simple momentary on switch as my random input device. When the user pushes the button to produce a new random number, the current value of TMR0 is read and converted into a decimal number to display on the LCD.

This technique is very simple and, other than adding the switch, doesn't require any extra hardware. High-end randomizers produce their data from such

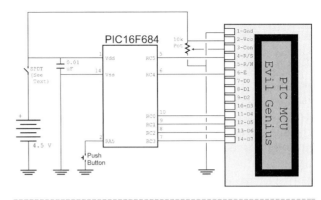

Figure 6-16   *Random circuit*

esoteric sources as the electrical noise inside a diode. Granted, the high-end randomizers can produce essentially limitless strings of random bits, but for most PIC microcontroller applications, the eight bits that are produced every time a user presses a button are more than adequate.

Experiment 48—Two-Bit LCD Display

PC/ Simulator

PICkit™ 1

starter kit

Parts Bin

1 PIC12F675

1 74LS174 hex D flip flop

1 1N914 (1N4148) silicon diode

1 Two-line, 16-column LCD

1 1k resistor

1 10k breadboard-mountable potentiometer

2 0.01 µF capacitor (any type)

1 Breadboard-mountable SPDT switch (E-Switch EG1903 recommended)

1 Three-cell AA battery

Tool Box

Needle-nose pliers

Breadboard

Wiring kit

A six-wire LCD interface is a great improvement over the full 11-wire original interface. It's still a lot of I/O pins, especially for the limited number in the PIC16F684 or the PIC12F675 (which also came with the PICkit™ 1 starter kit). What is needed is some way of converting a serial signal. A serial signal would be ideal, and a number of serial LCD interfaces are on the

market. These devices have three drawbacks. One is their cost: These interfaces can cost anywhere from $20 to $100. Secondly, they do not work at the speed of the microcontroller. Lastly, neither the PIC16F684 nor the PIC12F675 have an asynchronous serial interface. These issues do not eliminate the use of a serial LCD interface; they simply make its use more problematic.

A better solution is the two-wire synchronous serial interface shown in Figure 6-17. The shift register allows the minimal condition six-wire interface to be used with the LCD module. This circuit combines the most significant bit (which should only be set when all the other bits are loaded) and the data line to form the E clock. This method is obviously quite a bit slower than the method demonstrated in the previous experiment, but for most applications, the speed is not a major concern.

The actual operating waveforms are shown in Figure 6-18. First the contents of the shift register are cleared by shifting in six zeros and a high bit. Next the RS bit followed by the four data bits for the nibble are passed to the 74LS174, which is wired as a shift register. When the data is valid, the data line is strobed high and low to load and store the new character or instruction into the LCD. When I created the code for this experiment's circuit, I used the code from the previous experiment as a base.

For this experiment (see Figure 6-19), I used the PIC12F675 that came with the PICkit 1 starter kit. You could use a PIC16F684, but I wanted to use the PIC12F675 as an LCD enabled sensor controller in a (much) later experiment. By creating this experiment using the PIC12F675, I created the base that I would need for the later sensor experiments: A PIC16F684 (or any other PIC MCU) could be substituted in the

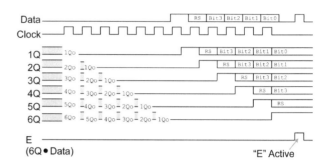

Figure 6-18   *Two-wire waveform*

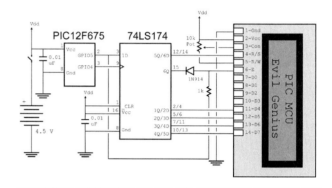

Figure 6-19   *Two-wire circuit*

circuit. You would simply have to change the Clk and the Data #define statements in the cLCD2Wire.c application with the pins and device that you are using, and then make sure the special function peripheral analog pin controls will allow these two pins to execute as digital I/O.

When I wired my prototype circuit, I used a long breadboard to allow a comfortable amount of space for the 74LS174 shift register. I also spent some time trying to keep the wires as neat as possible because I want to use this circuit in later experiments. If possible, put this breadboard circuit aside until you work up to the sensor experiments, which use this circuit as a base.

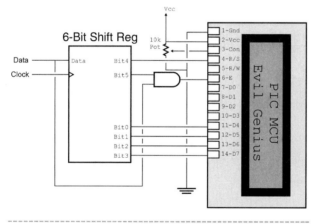

Figure 6-17   *Two-wire block diagram*

Experiment 49—Switch Matrix Keypad Mapping

PC/Simulator

PICkit™ 1

starter kit

Tool Box

Needle-nose pliers

Breadboard

Wiring kit

Parts Bin

1 PIC16F684

1 74LS174 hex D flip flop

1 1N914 (1N4148) silicon diode

1 Two-line, 16-column LCD

1 1k resistor

1 10-pin, 10k resistor SIP

1 10k breadboard-mountable potentiometer

2 0.01 µF capacitors (any type)

1 16-button switch matrix keypad with an eight- or nine-pin breadboard interface

1 Breadboard-mountable SPDT switch (E-Switch EG1903 recommended)

1 Three-cell AA battery pack

3 AA alkaline batteries

A switch matrix keypad can be a very useful tool for entering arbitrary data into an application or for application control. It can be so useful, in fact, that you will consider paying the often-exorbitant fee for a single keypad. New, they cost anywhere from $25 to $100 for a 4×4 switch matrix keypad. With a bit of hunting, you can probably find useable surplus switch matrix keypads for a few dollars. But, as they don't come with a datasheet, you will have to figure out how they are wired on your own. The task of *mapping*, or decoding, the keypad is surprisingly easy and can often be done using the same hardware as you will use for the final application that uses the keypad.

The keypad I chose to map for this experiment is the 16-pin keypad, shown in Figure 6-20, and has nine pins on the backside that can be pressed directly into a breadboard (which is why I chose it). The circuit used to map the keypad is based on the two-wire LCD interface shown earlier in this section, but instead of the PIC12F675, I used a PIC16F684 to take advantage of its 11 I/O pins. The circuit used in this application simply wires each pin to a pulled-up keypad pin with the two leftover pins connected to a two-wire LCD interface (see Figure 6-21).

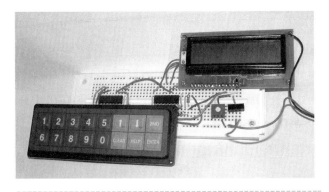

Figure 6-20 *Keypad mapping circuit consisting of a keypad connected to a PIC16F684 driving an LCD*

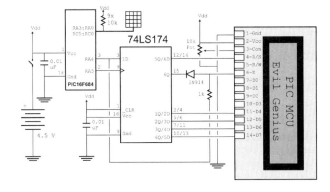

Figure 6-21 *Keypad decode*

The mapping software (cKeyDcode 2.c listed after this paragraph) is quite simple. It continually makes one pin a low output and scans the remaining pins to see if any are low (a key is pressed). If an input goes low, it is assumed to be a row, with the column being the output pin. These two pins are displayed on the circuit's LCD. When the key is lifted, the LCD values are cleared and the program resumes its scanning. I have not debounced the inputs, as multiple keypresses would be invisible to the user and not affect the operation of the application. I have not included the listing for cKeyDcode 1.c, as it is a copy of cLCD2Wire.c modified for the PIC16F684 and created as a base for cKeyDcode 2.c.

```c
#include <pic.h>
/*  cKeyDcode 2.c - Keypad Decode Utility

This Program Constantly scans the Keypad wired
to RA0-RA2 and RC0-RC5 by making one pin a low
output ("Column") and then scanning the pulled
up remaining pins ("Row") to see if any are low
and then displays the "Cathode" and "Anode" on
an attached 2 Wire LCD Interface.

This code is based on cKeyDcode 1.c

RA4 - Data
RA5 - Clock

RA0 - Pin 9 of Keypad
RA1 - Pin 8 of Keypad
RA2 - Pin 7 of Keypad
RC0 - Pin 6 of Keypad
RC1 - Pin 5 of Keypad
RC2 - Pin 4 of Keypad
RC3 - Pin 3 of Keypad
RC4 - Pin 2 of Keypad
RC5 - Pin 1 of Keypad

myke predko
04.12.05

*/

__CONFIG(INTIO & WDTDIS & PWRTEN & MCLRDIS &
UNPROTECT \
    & UNPROTECT & BORDIS & IESODIS & FCMDIS);

//                      0123456789012345
const char TopMessage[] = "Row:            ";
const char BotMessage[] = "Column:         ";

#define Clk  RA5       // Define the LCD Serial
                       //     Control Pins
#define Data RA4

const int Twentyms = 1250;   // Declare a
                             //     Constant for 20
                             //     ms Delay
const int Fivems = 300;
const int TwoHundredus = 10;

NybbleShift(int LCDOut, int RS) // Shift Out the
                                //     Nybble
{

int i;

  Data = 0;
  for (i = 0; i < 6; i++)      // Clear the Shift
                               //     Register
  {
   Clk = 1;                    // Just Toggle the
                               //     Clock
   Clk = 0;
  }  // rof

  LCDOut = LCDOut | (1 << 5) | ((RS & 1) * (1 << 4));
  for (i = 0; i < 6; i++)      // Shift Data Out
  {
   if (0 != (LCDOut & (1 << 5)))
    Data = 1;                  // Shift Out the
                               //     Highest Bit
   else
    Data = 0;
   LCDOut = LCDOut << 1;       // Shift Up the Next
                               //     Bit
   Clk = 1;                    // Clock the Bit into
                               //     the S/R
   Clk = 0;
  }  // rof

  NOP();
  Data = 1;                    // Clock the Nybble
                               //     int the LCD
  NOP();
  Data = 0;

}  // End NybbleShift

LCDWrite(int LCDData, int RSValue)
{                              // Send Byte to LCD

int i, j;

  NybbleShift((LCDData >> 4) & 0x0F, RSValue);
  NybbleShift(LCDData & 0x0F, RSValue);
                               // Shift out Byte

  if ((0 == (LCDData & 0xFC)) && (0 == RSValue))
   i = Fivems;                 // Set Delay Interval
  else
   i = TwoHundreds;

  for (j = 0; j < i; j++);  // Delay for Character

}  // End LCDWrite

main()
{

int i, j;

  PORTA = 0;              // Start with Everything Low
  CMCON0 = 7;             // Comparator Off
  ANSEL = 0;              // Turn off ADC
  TRISA4 = 0;             // Enable I/O Pins
  TRISA5 = 0;

// Initialize LCD according to the Web Page
  j = Twentyms;
  for (i = 0; i < j; i++);  // Wait for LCD to
                            //     Power Up
```

```
NybbleShift(3, 0);      // Start Initialization
                        Process
j = Fivems;             // Send Reset Command
for (i = 0; i < j; i++);

NybbleShift(3, 0);      // Repeat Reset Command
j = TwoHundredus;
for (i = 0; i < j; i++);

NybbleShift(3, 0);      // Repeat Reset Command
                        Third Time
j = TwoHundredus;
for (i = 0; i < j; i++);

NybbleShift(2, 0);      // Initialize LCD 4 Bit
                        Mode
j = TwoHundredus;
for (i = 0; i < j; i++);

LCDWrite(0b00101000, 0);  // LCD is 4 Bit I/F,
                          2 Line

LCDWrite(0b00000001, 0);  // Clear LCD

LCDWrite(0b00000110, 0);  // Move Cursor After
                          Each Character

LCDWrite(0b00001110, 0);  // Turn On LCD and
                          Enable Cursor

for (i = 0; TopMessage[i] != 0; i++)
  LCDWrite(TopMessage[i], 1);

LCDWrite(0b11000000, 0);  // Move Cursor to the
                          Second Line

for (i = 0; BotMessage[i] != 0; i++)
  LCDWrite(BotMessage[i], 1);

while(1 == 1)             // Finished - Scan the
                          Keypad
{
  for (i = 1; i <= 9; i++)
  {
    switch(i)             // Pull Down Each Pin to
                          Find "Columns"
    {
      case 1:
        TRISC5 = 0;
        break;
      case 2:
        TRISC4 = 0;
        break;
      case 3:
        TRISC3 = 0;
        break;
      case 4:
        TRISC2 = 0;
        break;
      case 5:
        TRISC1 = 0;
        break;
      case 6:
        TRISC0 = 0;
        break;
      case 7:
        TRISA2 = 0;
        break;
      case 8:
        TRISA1 = 0;
        break;
      case 9:
        TRISA0 = 0;
        break;
    }  // hctiws

for (j = 1; j <= 9; j++)
  switch(j)
  {
    case 1:
      if ((i != j) && (0 == RC5))
      {
        LCDWrite(0b10000101, 0);
        LCDWrite(j + '0', 1);  // Write Row
        LCDWrite(0b11001000, 0);
        LCDWrite(i + '0', 1);  // Write Column
        while (0 == RC5);
      }  // fi
      break;
    case 2:
      if ((i != j) && (0 == RC4))
      {
        LCDWrite(0b10000101, 0);
        LCDWrite(j + '0', 1);  // Write Row
        LCDWrite(0b11001000, 0);
        LCDWrite(i + '0', 1);  // Write Column
        while (0 == RC4);
      }  // fi
      break;
    case 3:
      if ((i != j) && (0 == RC3))
      {
        LCDWrite(0b10000101, 0);
        LCDWrite(j + '0', 1);  // Write Row
        LCDWrite(0b11001000, 0);
        LCDWrite(i + '0', 1);  // Write Column
        while (0 == RC3);
      }  // fi
      break;
    case 4:
      if ((i != j) && (0 == RC2))
      {
        LCDWrite(0b10000101, 0);
        LCDWrite(j + '0', 1);  // Write Row
        LCDWrite(0b11001000, 0);
        LCDWrite(i + '0', 1);  // Write Column
        while (0 == RC2);
      }  // fi
      break;
    case 5:
      if ((i != j) && (0 == RC1))
      {
        LCDWrite(0b10000101, 0);
        LCDWrite(j + '0', 1);  // Write Row
        LCDWrite(0b11001000, 0);
        LCDWrite(i + '0', 1);  // Write Column
        while (0 == RC1);
      }  // fi
      break;
    case 6:
      if ((i != j) && (0 == RC0))
      {
        LCDWrite(0b10000101, 0);
        LCDWrite(j + '0', 1);  // Write Row
        LCDWrite(0b11001000, 0);
        LCDWrite(i + '0', 1);  // Write Column
        while (0 == RC0);
      }  // fi
      break;
    case 7:
      if ((i != j) && (0 == RA2))
      {
        LCDWrite(0b10000101, 0);
        LCDWrite(j + '0', 1);  // Write Row
        LCDWrite(0b11001000, 0);
        LCDWrite(i + '0', 1);  // Write Column
        while (0 == RA2);
      }  // fi
      break;
```

```
     case 8:
      if ((i != j) && (0 == RA1))
      {
       LCDWrite(0b10000101, 0);
       LCDWrite(j + '0', 1);  // Write Row
       LCDWrite(0b11001000, 0);
       LCDWrite(i + '0', 1);  // Write Column
       while (0 == RA1);
      }  // fi
      break;
     case 9:
      if ((i != j) && (0 == RA0))
      {
       LCDWrite(0b10000101, 0);
       LCDWrite(j + '0', 1);  // Write Row
       LCDWrite(0b11001000, 0);
       LCDWrite(i + '0', 1);  // Write Column
       while (0 == RA0);
      }  // fi
      break;
     }  // hctiws
    LCDWrite(0b10000101, 0);
    LCDWrite(' ', 1);     // Clear Row
    LCDWrite(0b11001000, 0);
    LCDWrite(' ', 1);     // Clear Column

    TRISC = 0b111111;    // Restore Everything
                          High for Next
    TRISA = 0b001111;    // Scan
   }  // rof
  }  // elihw
 }  // End cKeyDcode 2
```

After running the program, I created a table that maps the rows and columns for each of the keypad's pins (see Table 6-1). The numbers after "Row" and "Column" indicate the keypad pin numbers, not some definition of row and column.

As I said previously, I used a keypad with a nine-pin interface—pin 9 is never listed as a row or column. Testing the keypad with a digital multimeter (DMM), I discovered that pin 9 is an additional row pin that gets pulled low when the "2nd" key is pressed. This feature allows "2nd" to be pressed along with the number keys to indicate a hexadecimal number. I will use this keypad (and the information provided here) as part of a sensor control module, and, instead of using pin 9, I have come up with a method of entering in hexadecimal values by recording that "2nd" was pressed.

Table 6-1
Sample Keypad Decoding Matrix (Numbers beside "Row" and "Column" are the keypad's pin numbers)

	Column 1	Column 2	Column 3	Column 4
Row 8	"2nd"	Down Arrow	Up Arrow	ENTER
Row 7	9	6	3	HELP
Row 6	8	5	2	0
Row 5	7	4	1	CLEAR

When I see an interesting keypad or switch matrix keyboard (often built for old home computer systems), the ease by which it can be decoded is what leads me to finally choose it. In this experiment, I showed you how to map a keypad; the process of mapping a keyboard is exactly the same—it simply requires more I/O pins. Although you might be tempted to map the keypad using a DMM, why don't you try out this application with something like a PIC16F877A? This chip (supported by the PICC Lite™ compiler, so the ckeyDcode 2.c application can be used as a base) provides up to 33 I/O pins, which means that you will be able to map keyboards with up to 31 I/O pins with two pins left over for the two-wire LCD interface.

For Consideration

Although it is going beyond the scope of this book, I did want to present you with the concept of interrupts and how they can perform some pretty amazing tricks. Interrupts are special software subroutines that execute when a hardware event (e.g., a timer overflowing, a pin changing state, data being received, or a conversion operation being completed) takes place. These subroutines are usually written to be independent of the mainline code, with flag bits or data variables passing the results to the mainline code. Interrupts can reduce your application code requirements by up to 80 percent and variable requirements by 30 percent or more. It can be difficult structuring an application to use interrupts and making sure they do not negatively affect the operation of the application.

The execution flow of the interrupt subroutine, or handler, is shown in Figure 6-22 and starts with a hardware event requesting a response from the interrupt handler. When the processor responds to the request, the current program counter is saved and the state flip flops in the processor are set to indicate that an interrupt is being handled. Execution control is then passed to the subroutine handler, and the registers required for all program execution (called the *context registers*) are saved before the registers are changed for the interrupt handler. The interrupt handler itself should execute as quickly as possible to avoid interrupting the flow and responsiveness of the mainline and to make sure it isn't executing when another interrupt request is coming in, which could result in the second interrupt being missed.

I must emphasize that interrupts are *requests*; the application code can choose to not respond to them immediately or even simply to handle them as part of the mainline code. This range of variability must be accommodated in the application to make sure inter-

For Consideration

rupt requests are not lost or handled in the wrong order, which can happen quite easily. Once the interrupt handler has finished executing, the context registers are restored to their pre-interrupt values, the interrupt request hardware is reset, and the execution returns to the mainline, as if nothing had happened.

Despite these warnings, interrupts can be added to an application quite easily as shown in cPKLEDInt.c. This application is another of the PICkit 1 starter kit LED sequence programs in which TMR1 is allowed to run at clock speed (4 MHz) with a divide-by-8 prescaler. Every time TMR1 overflows, which happens after a little over a half second, an interrupt request is made and responded to by turning off the current LED and turning on the next one in sequence.

```
#include <pic.h>
/* cPKLEDInt.c - Roll Through PICkit LEDs Using
TMR1 Int

This Program will roll through each of the 8
LEDs built
  into the PICkit PCB.  When the T1 interrupt is
active.

The LED values are:

LED    Anode    Cathode
D0     RA4      RA5
D1     RA5      RA4
D2     RA4      RA2
D3     RA2      RA4
D4     RA5      RA2
D5     RA2      RA5
D6     RA2      RA1
D7     RA1      RA2

myke predko
04.09.10

*/

    __CONFIG(INTIO & WDTDIS & PWRTEN & MCLRDIS &
UNPROTECT \
    & UNPROTECT & BORDIS & IESODIS & FCMDIS);
```

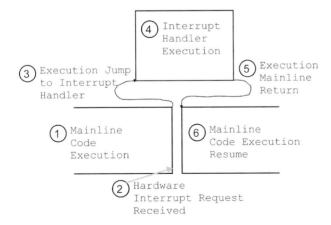

```
int k = 0;

const char LEDValue[8]  = {0b010000, 0b100000,
                           0b010000,0b000100,
                           0b100000, 0b000100,
                           0b000100, 0b000010};
const char TRISValue[8] = {0b001111, 0b001111,
                           0b101011, 0b101011,
                           0b011011, 0b011011,
                           0b111001, 0b111001};

void interrupt tmr1int(void)
{                           // Respond to Timer 1
                               Interrupt

   TMR1IF = 0;              // Reset Timer Interrupt
                               Request

   PORTA = LEDValue[k];     // Point to the Current
                               LED
   TRISA = TRISValue[k];

   k= (k + 1) % 8;          // Increment k within
                               range of 0-7

}  // tmr1int


main()
{
    PORTA = 0;
    CMCON0 = 7; // Turn off Comparators
    ANSEL = 0;                   // Turn off ADC
    TMR1L = TMR1H = 0;           // Reset Timer 1
    T1CON = 0b00110101;         // Enable TMR1

    PEIE = 1;    // Enable Peripheral Interrupts
    GIE  = 1;    // Enable Global Interrupts
    TMR1IE = 1; // Enable Timer 1 Interrupts

    while(1 == 1);              // Loop Forever

}  // End cPKLEDInt
```

The main interrupt operation control is the INT-CON register's *global interrupt enable* (GIE) bit, which, when set, allows interrupt requests from the different hardware sources in the PIC MCU. For every interrupt source, there is a register that ends in "IE" (for *interrupt enable*). And when this bit is set, an interrupt will be requested and the interrupt subroutine will be called. In the PIC MCU, there can only have one interrupt subroutine, and in the subroutine, the different *interrupt request flag* (IF) bits are polled to see the source of the interrupt. In cPKLEDInt.c, there is only one interrupt enable bit set (TMR1IE for TMR1 overflow interrupt requests), so I don't bother polling the register.

The PICC Lite compiler makes working with interrupts quite a nice experience; it places the interrupt subroutine handler code at the correct address within the PIC MCU and handles the context register, saving and restoring for you. When you are ready to start

Figure 6-22 *Interrupt flow*

For Consideration

trying out interrupts in your own application, I highly recommend using the PICC Lite compiler before assembly language.

Comparing the application hex file size of cPKLED.c and cPKLEDInt.c, you'll see that cPKLED.c is 130 instructions and uses 10 file register bytes, while cPKLEDInt.c is 102 instructions (a 20 percent reduction in size) and uses nine file register bytes. What is really amazing about this application is that the mainline, consisting of the single "while (1 == 1);" statement now, could be used for another task because the TMR1 interrupt driven subroutine is completely independent of the mainline code. And because it

requires approximately 38 instruction cycles out of every 524,288, only 0.0075 percent of the available instruction cycles are lost. You can have a complete application running while the LEDs are active and not even notice a drop in performance. This type of improvement in program parameters is pretty typical when interrupts are added to an application. When you are comfortable with programming the PIC MCU in C and assembler, you will be very ready to start to implement interrupts in your own projects. You'll learn how they can make a program more efficient and (ironically) easier to write because complex hardware interfaces can be independent of the mainline code.

Sample C Microcontroller Applications

A certain perception is held by many hobbyists that practical microcontroller applications *have* to be coded in assembler. The reason for this feeling is based on the expected superiority of assembly language applications in terms of speed and size. These expectations as to the superiority of assembly language really don't hold any water; modern compilers, such as PICC Lite™ compiler, produce machine code that is often just as efficient as that using assembly language for a given operation or statement. And no reason can be given to use assembly language except in situations where custom interfaces are required and not available in prewritten libraries. Being able to write applications in C is primarily dependant on the ability of the programmer to design application code that is properly organized.

For some reason, properly organizing the microcontroller application seems to be quite difficult, when in fact they usually only need to use the following template:

```
main()            // Basic Microcontroller
                  Application Code Organization
{

// Variable/Peripheral Initialization
// (Timer) Delay Initialization (if required)

    while(1 == 1)                // Main Loop
    {

//   Perform Interface Operation

//   (Timer) Delay before the Next Loop/Interface
     Operation

    }  // elihw
}  // End Sample C Based Application
```

There really isn't much to this type of application, although I should point out two very important assumptions:

- The output response to a given set of inputs (or internal state values) takes essentially zero time —the outputs are based on a calculation of the inputs.

- There is no chance that an input state will be missed.

As a rule of thumb, you can assume that "essentially zero time" means less than 5 percent of the next loop delay. Any longer than this (especially if it is variable), then you will find that the application will respond erratically and certainly not the way you expect.

If the application's inputs are asynchronous and there is a chance that something will be missed, then you will have to modify the application organization to poll the inputs while the next loop delay is executing:

```
main()            // Microcontroller Asynchronous
                     Input Application
{                 // Code Organization

// Variable/Peripheral Initialization
// (Timer) Delay Initialization (if required)

    while(1 == 1)                // Main Loop
    {

// Perform Interface Output Operation

       for (i = 0; i < Dlay; i++) // Next Loop
                                     Delay
       {
// Input Poll/Processing
       }  // rof

    }  // elihw
}  // End Sample C Based Application
```

This organization of the application code allows asynchronous inputs to be recognized without affecting the output timing. The reason for maintaining the output delay is to allow any output interface circuitry to complete their operation and any devices connected to them to accept the complete output from the microcontroller and respond appropriately. For proper operation, I recommend using a timer instead of a for loop as I have shown previously, and as the input is polled, the timer value should also be polled.

The two code samples shown here should not be surprising; I have been using them for many of the experiments' code and will use them as a base for the applications contained in this section. In most practical applications, a great number of *dead time* execution cycles exist that are not required for application execution. And the code samples presented here take advantage of this characteristic by organizing the input and output operations to best implement the application. By correctly organizing your applications execution, you will have a simple base that can be easily modified into providing the required functions. Most importantly, these sample applications can be written in either C or assembler with equal execution efficiency.

You probably have noticed that I have not included a list of parts and tools for the experiments in this section. This is deliberate, as each experiment is unique and independent of the others in the section. I have included the list of parts and tools required for each experiment, however, so you can build them on your own.

Experiment 50—Pumpkin LED Display

PC/ Simulator

PICkit™ 1

starter kit

Parts Bin

1	PIC16F684
1	14-pin socket
6	High-intensity orange LEDs
1	0.01 µF capacitor (any type)
1	120V 8-pin SIP resistor
1	SPST switch
1	Prototyping PCB (see text)
1	Three-cell AAA battery pack
3	AAA batteries

Tool Box

DMM
Needle-nose pliers
Soldering iron
Solder

I am breaking new ground in this book. In all my previous books, I show how to take advantage of *linear feedback shift registers* (LFSRs) to create a seemingly random display for the holidays. In this book, I am going to use LFSRs to create a random-seeming "candle" (see Figure 7-1) to put inside pumpkins at Halloween. The project idea was taken from Mark McCuller's "Pumpkin Light LED" project presented in October 2004's *Design News* and uses resistor- and capacitor-based reflex oscillators to flash LEDs on and off. I felt that the original would be quite complex to assemble and a PIC® MCU application would allow me to decrease significantly the number of parts required (and the work required to assemble it). I daresay my version is somewhat cheaper to build as well. Despite the project's simplicity, it actually works quite well and won't burn and dry up the inside of your pumpkins the way a regular candle does.

The Pumpkin circuit uses a 4.5V power supply (provided by three AA batteries) that serves as the base for a small, cut-down prototyping PCB to which a switch, a

PIC16F684, a 0.01 µF capacitor, a 120Ω SIP resistor, and the six high-intensity orange LEDs are soldered.

Figure 7-1   *Pumpkin artificial candle on a three-cell AA battery pack*

This circuit (see Figure 7-2) should not be a surprise to you, and is very similar to the previous experiments in this book. I was able to solder together the circuit neatly on a cut-down prototyping PCB. To further simplify the assembly work, I used an eight-pin, 120Ω *single-inline package* (SIP) resistor, which contains seven resistors, each connected to a common point.

To mimic the flickering candle, I wanted the LEDs to turn on and off seemingly randomly, for random periods of time (ranging from about one-tenth of a second to one or two seconds). The most obvious way of implementing this is by an LFSR, and I decided on using a six-bit LFSR for the six LEDs on PORTC of the PIC16F684 and a four-bit LFSR to specify the length of time the LEDs would be in a specific state. I expected the code to be very simple and similar to some of the earlier experiments. However, before blindly writing the LFSR code and attempting to debug it in the application circuit, I decided I would simulate it first.

The LFSRs for this application each have a single tap, which is XORed with the most significant bit and fed into the input of the first bit of the shift register. To test and demonstrate their operation, I created the application cLFSR.c:

```
#include <pic.h>
/*  cLFSR.c - Test LFSR Taps

This program Tests 6 and 4 bit LFSR Taps to make
sure they Perform a
  Maximum Number of times ((2 ** n) - 1)

myke predko
04.11.17

*/

int fourBitLFSR, sixBitLFSR;
int fourBitCount, sixBitCount;

main()
{

  fourBitLFSR = 1;    // Start at 1
  fourBitCount = 1;   // Keep Track of State #

  fourBitLFSR = ((fourBitLFSR << 1) & 0x0F) +
               ((fourBitLFSR >> 3) ^
               ((fourBitLFSR >> 2) & 1));
```

```
  while (fourBitLFSR != 1)
  {
    fourBitLFSR = ((fourBitLFSR << 1) & 0x0F) +
                 ((fourBitLFSR >> 3) ^
                 ((fourBitLFSR >> 2) & 1));
    fourBitCount = fourBitCount+ 1; // Keep Track
                                    of State #
  } // elihw

  sixBitLFSR = 1;    // Start at 1
  sixBitCount = 1;   // Keep Track of State #

  sixBitLFSR = ((sixBitLFSR << 1) & 0x3F) +
             ((sixBitLFSR >> 5) ^
             ((sixBitLFSR >> 4) & 1));
  while (sixBitLFSR != 1)
  {
    sixBitLFSR = ((sixBitLFSR << 1) & 0x3F) +
               ((sixBitLFSR >> 5) ^
               ((sixBitLFSR >> 4) & 1));
    sixBitCount = sixBitCount+ 1;    // Keep Track
                                     of State #
  }  // elihw

  while(1 == 1);

}  // End cLFSR
```

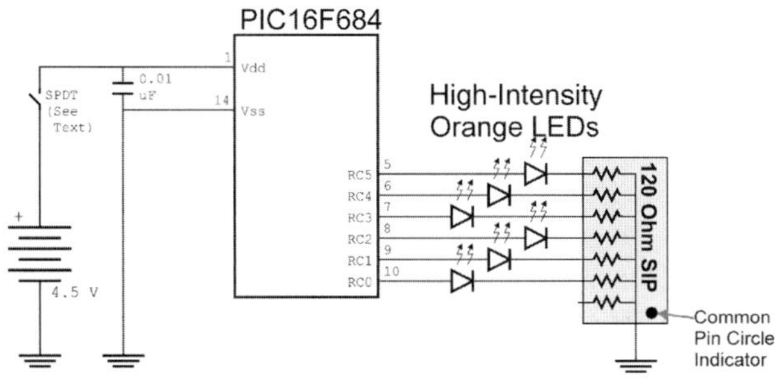

The two *count* variables are used to determine how many iterations of the LFSR would execute before the value returned to 1, at which time it would start all over again. When the LFSR taps are properly placed, there should be $2^n - 1$ iterations (15 for the four-bit LFSR and 63 for the six-bit LFSR). Amazingly enough, the two taps (those at the shift register's second to last bit) produced perfect results, and they were integrated into the cPumpkin.c application code:

```
#include <pic.h>
/*  cPumpkin.c - Randomly Light LEDs on PORTC

This Program will run continuously turning on
LEDs for varying Lengths of time to simulate a
candle in a Pumpkin.  Six Yellow and Orange High
Intensity LEDs should be used with this project.

Hardware Notes:
PIC16F684 Running at 4 MHz
RC5:RC0 - 6 LEDs

myke predko
04.11.17

*/

__CONFIG(INTIO & WDTDIS & PWRTEN & MCLRDIS &
UNPROTECT \
     & UNPROTECT & BORDIS & IESODIS & FCMDIS);


int i, j, k;
int fourBitLFSR, sixBitLFSR;

main()
{

  fourBitLFSR = 1;          // Start at 1
  sixBitLFSR = 1;           // Start at 1

  PORTC = 0;
  CMCON0 = 7;               // Turn off Comparators
```

PIC16F684

High-Intensity Orange LEDs

120 Ohm SIP

Common Pin Circle Indicator

4.5 V

Figure 7-2   *Pumpkin circuit*

```
ANSEL = 0;          // Turn off ADC
TRISC = 0;

while(1 == 1)       // Loop Forever
{
  for (k = 0; k < fourBitLFSR; k++)    // Delay
                                        0.1s x 4
                                        Bit LFSR

    for (i = 0; i < 255; i++)
      for (j = 0; j < 24; j++);

      PORTC = sixBitLFSR | fourBitLFSR; // Maximize
                                        # of
                                        LEDs on

  fourBitLFSR = ((fourBitLFSR << 1) & 0x0F) +
                ((fourBitLFSR >> 3) ^
                ((fourBitLFSR >> 2) & 1));

  sixBitLFSR = ((sixBitLFSR << 1) & 0x3F) +
               ((sixBitLFSR >> 5) ^
               ((sixBitLFSR >> 4) & 1));

} // elihw
} // End cPumpkin
```

The code is quite simple and straightforward except for two places I would like to bring to your attention. The first is in the delay loop: I changed the end values in a 500 ms delay so that it was 100 ms, and then I added another for statement to multiply the delay by the value of the four-bit LFSR. Similarly, rather than just pass the six-bit LFSR to PORTC, I decided to OR the two LFSRs together to get the maximum number of LEDs on at any given time.

The application works quite well and for a surprisingly long time. I must caution you against trying to stare into the high-intensity orange LEDs. They are bright enough to give you a headache and produce spots before your eyes.

Experiment 51—Reaction-Time Tester

PC/ Simulator

PICkit™ 1

starter kit

Parts Bin

1	PIC16F684
1	14-pin socket
1	10-LED bargraph display
1	10 μF electrolytic capacitor
1	0.01 μF capacitor (any type)
2	10k resistors
10	100Ω resistors
2	SPST switches
1	Prototyping PCB (see text)
1	Two-cell AAA battery pack
2	AAA batteries

Tool Box

DMM

Needle-nose pliers

Soldering iron

Solder

Wire-wrap wire

Weldbond adhesive

A gimmick that became quite popular when I came of drinking age was the reaction-time, or sobriety tester, that consisted of a simple circuit that would run a series of LEDs until a button was pressed. The reaction time being tested was that of an individual who had been drinking, and of course, the reaction time would vary according to how much he or she had to drink. The circuit presented here can be built for just a few dollars and an hour's worth of time.

The reaction-time tester circuit (see Figure 7-3) probably looks similar to the circuit used in the previous experiment. The difference between the two circuits is simply a 10-LED bargraph display and two buttons. You should also note that there isn't a power switch built into the application; I take advantage of the low-power sleep mode of the PIC16F684 microcontroller as well as its ability to run on the power provided by two AA batteries. The button connected to

RA2 will start the test, and the button at RA3 is the user respond button, which is to be pressed when the user sees the LEDs start to sequence on and off.

The operation of the circuit is quite simple, after power is applied (i.e., when RA2 is pressed), the circuit delays at some random interval (using the current TMR0 value) between one and two seconds. If RA3 is pressed during this time, the test stops and the two extreme LEDs start flashing. After the random interval, the LEDs are sequenced on until RA3 is pressed. If RA3 is pressed during the sequence, the operation stops at the currently displayed LED. If the user doesn't respond, then the last LED starts to flash. The flashing LED(s) or just one LED will stop, and the PIC MCU will go into sleep mode after 1 minute unless RA2 is pressed and another game is started.

When I built my prototype circuit (see Figures 7-4 and 7-5), I used a basic prototyping PCB that had some generic signal and power traces, and then filled in with point-to-point wiring for the remainder. When I was finished, I used a bead of Weldbond adhesive to hold down the wires. Before writing the application, I created two test programs. The first, which tests the two buttons, is cReact 1.c:

```
#include <pic.h>
/*  cReact 1.c - C Program to Test LED Buttons
on Reaction Tester

This Program is a modification of "cButton"

RA2 - Button Connection (Start/Power On)
RA3 - Button Connection (Respond)
RA0 - LED10
RA1 - LED9
RA4 - LED8
RA5 - LED7
RC0 - LED6
RC1 - LED5
RC2 - LED4
RC3 - LED3
RC4 - LED2
RC5 - LED1

myke predko
04.11.28

*/

__CONFIG(INTIO & WDTDIS & PWRTEN & MCLRDIS &
UNPROTECT \
    & UNPROTECT & BORDIS & IESODIS & FCMDIS);

int i, j;

main()
{
  PORTA = 0x3F;      // All Bits are high because
                        LEDs
  PORTC = 0x3F;      // are negative active
  CMCON0 = 7;        // Turn off comparators
  ANSEL = 0;         // Turn off ADC
  TRISA = 0b001100;  // RA2/RA3 are inputs
  TRISC = 0;         // All of PORTC are outputs
```

```
  while(1 == 1)      // Loop forever
  {
    RA1 = RA2;       // Power On switch (to LED9)
    RC5 = RA3;       // Respond switch (to LED1)
  }  // elihw
}  // End cReact 1
```

The second program, "cReact 2.c" tests the sequence operation of the LEDs and was based on "cPKLED.c":

```
#include <pic.h>
/*  cReact 2.c - C Program to Test LED
Connections on Reaction Tester

This Program is a modification of "cPKLED"

RA2 - Button Connection (Start/Power On)
RA3 - Button Connection (Respond)
RA0 - LED10
RA1 - LED9
RA4 - LED8
RA5 - LED7
RC0 - LED6
RC1 - LED5
RC2 - LED4
RC3 - LED3
RC4 - LED2
RC5 - LED1

myke predko
04.11.28

*/

__CONFIG(INTIO & WDTDIS & PWRTEN & MCLRDIS &
UNPROTECT \
    & UNPROTECT & BORDIS & IESODIS & FCMDIS);

int i, j, k;

main()
{
  PORTA = 0x3F;      // All Bits are high
                        because LEDs
  PORTC = 0x3F;      // are negative active
  CMCON0 = 7;        // Turn off comparators
  ANSEL = 0;         // Turn off ADC
  TRISA = 0b001100;  // RA2/RA3 are inputs
  TRISC = 0;         // All of PORTC are outputs
```

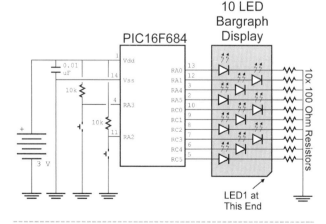

Figure 7-3   *Reaction-time circuit*

Figure 7-4 *The prototype reaction-tester circuit was built on a prototyping PCB*

```
k = 0;

while(1 == 1)              // Loop Forever
{

 for (i = 0; i < 255; i++) // Simple Delay Loop
  for (j = 0; j < 129; j++);

    PORTA = 0x3F;          // Turn off LEDs
 PORTC = 0x3F;

 switch(k)
 {
  case 0:               // LED1
   RC5 = 0;
   break;
  case 1:               // LED2
```

Figure 7-5 *Backside point-to-point wiring used on the prototype reaction-tester circuit*

```
   RC4 = 0;
   break;
  case 2:               // LED3
   RC3 = 0;
   break;
  case 3:               // LED4
   RC2 = 0;
   break;
  case 4:               // LED5
   RC1 = 0;
   break;
  case 5:               // LED6
   RC0 = 0;
   break;
  case 6:               // LED7
   RA5 = 0;
   break;
  case 7:               // LED8
   RA4 = 0;
   break;
  case 8:               // LED9
   RA1 = 0;
   break;
  case 9:               // LED10
   RA0 = 0;
   break;
 }  // hctiws

 k = (k + 1) % 10;

} // elihw
} // End cReact 2
```

After using cReact 1.c to verify the operation of the buttons and cReact 2.c to show the LEDs were wired correctly, I created cReact 3.c, which is the reaction tester.

In cReact 3.c, you'll see a couple of things that won't be obvious to you. The first is the group of nested for loops that also polls button and falls out if a button is pressed unexpectedly or out of turn. When you look at the loops, try to imagine their operation with both the button pressed and with it released. You can also simulate this code to see exactly how it works by using the synchronous stimulus function of MPLAB® IDE.

The second piece that will seem unusual is the sequence of instructions leading up to and following the SLEEP(); function. This is not a function, it is an assembler instruction that causes the PIC MCU to go into a low-power mode, or sleep, and it can be awakened either by cycling the application's power or causing an interrupt request on RA2. There is quite a bit of theory behind the sleep operation and the different modes that can be employed with it. Trust me; the "INTE = 1;/SLEEP(); NOP();" statement sequence will put the PIC16F684 into a low-power mode and will resume operation when the RA2 button is pressed.

Experiment 52—Rokenbok® Monorail/Traffic Lights

PC/Simulator

PICkit™ 1

starter kit

Parts Bin

1 PIC16F684

1 14-pin socket

2 Red LEDs

2 Yellow LEDs

1 10 μF electrolytic capacitor

1 0.01 μF capacitor (any type)

2 LDRs

2 10k resistors

2 100 Ohm resistors

1 SPST switch

1 Prototyping PCB

1 Three-cell AAA battery pack

3 AAA batteries

Tool Box

DMM

Needle-nose pliers

Dremel tool with abrasive cutter

Soldering iron

Solder

Wire-wrap wire

2 mm heat-shrink tubing

Weldbond adhesive

5-minute epoxy

Crazy glue

My daughter and I love to play with our Rokenbok remote-control building system. The toy consists of a number of remote-control construction vehicles that interact with building materials used to create different buildings and structures. To help enhance the experience, I wanted to modify a monorail/truck-level crossing so that dummy warning lights would flash when a train approached (see Figure 7-6). This seemed like a perfect project for a PIC MCU. I always find that hacking toys to be something of a challenge; the electronics built into the toys do not lend themselves to interfacing with more general-purpose electronics (generally the electronics selected were the lowest cost available and have their own special characteristics), structural features are difficult to cut through or modify without significantly changing the look of the toy, and the weight of the additional electronics can seriously degrade the performance of the toy. I can happily say that this modification went quite well, with the only major surprise being in the operation of the PIC MCU.

The theory behind the modification is quite simple. Two *light dependant resistors* (LDRs) were placed at either end of the crossing, by soldering them to cut down prototyping PCBs and gluing the PCBs into track pieces (see Figure 7-7). On power up, it is assumed that the train is not over the LDRs, and therefore they sample the ambient light eight times and average the result. When the train moves over the LDRs, it is assumed that the ambient light to them is cut off and the change in the LDR results in a change in the value read by the PIC MCU's ADC of greater than 25 percent. This causes the LEDs placed in the vehicle warning lights to flash for as long as either sensor is covered plus 10 seconds. Despite the simplicity of operation, the circuit works very well and is actually quite impressive.

Two nice things about this circuit are that the LDRs do not need to be wired to specific ADC inputs (either one will do) and, practically speaking, LED colors do not need to be matched. These advantages eliminated the need for testing each wire for a specific connection, and thus eliminated the difficult task of sorting out 10 two-foot wires in a small toy.

The installation of the LDRs is quite simple. The 2-foot-long pieces of wire were soldered to each of the LDRs' leads and covered in heat-shrink tubing. I used the three square black structural rods for the track and, using Krazy Glue, I attached the PCBs to the inside of

Figure 7-6 *Rokenbok monorail-level crossing with train sensors and flashing warning lights*

the middle square of the structural rod. At this position, the monorail and its trailer would cover both sensors. I was surprised to discover that Krazy Glue behaves as a solvent for the black plastic, literally welding in the small PCBs without need of another glue.

Red and yellow LEDs were installed in two warning lights that snap into stalks. I did so by drilling ³/₈-inch holes in the plastic on one side and ¹/₂-inch holes in the other to allow the flanges of the LEDs to fit through and the ends of the LEDs to poke out the other sides. Like the LDRs, the 2-foot lengths of wire were soldered to the LEDs, but with a few modifications. The cathodes of the two LEDs were joined with a 2.5-inch length of wire, and single cathode wire was used for the two LEDs. I then soldered a 100Ω resistor

to the anode of each LED that had a 2-foot piece of wire soldered to it and put heat-shrink tubing over the length of the resistor and its solder connections. These solder connections allowed me to easily identify which wire was used for which purpose. A simpler method of keeping track of the wiring might have been to use different color wires.

I threaded the wires through the structural rods and the bottom side of the crossing. Using a Dremel tool, I cut away ribs in the center of the crossing to allow the wires from one side to pass through without causing the crossing to rock when a vehicle or the train went over it. These wires were held in place using 5-minute epoxy.

With the toy modified and the wiring in place, I then created a prototyping PCB with the circuit shown in Figure 7-8. The 10k pull-up resistors on the LDRs will convert the LDRs into a voltage divider circuit that can be measured using the PIC MCU's ADC. This is actually quite a simple circuit, although you will find it quite difficult to wire using the different wires leading from the toy. When planning your PCB, make sure you keep the wires from interfering with the monorail train as it passes over the crossing.

To test the circuit, I created a simple program (called cRok 1.c) that simply flashed the LEDs for 30 seconds, and then I jumped right into the application (cRok 2.c):

```
#include <pic.h>
/*  cRok 2.c - Rokenbok Application

This Program will first sample the ambient light
on the two LDRs for 8 Samples, 50 ms apart.
With these Samples, the code will execute,
waiting for the sensors to be covered and then
uncovered.  When either is Covered (LDR Voltage
changes by 25% or more) then the LEDs flash.
When both are left uncovered, the lights flash
for an additional 10 seconds.

PIC16F684 Pins:
RA0 - LDR1
RA2 - LDR2
RC2 - LED1
RC3 - LED2

myke predko
04.11.30

*/

__CONFIG(INTIO & WDTDIS & PWRTEN & MCLRDIS &
UNPROTECT \
  & UNPROTECT & BORDIS & IESODIS & FCMDIS);


int i, j, k;
int LightTimer, FlashTimer;
int LDR0, LDR2;                 // LDR Average Values

const int fiftyms = 3333;  // 50 ms Delay
```

Figure 7-7 *Monorail LDR soldered to a small piece of PCB and glued inside the black monorail track (one on each side of the crossing)*

```
main()
{

   PORTA = 0;
   PORTC = 0;
   CMCON0 = 7;                    // Turn off Comparators
   ANSEL = 0b00000101;            // ADC on RA0 & RA2
   ADCON0 = 0b00000001;           // Turn on the ADC
                // Bit 7 - Left Justified Sample
                // Bit 6 - Use VDD
                // Bit 4:2 - Channel 0
                // Bit 1 - Do not Start
                // Bit 0 - Turn on ADC
   ADCON1 = 0b00010000;           // Select the Clock as
                                  Fosc/8
   TRISC = 0b110011;              // RC2 & RC3 are LED
                                  Drivers
   LightTimer = 0;                // Nothing Flashing Yet
   FlashTimer = 0;

   for (j = 0; j < 5; j++)  // Delay to Let Power
                            Settle
    for (i = 0; i < fiftyms; i++);
    LDR0 = 0;
   for (j = 0; j < 8; j++) // Get RA0 Light
                           Average
   {
    GODONE = 1;               // Sample RA0
    while (!GODONE);
     LDR0 = LDR0 + ADRESH;
    for (i = 0; i < fiftyms; i++);
   } // rof
   LDR0 = LDR0 / 8;               // Get Average Reading

   ADCON0 = 0b00001001;           // Sample on RA2
   LDR2 = 0;
   for (j = 0; j < 8; j++) // Get RA0 Light Average
   {
    GODONE = 1;               // Sample RA0
    while (!GODONE);
     LDR2 = LDR2 + ADRESH;
    for (i = 0; i < fiftyms; i++);
   } // rof
   LDR2 = LDR2 / 8;               // Get Average Reading
   for (i = 0; i < fiftyms; i++);

    j = 0;                        // Use j for Channel
                                  Select
   while(1 == 1)                  // Loop Forever
   {

    if (j < 2)                    // RA0 Read
    {
     ADCON0 = 0b00000001;  // Sample on RA0
     for (i = 0; i < fiftyms; i++);
      GODONE = 1;              // Do ADC Read
     while (!GODONE);
      k = (ADRESH * 100) / LDR0;
     j = j + 1;                // Increment Sample
    }
    else                         // Sample on RA2
    {
     ADCON0 = 0b00001001;  // Sample on RA2
     for (i = 0; i < fiftyms; i++);
      GODONE = 1;              // Do ADC Read
     while (!GODONE);
      k = (ADRESH * 100) / LDR2;
     j = (j + 1) % 4;
    } // fi

    if ((0 == (j & 1)) && ((k > 125) || (k < 75)))
    {                  // Start/Continue Flashing
```

```
     LightTimer = 10 * 20; // Flash for 10 Seconds
     if (0 == (PORTC & 0b001100))
     {          // If Not Flashing, Start
      PORTC = 0b001000;
      FlashTimer = 10;
     } // fi
    } // fi

    if (LightTimer != 0)    // Still Flashing?
    {
     LightTimer = LightTimer - 1;
     if (0 == LightTimer)
     {          // Finished Flashing
      FlashTimer = 0;        // Finished Flashing
      PORTC = 0;
     } // fi
    } // fi

    if (LightTimer != 0)    // Change Flashing?
    {
     FlashTimer = FlashTimer - 1;
     if (0 == FlashTimer)
     {                    // Change Flashing
                          Lights
      FlashTimer = 10;      // Reset Flash Timer
      PORTC = PORTC ^ 0b001100;
     } // fi
    } // fi
   } // elihw
} // End cRok 2
```

When I created cRoc 2.c, I expected it to work with minor tweaking. I was surprised to discover that when power was applied, the LEDs flashed continually. This is probably the most dreaded situation imaginable; I had a problem with a hardware device that was not supported by the simulator (which had worked correctly when I simulated it), and I had not designed any way of passing data out of the application. After about 3 hours of working at characterizing the problem, I discovered that to get an accurate reading for comparison to the average value, the ADC needed a dummy read after changing channels. When you look at this code, you'll see that I actually loop through each ADC's channel twice before I compare the result to the averaged amount for the channel.

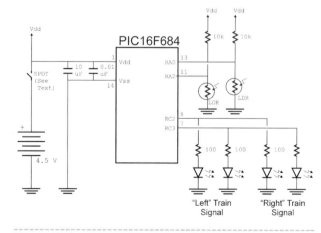

Figure 7-8 *Train crossing circuit*

To find the problem, I first simulated the application to make sure it worked as I expected it to. You will find that the simulator executes straight through the ADC operations with the GODONE bit being immediately reset after it is set. Using the Watch window register value update feature (double-click on the register's value to bring up an edit window), I tested the application for different ADC return values and made sure that the code worked properly. Next, I commented out all the code for one of the two channels and I found the application started working, until I returned the commented out code. Next, by flashing the LEDs, I output the value read versus the value average and discovered they were different. Finally, I tried to do a dummy read of the ADC channel before the live read, which was checked against the averaged value.

The previous paragraph does not convey the frustration and annoyance I felt in working through the problem. But, by working step by step, ensuring the program worked in the simulator as I expected, checking my assumptions rigorously, and not presuming the root cause of the problem beforehand, I was able to debug the application. This is a useful sequence you can use when you are faced with an application that doesn't work or behaves in a manner that is totally unexpected.

Experiment 53—Seven-Segment LED Thermometer

PC/ Simulator

PICkit™ 1

starter kit

Parts Bin

1 PIC16F684

1 74HCT138 three to eight decoder

1 14-pin socket

4 2N3906 bipolar NPN transistors

1 Thermistor (Radio Shack 271-110 recommended)

2 0.01 μF capacitors (any type)

8 100Ω resistors

4 330Ω resistors

1 10k, 1-percent tolerance resistor

1 SPST switch

1 Two-cell AA battery clip

Tool Box

DMM

Scientific calculator

Needle-nose pliers

Soldering iron

Solder

Wire-wrap wire

Weldbond adhesive

Probably the most popular PIC MCU project I have ever designed was a three-digit digital thermometer for the PIC16C84 that was written in assembly language. For some reason this circuit and software was extremely popular and was used as the basis for a wide variety of different applications, including monitoring the temperature of a chicken incubator! The original digital thermometer used a timed I/O pin resistance measurement of an RC network. Two input pins were used to calibrate the thermometer output, and the calibration value was saved in the data EEPROM of the PIC MCU. The original circuit works quite well, but I wanted to eliminate the need for the calibration value

as well as write the application code in a high-level language that did not obfuscate the operation of the thermometer. To meet these objectives, I used the PIC16F684's ADC along with the PICC Lite compiler language and its floating-point number capabilities.

The circuit (see Figure 7-9), although fairly complex, is quite straight forward. One common anode, seven-segment LED display is active at any one time, and the application sequences through each LED display 50 to 64 times per second, giving the appearance that all four are active at the same time. The seven-segment LED displays are powered by PNP transistors

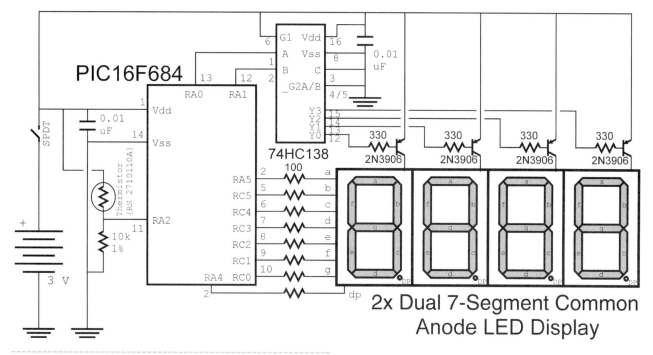

Figure 7-9   *Digital thermometer circuit*

that are, in turn, driven by an eight-bit decoder. This may seem like an ungainly method of driving the seven-segment displays, but it is required due to the lack of available pins from the PIC16F684 to drive the segment common pins.

I built my prototype on a fairly small PCB (see Figure 7-10) and used point-to-point wiring to create the circuit as shown in Figure 7-11. There are a lot of wires to solder, and when you have completed the circuit, I suggest you test it with cThermo 1.asm. This will initialize the PIC16F684 and display the ADC value for the resistor/thermistor voltage divider. I used the Radio Shack thermistor because it is reasonably priced and the back of the package lists the thermistor's resistance for different temperatures, which is nice to have when debugging software.

Figure 7-10   *Digital thermometer top with thermistor by the on/off switch*

```
#include <pic.h>
/*  cThermo 1.c - Display the ADC Value

This program will display the current PIC16F684
ADC Value on four 7 Segment Common Anode LED
Displays.

The Code/Wiring is based on "c4x7Seg.c" Noting
that Negative Active LED Wiring is used.

Hardware Notes:

RA5 - Segment a
RC5 - Segment b
RC4 - Segment c
RC3 - Segment d
RC2 - Segment e
RC1 - Segment f
RC0 - Segment g

RA0 - Right 7 Segment Display
RA1 - Left  7 Segment Display

myke predko
04.12.23

*/

__CONFIG(INTIO & WDTDIS & PWRTEN & MCLRDIS &
UNPROTECT \
    & UNPROTECT & BORDIS & IESODIS & FCMDIS);
```

Figure 7-11 *Digital thermometer backside showing point-to-point wiring*

```
int i, j;
int DisplayValue, DisplayDigit, DisplayLED;
int ADCState = 0;

const char LEDDigit[] = {
// RRRRRRR - PIC16F684 Pin
// ACCCCCC
// 5543210
// abcdefg - LED Segment
    0b0000001,                    // "0"
    0b1001111,                    // "1"
    0b0010010,                    // "2"
    0b0000110,                    // "3"
    0b1001100,                    // "4"
    0b0100100,                    // "5"
    0b0100000,                    // "6"
    0b0001111,                    // "7"
    0b0000000,                    // "8"
    0b0001100,                    // "9"
    0b0001000,                    // "A"
    0b1100000,                    // "b"
    0b0110001,                    // "C"
    0b1000010,                    // "d"
    0b0110000,                    // "E"
    0b0111000};                   // "F"

main()
{

    PORTA = 0;
    PORTC = 0;
    CMCON0 = 7;            // Turn off Comparators
    ANSEL = 1 << 2;        // RA2/(AN2) is an Analog
                           // Input
    ADCON0 = 0b10001001;   // Turn on the ADC
            // Bit 7 - Right Justified Sample
            // Bit 6 - Use VDD
            // Bit 4:2 - Channel 2
            // Bit 1 - Do not Start
            // Bit 0 - Turn on ADC
    ADCON1 = 0b00010000;   // Select the Clock as
                           // Fosc/8
    TRISA = 0b001100;      // RA5/RA1/RA0 are Outputs
    TRISC = 0b000000;      // All Bits of PORTC are
                           // Outputs
```

```
    DisplayValue = 0;    // Start Displaying at
                         // 0x0000
    DisplayLED = 0;      // Display the 1s first
    j = 0;               // Start Counting at Zero

    while(1 == 1)        // Loop Forever
    {
      DisplayDigit = (DisplayValue >> (DisplayLED * 4))
                 & 0x0F;
      PORTA = ((LEDDigit[DisplayDigit] >> 1) & 0x020)
             + (1 << 4) + (3 - DisplayLED);
      PORTC = LEDDigit[DisplayDigit];

      DisplayLED = (DisplayLED + 1) % 4; // Next
                                         // Digit

      NOP();             // Used for 5 ms Timing
      for (i = 0; i < 327; i++);         // 5 ms
                                         // Delay
                                         // Loop
      NOP();             // Used for 5 ms Timing

      j = j + 1;         // Increment the Counter?
      if (50 == j)       // 1/4 Second Passed?
      {
        j = 0;           // Reset for another 1/4
                         // Second
        switch(ADCState)
        {
          case 0:        // Start ADC Operation
            GODONE = 1;
            ADCState = 1;
            break;
          case 1:
            ADCState = 0;
            DisplayValue = (ADRESH << 8) + ADRESL;
            break;
        }  // hctiws
      }  // fi
    }  // elihw
}  // End cThermo 1
```

When cThermo 1.c is running, you can test the operation of the ADC and the display by breathing on the thermistor or by putting it in the sun or in a refrigerator. After you have verified the operation of the display and thermistor, I suggest that you coat the back of the PCB and all the wires with Weldbond or epoxy to hold them down and prevent them from being damaged.

The next program demonstrating the operation of the circuit is cThermo 5.c, which displays the current resistance of the thermistor (in 10Ω increments). I found that the resistance displayed was almost exactly the resistance specified on the back of the thermistor's package.

The thermistor and 10k, 1 percent resistor form a voltage divider, which is read by the PIC MCU's ADC. The formula for the 10-bit ADC value (assuming that the full range is 1,023) is as follows:

$$ADC / 1{,}023 = 10k / (10k + R_{Thermistor})$$

Rearranging this formula, $R_{Thermistor}$ can be calculated as:

$$R_{Thermistor} = [(1,023 \times 10k) / ADC] - 10k$$

This is the formula used in cThermo 5.c to calculate the thermistor resistance value before it is converted to individual decimal digits and displayed. I broke the formula into different pieces and put them in a state machine to make sure that the floating-point operations would not delay the operation of the TMR0-based digit display code.

cThermo 5.c was my first experience with floating-point numbers and variables in the PICC Lite compiler, and I must say it was an educational one. In the code for cThermo 2.c through cThermo 4.c, I tried a number of different approaches to solve the problem and discovered something that should have been obvious. The functions that provide floating-point operators in the PIC MCU take up a lot of instructions. I found that to best implement the PICC Lite compiler code with floating-point operators, I should restrict the number of operations to three.

In the previous code, you will see the three operators:

- Floating-point-to-integer conversion using the "(int)" type cast
- Floating-point subtraction
- Floating-point division

To convert the thermistor resistance to a temperature, I found that the resistance varies negatively with the temperature. This was confirmed by a table on the datacard that came with the RadioShack thermistor. This means that as the temperature of the thermistor goes up from a nominal temperature of 25°C, the resistance goes down. Similarly, when the temperature of the thermistor drops, its resistance increases. With a little bit of experimentation with a calculator, I discovered that the resistance varied by 1.039 percent for every degree Celsius of temperature change. This relationship seemed to be true for +/- 40°C from 25°C, which was within the range of temperatures for a thermometer to be used around the home.

When you order a specific thermistor, you can find this information in its datasheet, otherwise you will have to come up with some kind of way of experimentally finding this information. This is not easy to do. The temperature versus resistance information printed on the packaging is why I selected the RadioShack part. When you look at the datasheets, you'll see that the percent resistor change per degree Celsius is not constant across all temperatures, and the manufacturer will often include a formula for a temperature range of several hundred degrees Celsius.

To keep the application reasonable, I just kept the operating range to +/- 40°C with a center point of 25°C and came up with the voltage-to-temperature function for the thermistor and 10k, 1 percent precision resistor voltage divider:

$$Voltage_{RA2} = Vdd \times 10k/(10k + R_{Thermistor})$$

Because I used the full 10 bits of the ADC for this application, the ADC value could be converted to a voltage using the following equation:

$$Voltage_{RA2} = Vdd \times ADC/1,023$$

And combining them, I could solve $R_{Thermistor}$ from the ADC value:

$$Vdd \times ADC/1,023 = Vdd \times 10k/(10k + R_{Thermistor})$$

$$R_{Thermistor} = [(1,023 \times 10k)/ADC] - 10k$$

This calculation is carried out in states 65 and 66 of cThermo 5.c.

The next step was to convert this resistance into a temperature change from 25°C. Knowing that the thermistor's resistance changed by a set percent for every degree Celsius change in temperature, the resistance of the thermistor could be expressed as:

$$R_{Thermistor} = 10k \times (1.039^{25°C - T})$$

Plugging this value into the ADC equation for $R_{Thermistor}$, the temperature difference from 25°C could be written out as:

$$[(1,023 \times 10k)/ADC] - 10k = 10k \times (1.039^{25°C - T})$$

Removing the common 10k term, the equation becomes:

$$(1,023/ADC) - 1 = 1.039^{25°C - T}$$

Now, depending on how much mathematics background you have, this may seem insolvable. But, by knowing the following:

$$Num^{Power} = e^{ln(Num) \times Power}$$

where e is the base of the natural logarithm (ln), and by using this relationship on the previous equation, along with a bit of algebra, you can find:

```
Temp = 25°C - {ln[(1,023/ADC) -
1]/ln(1.039)}
```

Even if you understand exactly what I've done here and have had the necessary instruction in algebra, relations, and functions, I'm sure your brain is hurting. I suggest that you stand up and take a break. I know I had to. The temperature formula is deceptively simple: It took me four days to work through the circuit and equations to come up with this equation, which is specific to the PIC16F684's 10-bit ADC and independent of Vdd. Calculating the temperature using a formula that is independent of Vdd allows the circuit to work with different voltage power supplies without having to change the software. It is actually quite an important feature of the formula.

It seemed like a cruel joke to me, but the worst part of developing the application was yet to come. Despite having a seemingly simple conversion formula, I found that I had a fairly serious size problem; the code seemed to be 100 to 110 instructions too large for the 1024 instructions available to the PIC16F684 PICC Lite compiler.

The problem with the code size was the need for including the C natural logarithm function (known as *log* in C, not *ln* as is normally used in mathematics). This function seems to take over 400 instructions, and I could only allow 300 or so. I was very surprised at this situation as I believed that the natural logarithm (and exponent) functions were required for the floating-point division operator. Further complicating the situation was the fact that the MPLAB IDE simulator continually flagged a subroutine return underflow before executing the application code.

It took me another three days, but I was able to get the application working for the previous formula. By carefully managing integer variable sizes and looking for opportunities to share application code as much as possible, I was able to shoehorn the application in, with a dozen or so instructions to spare, and it seemed to work. cThermo 6a.c is a modification of the final result, and I found that the digits would occasionally flash noticeably. I made the assumption, first, that the log (C

natural logarithm function) took longer than the 4,000 cycles (4 ms) available to it between LED display updates, and then I enabled the 8 MHz clock and doubled the number of cycles between LED display updates so the 4 ms delay stayed constant.

Looking at the work that went into this experiment, along with the trials and tribulations I had to endure, you're probably asking yourself if the effort was worth it. This is an especially germane concern, considering that I could have spread it out into three experiments: driving four displays, a seven-segment LED Ohmmeter, and the final digital thermometer. I would have to say yes, because it gave me some practical experience with floating-point values in the PIC MCU and, despite all the problems I encountered, the voltage-to-temperature formula is fairly easy to observe and check visually.

I was able to come up with the following rules when working with floating-point operations in the PIC MCU:

- Limit the number of floating-point operations you are going to use to three: subtraction, division, and floating-point-to-integer casting.

- Do not use any of the logarithm or trigonometric functions in a PIC MCU where only 1k of program memory space is available unless you absolutely have to.

- If you are running display loops with floating-point operations executing inside the delays, recognize that the floating-point operations take a lot of time, and plan accordingly.

- Do not obfuscate your floating-point calculation code in order to make them more efficient. If you are going through the effort of including floating point in your application, make sure the theory behind the floating-point operations is sound and the formulas and algorithms you are implementing are reduced as far as reasonably possible. Try also to avoid many of the conversions and leaps that are necessary when fitting integer operations into a task that requires data manipulation.

Experiment 54—PIC MCU "Piano"

PC/Simulator

PICkit™ 1

starter kit

Parts Bin

1 PIC16F684

1 14-pin socket

8 1N914 silicon diodes

1 0.47 μF capacitor (any type)

1 0.01 μF capacitor (any type)

6 10k resistors

1 Piezo speaker

1 SPST switch

1 Three-cell AA battery clip

3 AA batteries

1 Prototyping PCB

Tool Box

DMM

Needle-nose pliers

Soldering iron

Solder

Wire-wrap wire

Indelible ink marker

Weldbond adhesive

In Figure 7-12, you will see that I have arranged 10 momentary on push buttons in an arrangement that is similar to a piano's keys. This probably doesn't seem like such an amazing application; the PIC16F684 has 12 pins and 10 of them could be easily connected to individual buttons, leaving two pins for driving out the musical signal. In the previous section, I introduced you to switch matrix keypads. So you might think that the 10 keys are wired as a switch matrix, resulting in only seven (five columns of two rows) pins being used. Actually, neither method is used. Instead, the wiring used in this application requires only six wires to interface to the 10 buttons. If you have a small number of button inputs for an application (and you have designed it to work with only one button at a time pressed), you will find cases where this is the most efficient method of wiring your application's button input.

The method used to provide the 10-button input can be seen in the piano's schematic (see Figure 7-13). For six of the natural keys, there is a traditionally pulled-up momentary on button. However, for the four sharp keys, a momentary on button connects the adjacent natural keys to ground through a silicon diode. When the sharp keys are pressed, current will flow through the diodes, pulling each of the adjacent natural keys low. Therefore, to sense when a sharp key is pressed, the two adjacent natural keys have to be polled.

This method can be used in a number of different applications. Remember that you are not limited to wiring the in-between momentary on buttons to just two standard pulled-up buttons; you can use this method with really any number of standard pulled-up buttons. In doing so, you will discover that locating the multiple diodes for each button in between can be a problem. I am bringing this up because it was a bit of an issue for me. However, I was able to find a

Figure 7-12   *PIC16F684-based 10-note "Piano"*

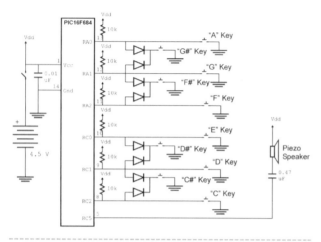

Figure 7-13   *Piano circuit*

Figure 7-14   *Backside of PIC16F684-based piano showing how the diodes are wired underneath the buttons*

prototyping PCB that would allow multiple connections at a single point, and by experimenting with how I placed the pull-up resistors and the diodes, I was able to come up with a scheme that was quite easy to wire and did not use up an excessive amount of space. In Figure 7-14, I show how the diodes are wired to the backside of the PCB underneath the natural key buttons.

cPiano.c is the application code I came up with for this experiment and it delays 50 ms and then polls the six input pins to see if any buttons are pressed. If a pin is pressed, then a PWM period and duty cycle is saved in the TMR2/CCP hardware and pin RC5 is put into output mode for the PWM signal to pass the sound to the Piezo speaker. To come up with the delay periods, I started with a table of the different note frequencies. I then converted them into microsecond periods and divided them by four so they could be loaded into the PR2 register for driving out the different frequencies. You'll find that the final output may be a bit flat or sharp; this is due to rounding errors that came with converting the periods into integers that TMR2 could work with.

```
#include <pic.h>
/*  cPiano.c - 10 Note "Piano"

This application monitors 10 piano keys via 6
I/O pins and plays the appropriate note as long
as the key is pressed for 50 μs increments.  The
PWM output is used for the notes with a 4x
prescaler.

The Keys are defined as:
Middle C - RC2     - 1,046 Hz  -  956 μs - PP = 240
C#       - RC1/RC2 - 1,108 Hz  -  902 μs - PP = 226
D        - RC1     - 1,174 Hz  -  852 μs - PP = 212
D#       - RC0/RC1 - 1,248 Hz  -  802 μs - PP = 200
E        - RC0     - 1,318 Hz  -  758 μs - PP = 188
F        - RA2     - 1,396 Hz  -  716 μs - PP = 180
F#       - RA1/RA2 - 1,480 Hz  -  676 μs - PP = 168
G        - RA1     - 1,568 Hz  -  638 μs - PP = 160
```

```
G#       - RA0/RA1 - 1,662 Hz  -  602 μs - PP = 150
A        - RA0     - 1,760 Hz  -  568 μs - PP = 142

Hardware Notes:
PIC16F684 Running at 4 MHz with Internal
Oscillator
RC5/P1A - PWM
RC2:RC0 - Sound Input Bits Noted Above
RA2:RA0 - Sound Input Bits Noted Above

myke predko
05.01.22

*/

__CONFIG(INTIO & WDTDIS & PWRTEN & MCLRDIS &
UNPROTECT \
  & UNPROTECT & BORDIS & IESODIS & FCMDIS);

int j;
char KeyValue;
char FreqOut;

main()
{
  CMCON0 = 7;              // Turn off Comparators
  ANSEL = 0;               // Turn off ADC
  T2CON = 0b00000101;      // TMR2 Has 4x Prescaler
  CCP1CON = 0b00001100;    // Enable PWM Output

  while(1 == 1)            // Loop Forever
  {
   NOP();                  // 50ms Delay
   for (j = 0; j < 3333; j++);
    NOP();

   KeyValue = ((PORTC & 7) << 3) + (PORTA & 7);

   if (0 == (KeyValue & (1 << 0)))
    if (0 != (KeyValue & (1 << 1)))
     FreqOut = 142;         // A
    else
      FreqOut = 150;        // G#
   else if (0 == (KeyValue & (1 << 1)))
    if (0 != (KeyValue & (1 << 2)))
```

```
  FreqOut = 160;        // G
else
  FreqOut = 168;        // F#
else if (0 == (KeyValue & (1 << 2)))
  FreqOut = 180;        // F
else if (0 == (KeyValue & (1 << 3)))
  if (0 != (KeyValue & (1 << 4)))
    FreqOut = 188;      // E
  else
    FreqOut = 200;      // D#
else if (0 == (KeyValue & (1 << 4)))
  if (0 != (KeyValue & (1 << 5)))
    FreqOut = 212;      // D
  else
    FreqOut = 226;      // C#
else if (0 == (KeyValue & (1 << 5)))
  FreqOut = 240;        // C
else                    // Nothing Pressed
  FreqOut = 0;

if (0 == FreqOut)       // Nothing Pressed
  TRISC = 0x3F;         // Turn OFF Output
else                    // Setup PWM
{
  PR2 = FreqOut;        // Output 50% Duty Cycle
  CCPR1L = FreqOut / 2; // PWM Signal
```

```
  TRISC5 = 0;           // Output Signal
  } // fi
 } // elihw
} // End cPiano
```

I should point out that this circuit is not limited to only 10 pins. For my application, the limiting criterion was space for buttons on the prototyping PCB. Using this method, if I had used all 11 available pins on the PIC16F684, I could have supported up to 22 keys. This is a good method to remember when you have a fair number of input buttons to poll but don't want to go through the hassle of the switch matrix code. Generally, this scheme requires one-half the buttons as I/O pins, and it can be extended far beyond the 10 buttons presented in this experiment. As a point of reference, it seems that 20 is the number of buttons where a switch matrix keypad is more efficient (in terms of I/O pins) than the method presented here.

Experiment 55—Model Railway Switch Control

PC/Simulator

PICkit™ 1

starter kit

Parts Bin

1 PIC16F684

1 14-pin socket

2 TRIACs (Radio Shack 276-1000 recommended)

1 5.1V, 1 Amp Zener diode

1 1N914 silicon diode

1 330 μF electrolytic capacitor

1 0.01 μF capacitor (any type)

1 220Ω, 1-watt resistor

1 10k resistor

1 SPST switch

1 Two-position terminal block

1 Three-position terminal block

1 Prototyping PCB

Tool Box

DMM

Oscilloscope

Needle-nose pliers

Soldering iron

Solder

Wire-wrap wire

HO scale remote train switch

Train power supply

Compared to other hobbies, model trains have changed very little since I was a kid. When I go to a local train store, I'm always depressed to see that power supplies are still based on a Variac (variable transformer) train power control and a transformer for 18-Volts AC accessory power. I realize that the DCC system does bring the hobby into the twenty-first century, but these control units as well as the remote units are very expensive. What is needed is a low-cost way of providing computerized control of trains and the layouts so that hobbyists of moderate means can experiment with computer control of their layouts. In this experiment, I will show how a PIC microcontroller can be used to control a switch using the accessory power from the standard train power supply.

Before going further, I want to note that this circuit is powered by *alternating current* (AC). The AC used in this experiment is the benign 18-VAC accessory power available from a hobby train power supply. The circuit's power supply is not designed for use with household AC (110 or 220 volts depending on where you live), which can burn or electrocute you. Although the methods used to design the power supply are the same, and I show how I calculated the values for this experiment, the component values are not valid for household AC. If you want to interface your application with household AC power (or any AC voltages other than the 18 volts used in this experiment), make sure you consult with an engineer or electrician before connecting the circuit to the power supply. Be sure you have selected correct values and have wired the circuit according to your local code.

The control circuit is quite simple (see Figure 7-15), and I was able to wire it on a small prototyping PCB (see Figure 7-16). The PIC device is powered by a 5.1-volt Zener diode with a 220Ω resistor and standard resistor to rectify and regulate the current. The 330 μF capacitor ensures that the voltage is smooth. The PIC MCU continuously polls the SPST switch, and if it changes state, one of the two TRIACs is driven for a few milliseconds to change the solenoid in the track switch. If you are wondering, the application could be easily expanded to handle multiple switches. Although due to some of the issues that I had, there are some considerations that you should be aware of that I will share with you at the end of the experiment.

To calculate the value for the components, I assumed I would run the PIC16F684 at 5 volts, so I used a 5.1-volt Zener diode (which has a constant 5.1 volts across it) and a silicon diode and resistor to rectify and limit the current flowing through the circuit. The RadioShack TRIACs I used require 25 mA of current to operate, so I wanted to have a maximum of 50 mA (25 mA for the TRIAC and 25 mA for the PIC MCU and other components). To find the correct resistor value, I used Kirchoff's and Ohm's Laws:

$$V_{AC} = V_{Zener} + V_{Diode} + V_{Resistor}$$

$$18 \text{ volts} = 5.1 \text{ volts} + 0.7 \text{ volts} + V_{Resistor}$$

$$V_{Resistor} = 12.2 \text{ volts}$$

$$R = V / I$$

$$= 12.2 \text{ volts} / 50 \text{ mA}$$

$$= 244\Omega$$

I used a 220Ω resistor in the circuit because it is a standard value. For its power rating, if 50 mA is passing through the resistor, 0.55 watts of power is being dissipated. To ensure that there wouldn't be any power issues, I used a 1-watt-rated resistor.

With the power supply circuitry selected, I built the circuit on a prototyping PCB using small terminal blocks to connect the circuit to the 18-volt power supply and to the train switch using the wires that come with the train switch. (For this, I bought an ATL Remote Right-Hand Switch, item number 851.) Before burning a PIC MCU with code, I tested its operation by shorting pins RC4 and RC5 to ground to make sure the TRIACs (which are simply AC switches, controlled by current) would change the switch solenoid's state. In my application, I kept the I/O pins driving the TRIACs in input mode until the solenoid was to be driven.

When I tried the PIC16F684 in circuit, I found that the circuit worked fine to change my switch to one state, but had problems with the other state. In this state, the switch would start chattering, and the solenoid got very hot. When I put an oscilloscope probe on the I/O pins, I discovered that both of them were becoming active. Further investigation revealed that the solenoid state seemed to require so much power that the 18 volts of AC was reduced to zero. After considering the problem, I decided the simplest solution to the problem was to use the PIC16F684's EEPROM data memory to save the state of the switch in case the PIC MCU was reset. This was a fairly minor modification to the original code, and I called it cTrain 2.c:

```
#include <pic.h>
/*  cTrain 2.c - Control AC Driven Train Control

This program is a modification of "cTrain" to
take advantage of the EEPROM Data Memory to
record the last setting of the PIC switch

RA3 - Button Connection
RC4 - Train TRIAC1
RC5 - Train TRIAC2

myke predko
04.12.30
```

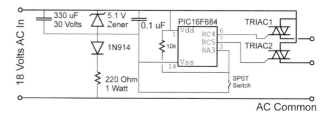

Figure 7-15 *Train*

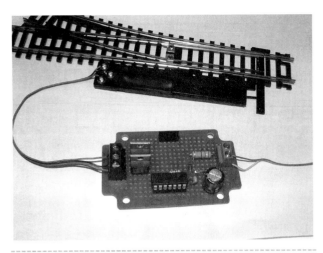

Figure 7-16 *Simple hobby train switch control circuit that replaces standard mechanical switch block*

```
*/

__CONFIG(INTIO & WDTDIS & PWRTEN & MCLRDIS &
UNPROTECT \
  & UNPROTECT & BORDIS & IESODIS & FCMDIS);

int i, j;
int BState = 2;

main()
{

  PORTC = 0;          // Output Bits are Low
  CMCON0 = 7;         // Turn off Comparators
  ANSEL = 0;          // Turn off ADC
                      // No Outputs (Yet)
  for (i = 0; i < 18000; i++)  // 250 ms Initial
                                         Delay
  if ((EEPROM_READ(0) == 0) &&
      (EEPROM_READ(1) == 0xFF) &&
      (EEPROM_READ(2) == 0x55) &&
      (EEPROM_READ(3) == 0xAA))
   BState = EEPROM_READ(4);
  else                // Initialize the State
  {
   EEPROM_WRITE(0, 0);      // Put in Check
                              Pattern
   EEPROM_WRITE(1, 0xFF);
   EEPROM_WRITE(2, 0x55);
   EEPROM_WRITE(3, 0xAA);
   EEPROM_WRITE(4, 2);      // Nothing Set Yet
  } // fi

  while(1 == 1)       // Loop Forever
  {
   if ((0 == RA3) && (BState != 0))  // Switch
                                        Change
```

```
  {
   i = 0;
   while (( i < 1100) && (0 == RA3))
    if (0 != RA3)
     i = 0;              // Switch Bounce
    else
     i = i + 1;          // Increment Debounce
                            Counter
    BState = 0;          // Indicate State Active
    EEPROM_WRITE(4, 0);         // Record in EEPROM
    for (i = 0; i < 6666; i++);    // 100 ms
     TRISC4 = 0;         // Enable RC4 as Output
    for (i = 0; i < 6666; i++);   // 100 ms
     TRISC4 = 1;         // Finished Driving Solenoid
  }
  else if ((0 != RA3) && (BState != 1))
                         // Switch Change
  {
   i = 0;
   while (( i < 1100) && (0 != RA3))
    if (0 == RA3)
     i = 0;              // Switch Bounce
    else
     i = i + 1;          // Increment Debounce Counter
    BState = 1;          // Indicate State Active
    EEPROM_WRITE(4, 1);      // Record in EEPROM
    for (i = 0; i < 6666; i++);   // 100 ms
     TRISC5 = 0;         // Enable RC5 as Output
    for (i = 0; i < 6666; i++);   // 100 ms
     TRISC5 = 1;         // Finished Driving Solenoid
   } // fi
 } // elihw
} // End cTrain 2
```

When you build this application, you may find that you do not need to save the state of the solenoid in EEPROM data memory. I have tried this application on only one train switch. As the saying goes: YMMV (*your mileage may vary*). You may find that you require the EEPROM for both solenoid states or you might find that you don't require it all.

This application can be easily extended to multiple switches (using a single PIC16F684, you could control four switches), although I suspect that you will find a certain amount of variation between the switches, which could lead to some problems. To minimize the chance of power problems, make sure you engage only one solenoid at a time; if four switches are active at the same time, you should sequence through them with some delay between each solenoid activation. To ensure there are no power problems, you may want to pass the Vdd voltage supply through a diode and then have a large capacitor (47 μF to 100 μF) provide a temporary power supply if a solenoid reduces the accessory voltage to zero.

Experiment 56—PC Operating Status Display

PC/Simulator

PICkit™ 1

starter kit

Parts Bin

1 PIC16F684

1 14-pin socket

7 1N914 silicon diodes

1 10-LED bargraph display

10 470Ω resistors (see text)

1 47 μF electrolytic capacitor

1 0.01 μF capacitor (any type)

1 Prototyping PCB

1 DB-25M solder cup connector

Tool Box

DMM

Needle-nose pliers

Soldering iron

Solder

Wire-wrap wire

Weldbond adhesive

One PC peripheral that has always fascinated me is the remote status monitor that provides you with an alternative form of feedback regarding the operation of your computer in the form of an LCD display, a glowing ball, or a series of LEDs. In this experiment, I have created a simple 10-LED operations display controlled by the parallel port that can be added to a Windows PC in just a few hours. It's important to note that although the PIC MCU side of the experiment is fairly simple to build (and is very similar to the other experiments in this section), the PC software, despite *not* requiring a specialized device driver, is fairly sophisticated and probably cannot be replicated or modified easily.

The circuit (see Figure 7-17) consists of a PIC16F684 connecting to a PC's parallel port and has the ability to turn on a single LED because it's powered by the PC's parallel port. Power is provided by seven of the parallel port's eight output pins. These pins are pulled up internally by the parallel port and source 1 mA of current, which should be sufficient for running the PIC MCU and a single LED. The remaining I/O pin is used, along with the printer strobe bit, to latch in a bit of data. When I built my application (see Figure 7-18), I used a 470Ω SIP and a 470Ω resistor for limiting the LED current. You can use either this solution or 10 individual 470Ω resistors.

The application code (cPerfMon.c) waits for eight bits to be sent to it and then XORs the first four bits against the second four bits, and, if the result is all bits set, the specified LED turns on.

```
#include <pic.h>
/*  cPerfMon.c - Display Performance Value from
PC

This program connects to the parallel port of a
PC and will receive eight bits of data on RA3
(Clock) and RA4 (Data) and compare the first
four bits with the second and if they are
complementary, will display the BCD value on a
10 LED bargraph display.

This program was written to work with
"PerfMon.c".

Hardware Notes:
PIC16F684 Running at 4 MHz
RA3 - Printer Port Clock
RA4 - Printer Port Data
RC5:RC0 - Low 6 Bits of the Display
RA2:RA0 - Bits 8:6 of the Display
```

Figure 7-17   *PerfMon circuit*

```
RA5 - Bit 9 of the Display

myke predko
05.01.22

*/


__CONFIG(INTIO & WDTDIS & PWRTEN & MCLRDIS &
UNPROTECT \
   & UNPROTECT & BORDIS & IESODIS & FCMDIS);


char BitCount;
char BitValue = 0;
char BitComp = 0;

main()
{

  PORTA = 0x3F;          // Display Negative Active
  PORTC = 0x3F;
  CMCON0 = 7;            // Turn off Comparators
  ANSEL = 0;             // Turn off ADC
  TRISC = 0;             // All 6 Bits Outputs
  TRISA = 0b011000;      // RA4:RA3 Inputs

  OSCCON = OSCCON | (1 << 4);  // Run PIC MCU at
                               //    8 MHz

  while(1 == 1)          // Loop Forever
  {
   for (BitCount = 0; BitCount < 4; BitCount++)
   {                     // Get First 4 Bits
    while (RA3);         // Wait for Bit
     BitValue = (BitValue >> 1) + (RA4 * 8);
    while (!RA3);        // Make Sure Bit is High
   }  // rof

   for (; BitCount < 8; BitCount++)
   {                     // Get Second 4 Bits
    while (RA3);         // Wait for Bit
     BitComp = (BitComp >> 1) + (RA4 * 8);
    while (!RA3);        // Make Sure Bit is High
   }  // rof


   if ((BitValue ^ BitComp) != 0x0F)
   {                     // No Match, Wait 1 Bit
    while(RA3);          // Wait for Low
    while(!RA3);         // Wait for High
   }
   else                  // Data OK - Light
                         //    Appropriate LED
   if (BitValue < 6)
   {                     // 0 to 50%
    PORTC = 0x3F ^ (1 << BitValue);
    PORTA = 0x3F;        // High LEDs Off
   }
   else if (BitValue < 9)
   {
    PORTC = 0x3F;
    PORTA = 0x3F ^ (1 << (BitValue - 6));
   }
   else                  // 90%+
   {
    PORTC = 0x3F;
    PORTA = 0x1F;
   }  // fi

   BitValue = 0;         // Reset the Comparison
                         //    Values
   BitComp = 0;

  }  // elihw
 }  // End cPerfMon
```

cPerfMon.c has a simple, but quite effective method of error detection; if after eight bits, the XOR'd result is not 0x0F (all bits set), it then waits past the next bit and repeats the process with the next eight bits until the XOR'd result is 0x0F. With the PC software sending data once every tenth of a second, it will take up to three-quarters of a second for the PIC circuit to become synchronized with the PC output.

Although the PICC Lite compiler code for the PIC16F684 is quite simple, the PC code to produce the dialog box shown in Figure 7-19 and monitor the amount of main memory that has been allocated by the different applications running in the PC will be very complex the first time you look at it. An important attribute of this application is that it does not require any special printer device drivers. If you have enough experience with the PC, you will know that most printers require specialized device drivers to provide bidirectional communications or even status monitoring to suspend sending data until the current buffered data has been printed. In this application, I simply open the LPT file device and send the eight bits of data as one bit of a byte (the other seven bits are high to ensure the hardware is powered by the parallel port). This is possible because the parallel port is wired as Loopback port. As such, it appears to the basic printer software as a "dumb" ASCII printer and will have data sent to it at full speed (roughly one byte every 50 μs).

The application code is written to be built under Cygwin and run under GCC, and it consists of five modules that are compiled and linked together. To debug the application, I used DDD, also running under Cygwin. The application and files should be buildable under Visual Studio. Using the open-source tools for development means that you can replicate the application and avoid the costs of using Visual C++ or Visual BASIC. If terms like GNU Project GCC C compiler, DDD debugger, and Cygwin Windows interface are Greek to you, don't worry; as you learn more about programming PCs and gain understanding of how to create and build Windows applications, these terms (and tools) will become more familiar to you.

The basic application itself was written in C (for GCC), is called PerfMon.c, and is listed below. This application can access either LPT1 or LPT2 or neither, and it remembers which one was active by using the PC's registry for storing the last used port. I originally wanted the application to display the PC's current load, but this turned out to be problematic because each version of Windows uses a different method of computing the current load. Instead, I display the current memory usage, which can also be a useful load indicator in a PC if it starts running very slowly. Along with the source code, ".rc" files are required that can

Figure 7-18   *The PerfMon application's circuit connects to and is powered by a PC's parallel port.*

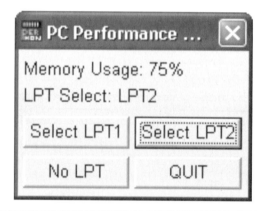

Figure 7-19   *PerfMon operating*

be found on the PICkit 1 Starter Kit's CD-ROM or at http://www.myke.com.

The application icons were created using Paintbrush and then converted into ico format. To build the entire application and be able to debug it using DDD, I used the following statements in the Cygwin X-Windows box:

```
windres %1.rc -O coff -o %1.res
gcc -mwindows -mno-cygwin -g -o %1.exe %1.c
%1.res -lgdi32 -luser32
```

Due to the untimely demise of my PC on which I normally use to create PC and PIC MCU software, I had to break up the PC and PIC MCU software development for this experiment onto two PCs. Amazingly enough, when the two pieces of software were brought back together, the application worked quite well, although occasionally the lit LED would change to an

unexpected value. The solution to this problem was to run the PIC MCU at 8 MHz, rather than at the standard 4 MHz, using the theory that the incoming data and changing bit values were too fast for the 4 MHz device. This change eliminated the problem with the errant LED, and the application has been running perfectly ever since.

Twice in this section, I encountered situations where the application worked well, but needed to be sped up to work perfectly. I want to make sure that you understand that in both cases, the problem was reasonably well defined and the clock doubling was used to address these issues. I mention this because I do not want you to think that by increasing the clock speed you can fix the problem every time you have an application that occasionally runs errantly. Before attempting any fix, you should have a theory behind the cause of the problem and what you expect to gain by the fix. If you don't, chances are you will not fix the problem or you'll end up with an entirely new problem requiring its own fix.

Introduction to PIC® MCU Assembly Language Programming

PC/ Simulator

Required Parts

1 PICkit-1 Starter Kit
1 PIC16F684

So far in this book, I have shown you the high-level C programming language and how to use it to program a PIC16F684 to perform various tasks. You should now be reasonably comfortable with programming the chip and coming up with applications that perform useful functions. You should also have used the simulator that is built into MPLAB® IDE to help you verify the operation of your program before you program a chip or to help you debug your program once you discovered your program doesn't work exactly as you thought it might. You are now ready for the next step on your journey: learning assembly language programming.

The computer program that converts an assembly language program into the .hex file and is then programmed into the PIC® microcontroller chip is known as an *assembler*. The assembler is analogous to the *compiler* program that converts high-level language statements into a .hex file. Unlike the PICC® Lite compiler, the PIC MCU assembler (known as MPASM® assembler) is built into the MPLAB IDE and is loaded automatically when you install MPLAB IDE.

Assembly language programming is usually looked upon with a great deal of trepidation; it is perceived to be more difficult to learn than a high-level language like C for programming applications and debugging failing applications. Another perception is that code written in assembly language is more efficient than code written in a high-level language. I would disagree with all of these statements. Assembly language programming is not more difficult to learn than a high-level language; it is merely different.

The difference is in the basic statements and in understanding how they are used to create applica-

tions. Once you are comfortable writing code in assembly language, you will find that it is just as much mental work as writing an application in a high-level language. You may find that an assembly language program takes more physical work (keying) than a high-level program, but this is due to the increased granularity of the assembly language instructions; each one performs a much smaller task than a high-level language statement. Additionally, I would say that debugging in assembly language is a lot easier than in a high-level language, because there is much less ambiguity about what the program is doing. You probably appreciate this statement knowing some of the strange things that can happen in C statements. Finally, an application that is poorly written in assembly language will not outperform an application that is poorly written in a high-level language.

The most important parameters of any program are its readability and the efficiency of the algorithms that are used to implement the required functions. If the program cannot be easily read, you will have problems completing it, getting it running, and debugging it. If no thought is put into the operation of the application, then it will not run appreciably faster. Regardless of the programming language, these parameters must always be considered when designing an application.

An excellent analogy to the differences between a high-level language (like C) and assembler can be illustrated by the commands needed to explain how to walk across the room and pick up an object. In a high-level language, the command statements will look something like the following: Turn right, Step forward six times, Bend over." Assembly language instructions for the same task would look like this: "shift weight on left leg, lift right leg two centimeters, move right leg 10 centimeters to the right, lower right leg." Both will get the job done, but the assembly language instructions work at a much lower level.

In this introduction, I have used the term *instructions* to describe the assembly language statements. At its most basic level, an instruction is a collection of bits that command the processor to carry out a simple task. For arithmetic operations, these bits specify the task,

where input parameter(s) come from, and where the result is stored. If multiple arithmetic operations are required to carry out a task, then multiple instructions are required. In addition to arithmetic instructions, a variety of other types of instructions exist that control the operation of the processor, the microcontroller in which it is built, and specify where and how the code is to execute. I will explain these instructions in this section. As I said previously, these instructions perform the same operations performed by the high-level statements, just in much smaller steps.

Experiment 57—The asmTemplate.asm File and Basic Directives

Before starting an assembly language program, I generally begin with a template similar to the one I introduced you to early in the book for C programming. This template is used to remind me of the basic information required in a program and the basic programming heading information needed to get a basic program running. The asmTemplate.asm file that I used for the programs presented in this book looks like:

```
  title   "asmTemplate - Assembly Language Coding
Template"
;
;   Explain the Operation of the Program.
;
;   "C" Equivalent Code:
;
;   // Put "C" Equivalent Function Here
;
;
;   Author
;   Date
;
  LIST R=DEC
  INCLUDE "p16f684.inc"


__CONFIG _FCMEN_OFF & _IESO_OFF & _BOD_OFF &
_CPD_OFF & _CP_OFF & _MCLRE_ON & _PWRTE_ON &
_WDT_OFF & _INTOSCIO

;  Variables
CBLOCK 0x20
;  Put Variable Names Here
;  If Variable Longer than 1 Byte, Put in ":#Bytes"
ENDC

  PAGE
org     0
 nop                     ; NOP Required for Debugger

;  Mainline Code
```

```
PAGE
;   Subroutines


  end
```

Let's go through the lines of this file to help you understand the different features of the template file. This will also serve to introduce you to some of the information that you will need to carry out assembly language programming.

The first line contains the *title* directive and will place the string in quotes on the first line of every listing page as a header. *Directives* are commands to the assembler that are used to control the assembly of the program. The title line that I use consists of the name of the file and a brief description of what it does; this helps me find a program with a specific function as well as identify the file in which it is contained.

The next series of lines are comments and use the semicolon (;) character to indicate that everything to the right is a comment and can be ignored. These are similar to C's double slash (//) comments. They explain the following:

- The operation of the program
- Analogous C code (if appropriate) to explain how the high-level function is to be implemented
- The author of the program and the date that it is written

The *list* directive is used to specify different assembly and listing commands to the assembler program, which then converts the assembly language program into the hex file of instructions to be programmed into the PIC MCU. If you look at the list directive parameters (shown at the end of this section), you will discover that there are quite a few, and many will seem like you should use them. However, the single parameter that I have specified as a list directive sets the default number base to 10 (decimal). (I feel that this is the only parameter that should be specified.) Adding additional list directive parameters can affect the operation of the final application, how it is programmed, and the difficulty to port the code to another PIC microcontroller part number.

The *include* directive loads an information file that adds additional directives and instructions to the program file. The pic16f684.inc file is found in the folders loaded during MPLAB IDE installation. These folders are used to define the various registers and functions of the PIC16F684, relieving you of the responsibility of having to do this (exactly as the "#include <pic.h>" directive did in the C programs). When you are developing more complex applications, you may find it useful to put common definitions and code in .inc files to avoid having to key repeatedly in the same information and provide a central repository of definitions and codes for all applications in your working current project.

The *__CONFIG* directive is used to specify the configuration word bits, just as the __config(directive did in C programming. Like the C version, each parameter is ANDed together to specify which bits are reset and which are set. Note that __CONFIG has two underscores and cannot start in the leftmost column. Also note that when the directive is listed here in the book, the parameters are generally printed on two or three lines, but they should all be in the same line. I have avoided putting this directive in applications that will not actually be programmed into a physical chip, because the double-line formatting can make it difficult to read and key correctly into the experiment's program if it is not necessary.

The *CBLOCK* and *ENDC* directives are used to declare program variables. I will explain how this is accomplished and, later in this section, I will note the points to take under consideration when declaring variables.

Next, the reset address (address 0x000) is specified using the *org* directive, and the application code starts after it. The assembler initializes its program counter to zero before starting to convert instructions into bit patterns, but it is customary to specify zero to ensure the memory location starts at the reset address. The *nop* (pronounced "no-op") string is actually an instruction, which commands the processor to do nothing over one instruction cycle. The value of this instruction will be explained later in the book; it is a lot more useful than you might imagine. Placing the nop at the reset address is necessary for using the MPLAB ICD2 debugger, and, although you may not have an MPLAB ICD2 debugger, it is a good idea to put this simple provision in for when it is needed (i.e., when the MPLAB ICD2 debugger is available).

Finally, the assembly language program can be written out in the file. Unlike in C programs, in assembly language I put subroutines *after* the mainline program. The assembler reads through the assembly language source file before starting to create the .hex file so that the address of labels (including subroutine labels) are identified and available when the .hex file is being created. This is different from a C compiler, which doesn't identify subroutine and function headers after they are used. I could place subroutines at the start of the program, but this would necessitate making the mainline of the program jump over the subroutines, which makes the program more difficult to read and follow.

I realize that in this assignment, I still have not given you enough information to start programming your own assembly language application. In the next few experiments, you will be able to start programming on your own.

Experiment 58—Specifying Program Memory Addresses

When I am programming in assembly language, I really don't care where the instructions are being placed in the PIC MCU's program memory. The MPASM assembler that is part of the MPLAB IDE does a very good job of managing program memory addresses and relieves the programmer of the responsibility for calculating the addresses themselves. By following a few simple conventions, you can write assembly language program without ever being concerned with the actual address of an instruction.

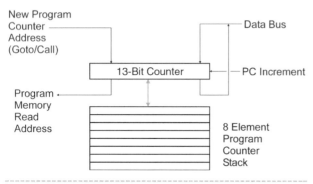

New Program Counter Address (Goto/Call) — Data Bus

13-Bit Counter — PC Increment

Program Memory Read Address

8 Element Program Counter Stack

Figure 8-1 *Program counter*

The program counter is a fairly sophisticated counter that keeps track of which instruction is to be executed next (see Figure 8-1). In normal operation, after an instruction is read in for decoding, the program counter is incremented causing the address output to point to the next instruction in the program. The value in the program counter can be changed four different ways:

- A new address can be imbedded in some instruction, and this address can be loaded into the program counter.

- The program counter can be changed algorithmically by the executing program.

- Certain instructions will increment the program counter when specific conditions are encountered.

- The program counter can be updated via a last in–first out (LIFO) stack.

As you become familiar with assembly language programming, you will become comfortable with using all four methods, but right now you should just concern yourself with changing the program counter using a value in an instruction.

The instructions that change the program counter to explicit values are the aptly named *goto* and *call* instructions, which change execution to the specified address or invoke the subroutine at the specified address, respectively. Looking at the goto instruction in its most basic form, its execution could be summarized as follows:

```
goto    0x????   ;  Program Counter = 0x0????
```

The value 0x0???? is known as an *immediate*, or *literal*, value or number. It is part of the program memory and cannot be changed during program execution. When people first started programming, this value was literally a number. Before entry into the computer system, the program would be written out on a sheet of paper with each address, and actual jump addresses would be entered manually into the program. The code snippet below, which toggles an output pin seven times, is written out with the addresses for each instruction to show how this is done:

```
0x01234   movlw    0x07     ;  Loop 7x
0x01235   bsf      0x05, 0  ;  PORTA, Pin 0 = 1
0x01236   bcf      0x05, 0  ;  PORTA, Pin 0 = 0
0x01237   addlw    0xFF     ;  Subtract 1 from
                            ;  Loop Counter
0x01238   btfss    0x03, 2  ;  If Zero Bit Set
                            ;  Skip Over Next
0x01239   goto     0x1235   ;  Else, Repeat Bit
                            ;  Pulse
```

This method of programming was very tedious and had the potential for many errors.

As time progressed, the ability to specify character strings for specific addresses was added to assemblers. By placing the label in the first column of the program file and ending it with a colon (:), the assembler could recognize it as a label and assign the address value to it. Then, when the label was encountered again, the assembler would substitute the address value for the label and avoid the need for the programmer to keep track of the address manually. The label also makes the program substantially easier to read. Using this ability, the snippet above becomes:

```
movlw    0x07     ;  Loop 7x - At Address
                  ;  0x01234
Loop:             ;  Label Indicating Where to
                  ;  Return Execution
bsf      0x05, 0  ;  PORTA, Pin 0 = 1
bcf      0x05, 0  ;  PORTA, Pin 0 = 0
addlw    0xFF     ;  Subtract 1 from Loop Counter
btfss    0x03, 2  ;  If Zero Bit Set Skip Over
                  ;  Next
goto     Loop     ;  Replace "Loop" in
                  ;  Instruction with Address
                  ;  Value (0x01235) when
                  ;  assembling program
```

Labels are strings of ASCII alphanumeric characters that start with a to z, A to Z, or _ and can have 0 to 9, a to z, A to Z, or _ for any of the remaining characters. Labels can be up to 255 characters in length (the maximum width of an MPASM assembler source file) and can optionally end in a colon. A label can be either on the same line as an instruction or on its own line. I recommend that labels are on their own line and always ended with a colon (as I have done with loop in the previous snippet). This will help you recognize labels in the program and not confuse them with instructions, directives, macros, or variables when you first read through an application.

This one simple ability to keep track of the address value for a label has been expanded to the three different methods shown in the following program:

```
      title   "asmAddress - Different Ways of
Specifying Addresses"
;
;   Demonstrate the 3 methods of specifying
;   program memory addresses in the PIC MCU using
;   MPASM assembler.
;
;   Hardware Notes:
;    PIC16F684 running at 4 MHz in Simulator
;
;
;   Myke Predko
;   04.09.10
;
   LIST R=DEC
  INCLUDE "p16f684.inc"

  PAGE

  org      0
   nop

   goto    Mainline
                ;   1st Method: Jump to Programmer
                ;   Specified Address

  org      47
                ;   User Specified Address with Label
Mainline:

   goto    $ + 1
                ;   2nd Method: Use Relative Address

Loop:
                ;   3rd Method: Jump to Label Address
   goto    Loop
                ;     Provided by MPASM assembler

   end
```

When you create the project for an assembly language program, you are going to do it in the same way you do it in a C language project. Before you attempt to link the assembly language source file to your project using Add File, click on "Project" and then on "Select Language Tool." You can then select "MPASM Assembler," followed by "OK" to specify that the project involves an assembly language program instead of a C program. Press Ctrl+F10 to build the assembly language program just as you would build a C language program.

The first method involves specifying a location for execution to jump to, putting in a label that is automatically given the address, and then executing a goto the label. The new address is specified by the org directive. This directive forces all subsequent instructions to be placed in program memory starting at this address. Instead of using the org directive, I could have set the goto address as a constant value or I could have just put the desired address in the instruction (e.g., goto 47). Although both of these methods seem simpler, there is the increased potential for either an incorrect address to be entered into the goto instruction and for the instructions at the address to be incorrect. By using the org directive to set the start of subsequent instructions at the label, the goto address will always be the instruction to start at the desired address, and the instructions themselves will start at the desired address.

The need for explicitly setting the starting address for a block of instructions is rare. The only situation I can think of where this is important for the PIC MCU is when you are setting the address of code that is used to implement the interrupt handler. The code addresses should be explicitly set if the application goes beyond the first code page in memory. Some people like to move blocks of code to specific areas of program memory to make debugging simpler. Because you will not be working with interrupts in the PIC MCU, and because the PIC16F684 does not have more than one page of program memory, there is no need for explicitly setting code addresses. The availability of the MPLAB IDE with the source code simulator eliminates any advantages of putting blocks of code in specific locations. You will find that it is faster to write an application with the different parts of the code butted, or concatenated, together.

The $ symbol in MPASM assembler returns the address of the instruction in which it is used. Jumping to the $ symbol turns the instruction into an endless loop. By adding or subtracting constant values to the $ symbol, going to addresses relative to the current address can be done quickly and easily, without the bother of trying to come up with a meaningful, unique label.

The third method is to use a unique label and goto it. This is generally the preferred method of most new programmers. The problem comes with trying to come up with unique labels for complex programs (the point I made previously). The conventions I tend to use for labels are as follows:

- There can only be one Loop and one Done label, and they are in the mainline.

- Within subroutines, I try to keep to one Loop, one Skip, one Done, and one End label suffix, with the label prefix being the name of the subroutine.

- Labels should be reasonably descriptive. Single-letter labels like a, n, and so on, make a program more difficult to read than if they describe their purpose.

If you have been doing some research about the PIC MCU, you would have discovered that the program memory *page size* is 2,048 instructions. Jumping between pages requires updating a register known as PCLATH before executing the goto instruction. This is not an issue when you are working with the PIC16 F684, which has a total of 2,048 instructions, but is something that you will have to understand if you were working with a PIC MCU, which has more than 2,048 instructions.

Experiment 59—Loading the WREG and Saving Its Contents

Unlike a high-level language in which data can flow through any variable, all the data in a PIC MCU assembly language program will always pass through the *working register*, which is known as *WREG*. In other microprocessors, this register is known as the *accumulator* and is the midway point when data is being transferred from one location to another, as well as one of the source points and a possible destination for mathematical operations. Despite the WREG's responsibility in assembly language programming, when you look at it, only three things can be done with it.

The first thing that can be done with the WREG is to load an eight-bit numeric value (known as a literal or immediate value) into it. This is typically done using the *movlw value* instruction, which stores value (or the least significant eight bits of the instruction) directly into the WREG (see Figure 8-2). This instruction executes without affecting any other resources in the PIC MCU.

Registers are the name given to a group (or a bank) of 128 addressable bytes, some of which can be used as variables in your application and others of which can access built-in hardware functions of the PIC MCU. The variable bytes are called *file registers*, and the hardware control function bytes are known as *special function registers*. These registers can be moved into the WREG using the "movf Register, d" instruction (see Figure 8-3).

The least seven bits of the movf instruction are the address of the register in the bank that is to be accessed. The value in this register is passed through a *zero check module* and then either passed back to the register or stored in the WREG. This action has caused me in the past to characterize the movf instruction as having a primary responsibility of testing whether or not the value of a register is equal to zero and a secondary responsibility of loading WREG with the contents of the register.

I think of the zero test module as a dotted AND bus (see Figure 8-4) with each bit being a control input to one of the pull-down transistor switches, and the bus output being the zero bit (called simply Z) of the STATUS register. If any of the bits are set, the line (and the Z bit in the STATUS register after the instruction executes) will be low.

In the description of the instruction, I noted that the contents of a register could be optionally stored into WREG or stored back into the register. The determination of where the register contents goes is made by the d or destination-bit parameter of the movf instruction. The destination-bit parameter is used by all instructions that move the contents of a register through WREG; it will become clearer as you work through the section.

In some references, you will see that if d is 1, the destination is the register, and if d is 0, the destination is WREG. In all of my code, and I adamantly suggest that you follow this convention, the letter *f* is used as the register destination and *w* is used as the WREG destination. In all of the Microchip files included here, f is equated to 1, and w is equated to 0, so the values are the same. But by using the letter codes, you should simplify the effort in remembering which number is used to initiate which action.

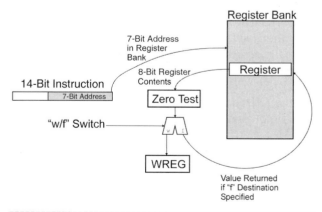

Figure 8-3   *Direct load (movf) instruction*

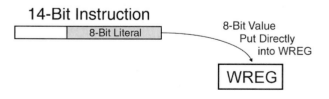

Figure 8-2   *Immediate load (movlw) instruction*

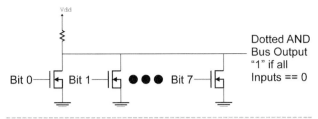

Figure 8-4   *Zero test module*

The final instruction, movwf Register, saves the contents of WREG into a register (see Figure 8-5). While the movlw uses literal addressing to store a specific value in WREG, the movf and "movwf" use direct addressing to transfer the contents of a register to or from WREG.

asmWREG.asm demonstrates the operation of the three instructions discussed in this experiment along with the clrw instruction. The clrw instruction performs the same operation as the movlw 0 instruction but also sets the zero flag (Z) of the STATUS register:

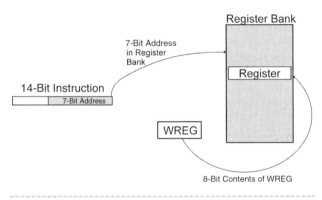

Figure 8-5   *Direct save*

```
        title   "asmWREG - Demonstrate Loading and
Saving WREG"
;
; This program demonstrates the "movlw",
; "clrw", "movf" and "movwf" instruction
; operation and how they are used with
;   WREG to move data within the PIC MCU.
;
;   Hardware Notes:
;    PIC16F684 running at 4 MHz in Simulator
;
;
;   Myke Predko
;   04.11.17
;
  LIST R=DEC
 INCLUDE "p16f684.inc"


  PAGE

 org       0
  nop                       ;   Required for MPLAB ICD2

  movlw     123             ;   Load WREG with Literal

  clrw                      ;   Clear/Load WREG with 0

  movlw     55              ;   Load WREG with Decimal
                            ;   Literal
  movlw     0x55            ;   Load WREG with Hex
                            ;   Literal
  movlw     b'00110111'     ;   Load WREG with Binary
                            ;   Literal
  movlw     'U'             ;   Load WREG with ASCII
                            ;   Character

  movwf     FSR             ;   Store WREG in a Register

  clrw                      ;   Clear WREG

  movf      FSR, f          ;   Set Zero Flag According
                            ;   to Contents of Register
  movf      FSR, w          ;   Load WREG with Register

  goto      $               ;   Finished, Everything
                            ;   Okay

 end
```

Experiment 60—Defining Variables

One of the most difficult concepts for programmers to understand when they are programming the PIC microcontroller in assembly language is that variables are not declared. To me, the term *variable declaration* means that memory is set aside for the variables used in the application and is kept separate from the application's code. In addition to specifying the memory used by the variables, often the application hex file will include initial values that are stored in the variables when the hex file is loaded. When you are specifying addresses for variables in the PIC MCU, that's all you're doing; you are not setting aside memory, and there is no ability to initialize the variables for specific purposes. In this experiment, I will present you with a way to specify variable addresses in the PIC MCU that is both efficient and resistant to user declaration errors.

The idea that PIC MCU assembly language variables are not declared as in a traditional system, with memory set aside for the variables in the hex file, may be confusing. You might try to specify the variable addresses manually using an equate (equ) directive. The *equ* directive is used to associate a label with a numeric value, and it could be used to declare variables in the following manner:

```
i equ 0x020    ;  Declare 8-Bit Counters
j equ 0x021
k equ 0x022    ;  Declare 16-Bit Counter
n equ 0x024    ;  Declare Following 8-Bit Counter
```

The biggest problem with this method should be immediately obvious—it's a lot of work! Not only is it work to write out each variable and make sure it has a unique address, but inserting or deleting variables requires updating addresses to make sure there aren't any holes. Another issue is how multibyte variables are declared and maintained. In the example declaration of the four variables above, it is easy to forget that k is 16 bits and takes up two bytes, and the next variable in sequence must start at an address two higher than the 16-bit variable.

In the mid-1990s, when I first started programming the PIC MCU, this was the accepted method of declaring PIC MCU assembly language variables. It certainly wasn't perfect and it was responsible for a lot of difficult-to-locate errors. A few years ago, Microchip added the *cblock* directive (with the endc directive ending) to MPASM assembler to make it easier to create sequences of numbers for a series of labels. The format for the cblock directive for declaring variables is:

```
cblock [Start Address]
Label1                  ; First Variable
Label2 Size in Bytes]   ; Second, multi-byte
                        ;  variable
Label3                  ; Third Variable
 endc
```

In the previous example, if [Start Address] was 0x020, Label1 would be associated with the constant value 0x021, Label2 with 0x022, and Label3, which follows the 16-bit variable Label2, will have the value 0x024. These values are assigned to the labels at assembly, or build, time without requiring any effort on the part of the programmer—including if you were to add or delete some values between builds.

To demonstrate the cblock directive as well as the equ directive methods of declaring variables in PIC MCU assembly language programs, I created asmDeclare.asm. In this program, I have created three single byte variables (the latter two on the same line and separated by a comma) and a multibyte variable. The equ

declared variable is at the same address as one of the cblock-declared variables (j) and, when written to, will overwrite the cblock-declared variable j.

```
   title   "asmDeclare - Defining variables in PIC
MCU Assembler"
;
;   This code demonstrates how variables are
;   declared in PIC MCU Assembler using the
;   "cblock" directive.
;
;   Hardware Notes:
;    PIC16F684 running at 4 MHz in Simulator
;
;
;   Myke Predko
;   04.09.04
;
  LIST R=DEC
 INCLUDE "p16f684.inc"

;  Variables
 CBLOCK 0x20
i                      ;  Single 8-Bit Variable
j, k                   ;  Two 8-Bit Variables
DataString:5           ;  Multi-Byte String Variable
 ENDC

Alternative_i EQU 0x21
       ; "Alternative_i", same address as "j"

  PAGE
;  Mainline of Declare

 org      0

   movlw   47    ;  Load "i" with Literal value
   movwf   i

   movlw   33    ;  Load "j" with Literal value
   movwf   j

   movlw   22    ;  Overwrite "j"
   movwf   Alternative_i

Loop:            ;  Finished, Loop Forever
  goto    Loop


  end
```

The body statements of the program are initialization statements for the variables, and the values that get stored in them can be displayed in the Watch window of the simulator. These are the same statements that were presented in the previous experiment, but applied to variables.

When you simulate asmDeclare.asm, you will discover that the variable j is overwritten when Alternative_i is written to. The reason for this is simple: Both variables are at the same address and a write to one variable will affect the other. This would not be the case if variables were truly declared in PIC MCU assembly language. An obvious conclusion to this experiment is that you should never mix cblock declared variables with equ declared variables. I would go further and say that variables should be declared using cblock only.

Along with declaring variables, the cblock directive can be used for declaring data structures and for enumerating labels for applications such as State Machines. Using cblock defined label values in these cases will allow simpler coding as well as eliminate much of the maintenance required of individual equ directive defined labels. The dynamic nature of the cblock directive (it is updated every build cycle) makes it ideal for use anywhere that changing numeric label values is required.

Experiment 61—Bitwise Instructions

Bitwise boolean operators are the most basic data processing instructions that can be performed in a processor. These instructions can be built out of standard gates quite easily and do not have the same operational issues that are present in addition and subtraction instructions. These instructions perform the basic logic functions over all eight bits of two parameters, one being WREG and the other being a parameter specified in the instruction.

Two-parameter bitwise instruction (i.e., AND, OR, and XOR) execution for direct addressing (those that end in wf) is shown in Figure 8-6. In this case, the contents of a register are operated on with the contents of WREG using one of the three Boolean operations. The destination of the result can be either in the WREG or back in the register using the d parameter of the instruction. The ability to return the value to the regis-

ter may seem to be unnecessary, but, as I will show in the next two sections, it can be extremely handy.

The data flow of bitwise Boolean operations that can be performed against the contents of WREG and a literal value (causing the instructions to end in lw) are shown in Figure 8-7. These instructions behave similarly to those with a direct address except that they have only one possible destination, WREG. In both cases, if the result of the operation is zero, then the zero (Z) STATUS bit is set, otherwise it is reset.

There is also a single-parameter bitwise instruction, called comf Register, d. This instruction complements (i.e., does not negate) each bit of the specified register and returns the result in either WREG or the source register. If you wanted to complement the contents of the WREG, you could use the following literal XOR instruction:

```
xorlw 0xFF      ;  Complement each bit of WREG
```

In this case, asmBitwise.asm demonstrates the operation of the different bitwise PIC MCU assembly language instructions.

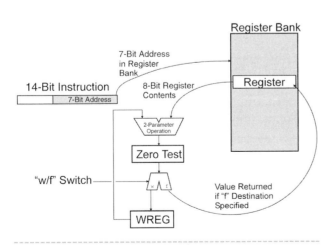

Figure 8-6   *Two-parameter direct*

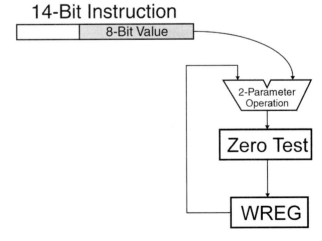

Figure 8-7   *Two-parameter literal*

```
        title   "asmBitwise - Demonstrate Bitwise
Instructions"
;
;    This program demonstrates the operation of
;    the Bitwise instructions on WREG and a File
;    Register including the operation of the Zero
;    Status Flag.
;
;    Hardware Notes:
;     PIC16F684 running at 4 MHz in Simulator
;
;
;    Myke Predko
;    04.11.17
;
     LIST R=DEC
     INCLUDE "p16f684.inc"

     CBLOCK 0x20        ;  Variable Declaration
     i, j
     ENDC

     PAGE

     org     0
     nop                ;  Required for MPLAB ICD2

     movlw   0x3F       ;  Initialize Variables
     movwf   i
     movlw   0xF3
     movwf   j

     movlw   123        ;  Initialize WREG
     andwf   i, w       ;  WREG = WREG & i
                        ;       = 0x7B & 0x3F
                        ;       = 0x3B
                        ;  Z = 0, i Unchanged

     movlw   123        ;  Initialize WREG
     andwf   i, f       ;  i = = WREG & i
                        ;       = 0x7B & 0x0F
                        ;       = 0x3B
                        ;  Z = 0, WREG Unchanged

     iorwf   j, w       ;  WREG = WREG | j
                        ;       = 0x7B + 0xF3
                        ;       = 0xFB
                        ;  Z = 0, j Unchanged
```

```
     comf    i, f       ;  i = i ^ 0xFF
                        ;    = 0x3B ^ 0xFF
                        ;    = 0xC4
                        ;  WREG Unchanged

     xorwf   i, f       ;  i = WREG ^ i
                        ;    = 0xFB ^ 0xC4
                        ;    = 0x3F
     xorwf   i, w       ;  WREG = WREG ^ i
                        ;       = 0xFB ^ 0x3F
                        ;       = 0xC4
     xorwf   i, f       ;  i = WREG ^ i
                        ;    = 0x3B ^ 0xC4
                        ;    = 0xFB

     goto    $          ;  Finished, Everything Okay

     end
```

Before going on, I would like you to consider the three XORs at the end of the program. I didn't explicitly summarize what they are doing. However, if you were to look through them, you would discover that they not only demonstrate the operation of the xorwf statement with different options, they also are a code snippet that you should keep handy. The following three instructions swap the contents of WREG with the contents of a register, and they are much more efficient than coming up with the equivalent assembly language statements for the traditional register swap:

```
     Temp = i;                    // swap "i" and
                                  "j"
     i = j;
     j = i;
```

When you perform traditional high-level programming, there aren't a lot of opportunities to use the bitwise instructions, but in assembly language programming, they are critical to your ability to create very efficient applications.

Experiment 62—Addition Instructions

PC/ Simulator

The PIC microcontroller's addition instructions are very similar to the bitwise instructions of the previous experiment: Data is set up in the WREG, it is arithmetically combined in some way, and the result is put back into either the WREG or a source file register. What makes this instruction different from the bitwise instructions is the production of two carry values that can be used as test values for controlling the execution path of the program.

The *carry* and *digit carry* STATUS register bits are normally referred to as C and DC, respectively, just as the zero bit is referred to as Z. These abbreviations are used in PIC MCU assembly language programming to indicate these bits and should never be used for variable or label names.

After passing the sum to the *Operation Result*, the carry bit of the addition operation is saved as bit 0 of the STATUS register and is used to indicate if the result of the addition operation is greater than 0x0FF. This bit is often used as an indicator for 16-bit (and greater) operations to indicate the results that addition to lower bytes has on the higher bytes.

The digit carry bit (bit 1 of the STATUS register) performs a similar operation but for the least significant four bits of the result. Digit carry is *not* the same as bit 4 of the sum. The digit carry is the carry result of bit 3 of the sum; it is difficult to illustrate this in diagrams (see Figure 8-8), but its operation can be illustrated clearly using a test program like asmAdd.asm.

When you simulate asmAdd.asm, I suggest that you display the contents of the STATUS register (and WREG) in a Watch window. The contents of the STATUS register should always be displayed as binary rather than the default hex. I realize that the state of the carry, digit carry, and zero bits, along with the contents of WREG are displayed at the bottom of the MPLAB IDE desktop, but still I find it useful to have them in the same Watch window as the other registers and variables used in the application.

```
title   "asmAdd - Demonstrate Addition
Instructions"
;
;   This program demonstrates the operation of
;   the addition instructions including the
;   operation of the Zero, Carry and Digit Carry
;   Status Flags.
;
;   Hardware Notes:
;    PIC16F684 running at 4 MHz in Simulator
;
;
;   Myke Predko
;   04.09.27
;
   LIST R=DEC
   INCLUDE "p16f684.inc"

 CBLOCK 0x20        ;  Variable Declaration
i, j
 ENDC

   PAGE

 org     0
   nop               ;  Required for MPLAB ICD2

   movlw   0x0F      ;  Initialize Variables
   movwf   i
   movlw   0xF0
   movwf   j

   clrw              ;  Clear WREG
                     ;  Z = 1

   addwf   i, w      ;  WREG = WREG + i
                     ;       = 0 + 0x0F
                     ;       = 0x0F
                     ;  Z = 0, C = 0, DC = 0

   addwf   i, w      ;  WREG = WREG + i
                     ;       = 0x0F + 0x0F
                     ;       = 0x1E
                     ;  Z = 0, C = 0, DC = 1
```

```
   addwf   i, w      ;  WREG = WREG + i
                     ;       = 0x1E + 0x0F
                     ;       = 0x2D
                     ;  Z = 0, C = 0, DC = 1

   addwf   j, w      ;  WREG = WREG + k
                     ;       = 0x2D + 0xF0
                     ;       = 0x(1)1D
                     ;       = 0x1D
                     ;  Z = 0, C = 1, DC = 0

   addlw   0xE3      ;  WREG = WREG + 0xE3
                     ;       = 0x1D + 0xE3
                     ;       = 0x(1)00
                     ;       = 0x00
                     ;  Z = 1, C = 1, DC = 1


   goto    $         ;  Finished, Everything Okay

 end
```

The first addition operation (zero plus 0x0F) has a nonzero result, and the three STATUS flags indicating the result of an addition instruction are all reset. I then add 0x0F to the value in WREG again and, as expected, the result is 0x1E. This second addition has a lower four-bit sum and a (digit) carry to the next larger four bits, so the DC STATUS bit is set. I should point out that coincidentally bit 4 of the sum is also set.

To prove that bit 4 being set is a coincidence, I repeated the addition of 0x00F to 0x1E in the WREG. In this case, the sum is 0x2D, but you will see that the digit carry flag is still set even though bit 4 of the sum is reset. The digit carry can be confounding and difficult to predict its value. For your initial PIC MCU assembly programming, I suggest that you ignore it and just focus on the carry and zero STATUS register bits for returning addition instruction information.

The final addition operations cause the sum in WREG to be greater than 0xFF, causing the carry flag to be set. Note that up to these instructions, the carry flag is reset because the results are always less than 0xFF. The first addition produces a result of 0x11D, and the most significant bit is used as the carry. This can be done because, unlike the digit carry, no other bits are part of the final sum.

The last addition value was chosen to produce a zero result, but when you simulate the program, you will notice that all three STATUS register bits will be set. Not only is the carry bit set because the sum is greater than 0x0100, but the zero bit is set (because the saved eight bits are all reset) *and* the digit carry bit is set (because the sum of the least significant four bits are greater than 0x00F). This may not seem important for addition, but this result will be very important in carrying out subtraction, as I will discuss in the next experiment.

Experiment 63—AddLibs: Strange Simulator Results

Before going on, I want to show you how the simulator will seemingly fail and produce an incorrect result for an arithmetic operation. The code for this experiment was taken from a high school quiz. The base code was given in the test, and the students were asked to verify what it did using the MPLAB IDE simulator. When the students entered the code the simulated result was totally unexpected.

I would like to ask you to load the following code, asmAdd 2.asm, into the MPLAB IDE and enable the simulator.

```
  title  "asmAdd 2 - Demonstrate Simulator
Operation"
;
;   This program Shows how the Simulator can
;   apparently display the wrong result to an
;   addition operation.
;
;   Hardware Notes:
;    PIC16F684 running at 4 MHz in Simulator
;
;
;   Myke Predko
;   04.12.09
;
  LIST R=DEC
  INCLUDE "p16f684.inc"

CBLOCK 0x20              ;  Variable Declaration
i
  ENDC

  PAGE

org      0
  nop                    ;  Required for MPLAB ICD2

  movlw    140           ;  Initialize Variable
  movwf    i
```

```
  addwf    i, w          ;  WREG = WREG + i
                         ;       = 140 + 140
                         ;       = 280
                         ;       = b'(1)00011000'
                         ;       = b'00011000'
                         ;       = 0x18
                         ;       = 24
                         ;  C = 1,  DC = 1,  Z = 0

end
```

After building the code, single-step through the code to the addwf instruction. After executing it, the simulator stops with a value of 0xB5 in WREG (my result), but as shown in the code's comments, the expected results are 0x018 or 24 (decimal).

What happened? The answer is no instruction is given after the addwf where the simulator can stop at, and therefore it executes a number of instructions in the simulated chip program memory before stopping. You can better see what is happening by clicking on "View" and then on "Program Memory." The instructions for the start of the application match the asmAdd 2.asm code, but the instructions afterward are all "addlw 0xff," and so it appears that the simulator has executed for 256 (0x100) instructions after executing the "addwf I, w" instruction before stopping.

This is actually a good example of how small programs work in a PIC MCU. Even though they take up a small fraction of the program memory available in the chip, the program memory is still present and they execute as addlw 0xFF instructions.

To avoid this issue, I tend to use an endless loop (goto $) although many others like to use a nop. Any instruction will do. I like the endless loop as it will keep the simulator from executing additional instructions, which can change the contents of registers or bit flags. These instructions will give the simulator an instruction to stop at rather than having the ambiguous situation of this application.

I realize that I have indicated the need for the instruction which ends the application, but the kids in the high school course were told this as well. It's easy to forget unless you have to work through it and try to understand what is happening, as I have in this experiment.

PC/ Simulator

Subtraction in PIC MCU assembly language will probably be the most consistently difficult thing you will have to work with. This is due to the nonintuitive way the instructions execute. You'll find that the data is handled backwards and the carry and digit carry flags will not work as you would expect. Many PIC MCU assembler programmers try to avoid working with the subtract instructions all together, instead creating similar capabilities with other instructions. In this experiment, I will show you how the PIC MCU subtraction instructions can be used in different situations.

The first point to remember about the subtract instructions is that they always add the negative contents of WREG to the instruction's parameter. The actual subtracting circuitry is really a modification to the addition circuitry with either the negative value of WREG or the unchanged value added to the second parameter (see Figure 8-8).

In grade school, you were probably told that subtraction was the same as adding the negative, that is:

$$A - B = A + (-B)$$

but when working with digital logic, this is not quite true. To negate a binary value (convert it to its complement in base 2), you complement the bits and increment the result:

$$-B = (B \char94 0xFF) + 1$$

When this 2's complement value is added to the second parameter, the eight-bit result is the same as you would expect for subtraction (again remembering that it is WREG taken away from the parameter). But the carry (C) and digit carry (DC) STATUS flags are not what you might expect. To simplify, don't worry about DC, but focus on predicting the state of the zero and carry flags based on different values (see Table 8-1). The carry flag after subtraction is often referred to as the negative borrow flag, because, when it is reset, the subtraction result is negative and a one should be borrowed from the next highest byte.

Further confusing the operation of subtraction is the apparent reversal of the instruction from what seems intuitive. In other processor assembly languages, when you see some instructions like the following:

```
mov   Acc, Value1
      ; Load Accumulator with the Subtractend
sub   Acc, Value2
      ; Subtract Value2 from Value1
```

you expect it to execute like this:

```
Acc = Value1-Value2;
```

But, in PIC MCU assembly language, the analogous instructions:

```
movf  Value1, w
      ; Load Accumulator with the Subtractor
subwf Value2, w
      ; Perform the Subtraction Operation
```

Table 8-1
Zero and Carry STATUS Bits After Subtraction Operation

WREG	Parameter	Results of Parameter + (-WREG)
0x02	0x01	WREG = 0xFF (-1), C = 0, Z = 0
0x02	0x02	WREG = 0, C = 1, Z = 1
0x02	0x03	WREG = 0x01, C = 1, Z = 0

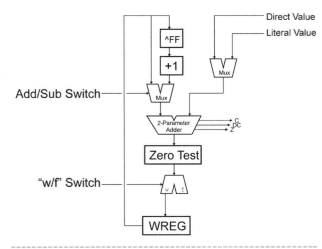

Figure 8-8   *Subtraction*

produce the actual operation:

```
WREG = Value2 - Value1;
```

This can be both confusing and frustrating when you are new to PIC MCU assembly language programming. There are a number of things that you can do to make the operations simpler. The first is to add the negative. Instead of trying to figure out how to use the sublw (subtract WREG from literal), you can add the negative. That is:

```
addlw       -47              ;    WREG = WREG - 47
```

Another method is to only use the subtraction instruction the same way, all the time. You could start with the basic subtraction statement:

```
A = B - C
```

and you can convert it to PIC MCU assembly language using the instruction sequence:

```
movf    C, w
subwf   B, w
movwf   A
```

In this sequence, if B or C is a literal value, then insert sublw and movlw in their places, respectively. If B is the same as A, the last instruction (movwf) could be eliminated and the subwf instruction changed to subwf B, f.

Finally, the method I recommend is to add the subtract instructions to your application as you think they will work, but write out what you expect the operations to do as I have in asmSubtract.asm. Then, when you are simulating the application, you can compare the actual results (both in WREG or the file register and in the STATUS register) to the expected values you marked in the application and on which you have based the application's operation.

```
    title  "asmSubtract - Demonstrate Subtract
Instructions"
;
;   The Subtract Instructions are a bit more
;   complex in operation than the addition
;   instructions:
;
;   REMEMBER, the Negative Contents of WREG are
;   Added to the instruction argument.
;
;   Hardware Notes:
;    PIC16F684 running at 4 MHz in Simulator
;
;
;   Myke Predko
;   04.11.19
;
  LIST R=DEC
  INCLUDE "p16f684.inc"
```

```
  CBLOCK 0x20     ; Variable Declaration
i, j
  ENDC

  PAGE

org     0
  nop             ; Required for MPLAB ICD2

  movlw   0x01    ; Initialize Variables
  movwf   i
  movlw   0x02
  movwf   j

  movf    i, w    ; Start with j - i
  subwf   j, w    ; WREG = j - WREG
                  ;      = 0x02 + (WREG ^ 0xFF) + 1
                  ;      = 0x02 + (0x01 ^ 0xFF) + 1
                  ;      = 0x02 + 0xFE + 1
                  ;      = 0x02 + 0xFF
                  ;      = 0x(1)01
                  ; WREG = 0x01, C = 1, DC = 1, Z = 0

  movf    j, w    ; Next i - j
  subwf   i, w    ; WREG = i - WREG
                  ;      = 0x01 + (WREG ^ 0xFF) + 1
                  ;      = 0x01 + (0x02 ^ 0xFF) + 1
                  ;      = 0x01 + 0xFD + 1
                  ;      = 0x01 + 0xFE
                  ;      = 0xFF
                  ; WREG = 0xFF, C = 0, DC = 0, Z = 0

  movf    i, w    ; Now, i - i
  subwf   i, w    ; WREG = i - WREG
                  ;      = 0x01 + (WREG ^ 0xFF) + 1
                  ;      = 0x01 + (0x01 ^ 0xFF) + 1
                  ;      = 0x01 + 0xFE + 1
                  ;      = 0x01 + 0xFF
                  ;      = 0x(1)00
                  ; WREG = 0x00, C = 1, DC = 1, Z = 0

  movlw   0x01    ; WREG = 0x10 - 0x01
  sublw   0x10    ; WREG = 0x10 - WREG
                  ;      = 0x10 + (WREG ^ 0xFF) + 1
                  ;      = 0x10 + (0x01 ^ 0xFF) + 1
                  ;      = 0x10 + 0xFE + 1
                  ;      = 0x10 + 0xFF
                  ;      = 0x(1)0F
                  ; WREG = 0x0F, C = 1, DC = 0, Z = 0

  movlw   0x10 ; WREG = 0x01 - 0x10
  sublw   0x0  ; WREG = 0x01 - WREG
                  ;      = 0x01 + (WREG ^ 0xFF) + 1
                  ;      = 0x01 + (0x10 ^ 0xFF) + 1
                  ;      = 0x01 + 0xEF + 1
                  ;      = 0x01 + 0xF0
                  ;      = 0xF1
                  ; WREG = 0xF1, C = 0, DC = 1, Z = 0

  goto    $       ; Finished, Everything Okay

end
```

Hopefully, I have not made the two subtract instructions scarier than they actually are. You might want to create an application like asmSubtract.asm and put in different values. But the best way to learn the subtract instructions is to use them in your own application and confirm they are working by using the simulator. By doing this you will be familiarizing yourself with the subtraction instructions work and becoming more proficient with the MPLAB IDE simulator.

Experiment 65—Bank Addressing

starter kit

So far, you have been introduced to how programs are structured, the Microchip PIC MCU include file, execution changes (gotos), register/variable declaration, and data processing. By all rights, you should be ready to start creating code for actual hardware applications. Unfortunately, the next topics don't build on each other as easily as the previous ones did, and to be able to code software to run in a physical chip you need the information in all the topics. To try and break this deadlock, I want to skip to what is normally one of the last topics, the PIC microcontroller register banks, and show you how, with the instructions and information you have been given so far, you can create your own applications.

Earlier in the section, I said that there are seven bits set aside in each instruction to accesses a register. They are the register's address. This means that each instruction can access up to 128 different byte-size registers. This isn't a bad amount of memory, but the PIC MCU designers wanted to be able to provide more, so they decided to *bank* the registers and provide up to four banks in the midrange PIC MCUs. Each bank consists of a number of common and unique special function registers (processor and hardware control and interface registers) and general purpose, or file, registers. These registers may or may not be shadowed between the PIC MCU's register banks, and there will always be the need to access registers in more than one bank in your application. The concept of register banks can be difficult to understand, but you must if you are to create PIC MCU applications.

The most vivid example I can give of the PIC MCU's banks is a spice rack; when I was a kid, a friend's mom had a spice rack her husband had made her. This rack stood on her kitchen counter and was able to hold many spice bottles. It was built to hold the spices on either side and was mounted on a small turntable. When a particular spice was required, the rack would be turned to expose the side that held that bottle of spice. This arrangement seemed to work well,

although the rack had to be turned repeatedly during the preparation of a meal because there was no way to prearrange the bottles so that a single dish would use spices from only one side of the rack.

The banks of registers in the PIC MCU are very similar in concept to this spice rack (see Figure 8-9). The different registers are arranged in such a way as to provide most of the useful functions on one bank (one side of the spice rack), but you may have to change the side that that is exposed during an application. The control for which side is exposed (or execute from) is the RP0 bit (bit 5) of the STATUS register. When this bit is reset, bank 0 is accessible; when this bank is set, bank 1 is accessible.

As can be seen in Figure 8-9, STATUS and a number of other registers can be accessed from either bank (like there was a hole cut in the spice rack so a bottle could be reached from either side). For transferring data between the two banks, the file registers in the address range of 0x70 to 0x7F are common between the two banks. Along with these *common, or shadowed, registers,* a number of other registers exist (both special function and file registers) that are unique to each bank. For most of your initial applications, you will be able to run almost exclusively out of bank 0. The only registers you will have to access are the TRIS registers and the ANSEL registers.

Changing the current bank can be accomplished by using the following code:

```
movf     STATUS, w
iorwf    1 << 5    ;  Set RP0 - Change to Bank 1
movwf    STATUS
```

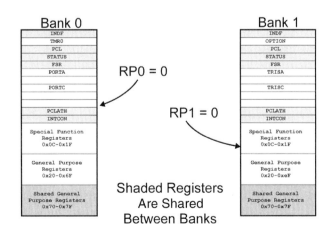

Figure 8-9   *PIC16F684 banks*

---

and

```
movf    STATUS, w
andwf   0xFF ^ (1 << 5)  ; Reset RP0 - Change
                         ; to Bank 0
movwf   STATUS
```

The *left shift operator* ($<<$) in the statements above, shifts 1 to the left by the specified number of bits. I use this method of specifying bit constants rather than using decimal, hex, or binary constants because it is very difficult to screw up. With constant values, you must always remember which digit, column, or value each bit value has. By using the left shift operator, remembering these values is not required. When setting or resetting individual bits, I recommend using this format for constant values, because it avoids the need to verify the value used for the specific bit.

When you are changing STATUS register bits, you cannot take advantage of the ability of the andwf and iorwf instructions to change all the bits of the STATUS register using code such as the following:

```
movlw   1 << 5  ;  Set RP0 - Change to Bank 1
iorwf   STATUS, f
```

The STATUS register's C, DC, and Z bits are affected by bitwise and arithmetic operations, so the PIC MCU designers decided to not allow statements that can change these flags to also write to the STATUS register simultaneously. The reason is simple: After the instruction, it is difficult to define which bit is valid. By forcing the user to read from and write to the STATUS register explicitly, there is no ambiguity in determining which value for which bit is to be used by the application code.

After changing to bank 1, if you were to access registers in that bank directly as you would in bank 0, say:

```
movf    TRISA, w
```

you would get the assembler message "Message[302] C: {Filename/Line Number}: Register in operand not in bank 0. Ensure that bank bits are correct." because the address specified in the PIC MCU include file gives TRISA the address 0x85. The number 0x85 is 133 decimal and uses eight bits, not the seven that is required for addressing the contents of the bank.

The simple solution to this problem would be to change the TRISA definition to 0x05, which seems correct because the TRISA register is at address 0x05 in bank 1. But I am going to suggest that the label is left as is and have the constant value XORed with 0x80, which will reset bit 7 of the address. The first is the index register uses it to differentiate between regis-

ters in bank 0 and bank 1 (it is an eight-bit value and can access each byte of the two 128-byte banks). This can be used in your applications to effectively optimize some code functions as well as to provide arrays in both banks that can be accessed easily. The second reason is that XORing every eight-bit address while executing in bank 1 will serve as a reminder to you to make sure the correct bank is executing. If the message comes up and you are in bank 0, then you should look over your code and add the correct bank switch commands. Similarly, if you are accessing a bank 0 register while in bank 1, by XORing the address with 0x80, you will get a warning message stating that bit 7 is set (telling you that you are accessing a bank 0 register while executing in bank 1).

With the information you now have, you can create a complete PIC MCU application and to demonstrate this. I created asmFlashNada.asm, which will flash "D0" in a PICKit™ 1 starter kit:

```
title   "asmFlashNada - PIC16F684 Flashing LED"
;
;  This Program Flashes the PICKit D0 LED
;  between RA4 (Positive) and RA5 (negative) at
;  approximately 2x per second without using
;  any advanced control instructions.
;
;  Hardware Notes:
;  PIC16F684 running at 4 MHz Using the Internal
;  Clock
;  Internal Reset is Used
;  RA4 - LED Positive
;  RA5 - LED Negative
;
;
;  Myke Predko
;  04.11.19
;
  LIST R=DEC
  INCLUDE "p16f684.inc"


  __CONFIG _FCMEN_OFF & _IESO_OFF & _BOD_OFF &
_CPD_OFF & _CP_OFF & _MCLRE_ON & _PWRTE_ON &
_WDT_OFF & _INTOSCIO

; Variables
CBLOCK 0x20
Dlay:2
ENDC

  PAGE
; Mainline

  org     0

  nop                      ; For ICD Debug

  clrf    PORTA            ; Initialize I/O Bits to
                           ; Off
  movlw   7                ; Turn off Comparators
  movwf   CMCON0
  movf    STATUS, w        ; Set the RP0 Bit (Bit 5)
  iorlw   1 << 5           ; to Execute in Page 1
  movwf   STATUS
  clrf    ANSEL ^ 0x80     ; All Bits are Digital
```

```
        movlw   b'001111'       ; RA4/RA5 are Digital Outputs
        movwf   TRISA ^ 0x80
        movf    STATUS, w       ; Clear RP0
        andlw   0xFF ^ (1 << 5)
        movwf   STATUS

Loop:                           ; Return Here after D0
                                ; Toggle
        clrf    Dlay + 1        ; High 8 Bits for Delay
        clrf    Dlay            ; Low 8 Bits for Delay
DlayLoop:
        movlw   1               ; Decrement the Inside
                                ; Loop
        subwf   Dlay, f
        movf    STATUS, w       ; Zero Flag Set?
        andlw   1 << 2          ; Check Bit 2
        addwf   PCL, f          ; Add to the Program
                                ; Counter
        goto    DlayLoop        ; Zero, Loop Around
        nop                     ; Space Change in
                                ; Address due
        nop                     ;  Zero Flag Being Set
        nop
        movlw   1               ; Decrement the Outside
                                ; Loop
        subwf   Dlay + 1, f
        movf    STATUS, w       ; Check for Zero
        andlw   1 << 2
        addwf   PCL, f
        goto    DlayLoop
        nop                     ; Space Change in Address
                                ; due
        nop                     ; Zero Flag Being Set
        nop

        movlw   1 << 4          ; Toggle RA4 (D0)
        xorwf   PORTA, f

        goto    Loop            ; Repeat


        end
```

There are really no new instructions in this application, at least none that you haven't been shown previously. There is however one part of the application that will require a bit of instruction. This is the idea of conditional jumps. No if, for, or while statements exist in PIC MCU assembly language, and so, to implement jumps that are conditional to the status of the previous instruction, I take advantage of one of the most wonderful features of the PIC MCU, its ability to access any register directly in the application code. To implement the conditional jump, I add the STATUS register's zero bit (bit 2) to the lower four bits of the program counter (the PCL register). The code works as follows:

```
Program Counter = Program Counter + 1 +
(STATUS & (1 << 2))
```

If the zero flag is reset (the previous instruction did not result in a zero), this code will simply continue on at the next instruction. If the zero flag is set, then the program counter will be loaded with the current instruction plus four (because bit 2 has a constant value of 4). I added the three nops, or no-operation, instructions to space out the code and make sure execution, when the zero flag of the STATUS register is set, takes place at the correct address. In the following experiments, I will show how conditional execution is implemented in a more conventional manner.

Tricks like the conditional jump using data movement and arithmetic operators are one of the reasons why I really like working with the PIC MCU. The PIC MCU, more than any other small processor architecture that I know, has been designed with the flexibility required to perform some amazing tricks that will help make your applications more efficient and help you to think through problems in different ways.

Experiment 66—Bit Instructions

PC/ Simulator

PICkit™ 1

starter kit

In the previous experiment, I discussed the concept of banks and the ability of the assembly language programs to access any register directly. I also showed how banks were changed and how bank 1 registers (including, most importantly, the TRIS registers) were addressed in an instruction so no message was produced by the assembler. Finally, the operation and access of the STATUS register was presented, and then a method of changing the program counter conditionally was presented. In short, I dumped a lot of new information on you.

The bit set (bsf) and bit reset (bcf) instructions presented in this experiment are somewhat less demanding but will be used as much in your assembly language programming. These two instructions will allow you to change the state of a register's bit in one instruction, without the need for ANDing or ORing the contents of the register. bsf and bcf are incredibly useful instructions and will reduce the size of your application code as well as make it easier to read.

In the previous experiment, when I discussed how to set and reset the RP0 bit of the STATUS register, I was also describing how the bsf and bcf instructions worked. Here are the bsf and bcf instructions:

```
;  bsf Register, Bit Operation
   movf    Register, w
   iorwf   1 << Bit           ;  Set Bit
   movwf   Register

;  bcf Register, Bit Operation
   movf    Register, w
   andwf   0xFF ^ (1 << Bit)  ;  Reset Bit
   movwf   Register
```

This description may be surprising, but this is exactly how they work. They do not, as you might expect, only set or reset the single appropriate bit, but they read in the entire register, modify the appropriate bit, and return the value to the register. I am pointing this out because there will be cases where, if some of a port register's bits already have the desired value and you want to change just one bit, the bcf and bsf instructions may not be the best instructions for the job. To get around this case, you should never use bsf or bcf in the port registers or other registers that can be written both by software and hardware.

To demonstrate the operation of the bsf and bcf instructions, I modified asmFlashNada.asm by replacing the instructions that could use bsf and bcf with them. It is not a major change to the application, but it does shorten it a bit. Like the previous experiment, asmFlashBit.asm can be run on a PIC16F684 in the PICKit 1 starter kit.

```
   title   "asmFlashBit - PIC16F684 Flashing LED"
;
;   This Program Flashes the PICKit D0 LED
;   between RA4 (Positive) and RA5 (negative) at
;   approximately 2x per second with only using
;   the Bit Set and Reset Instructions
;
;   Hardware Notes:
;    PIC16F684 running at 4 MHz Using the Internal
;    Clock
;    Internal Reset is Used
;    RA4 - LED Positive
;    RA5 - LED Negative
;
;
;   Myke Predko
;   04.11.19
;
   LIST R=DEC
```

```
   INCLUDE "p16f684.inc"

   __CONFIG _FCMEN_OFF & _IESO_OFF & _BOD_OFF &
_CPD_OFF & _CP_OFF & _MCLRE_ON & _PWRTE_ON &
_WDT_OFF & _INTOSCIO

;  Variables
CBLOCK 0x20
Dlay:2
ENDC

   PAGE
;  Mainline

   org     0

   nop                              ; For ICD Debug

   clrf    PORTA                    ; Initialize I/O Bits to
Off
   movlw   7                        ; Turn off Comparators
   movwf   CMCON0
   bsf     STATUS, RP0              ; Replaces:
                                    ; movf    STATUS, w
                                    ; iorlw   1 << 5
                                    ; movwf   STATUS
   clrf    ANSEL ^ 0x00             ; All Bits are Digital
   bcf     TRISA ^ 0x00, 4          ; Replaces:
                                    ; movlw   b'001111'
   bcf     TRISA ^ 0x00, 5          ; movwf   TRISA ^ 0x080
   bcf     STATUS, RP0              ; Replaces:
                                    ; movf    STATUS, w
                                    ; andlw   0xFF ^ (1 << 5)
                                    ; movwf   STATUS

Loop:                              ; Return Here after D0
Toggle
   clrf    Dlay + 1                 ; High 8 Bits for Delay
   clrf    Dlay                     ; Low 8 Bits for Delay
DlayLoop:
   movlw   1                        ; Decrement the Inside
                                    ; Loop
   subwf   Dlay, f
   movf    STATUS, w                ; Zero Flag Set?
   andlw   1 << 2                   ; Check Bit 2
   addwf   PCL, f                   ; Add to the Program
                                    ; Counter
   goto    DlayLoop                 ; Zero, Loop Around
   nop                              ; Space Change in
Address due
   nop                              ; Zero Flag Being Set
   nop
   movlw   1                        ; Decrement the Outside
Loop
   subwf   Dlay + 1, f
   movf    STATUS, w                ; Check for Zero
   andlw   1 << 2
   addwf   PCL, f
   goto    DlayLoop
   nop                              ; Space Change in
Address due
   nop                              ; Zero Flag Being Set
   nop

   movlw   1 << 4                   ; Toggle RA4 (D0)
   xorwf   PORTA, f

   goto    Loop                     ; Repeat

   end
```

Experiment 67—Bit Skip Instructions

PC/ Simulator

PICkit™ 1

starter kit

When you first look through the PIC16F684 instruction set, you will find execution address change goto instructions as well as subroutine call instructions. But nowhere will you find conditional gotos, branches, or other conditional execution address change instructions that you traditionally find in microprocessor and microcontroller architectures. What the PIC16F684 (and indeed all PIC MCU family microcontrollers) has instead is the ability to skip the next instruction based on whether or not a register bit is set or reset. The skipped instruction could be a goto address or it could be a single instruction that executes when the condition is false (and is not skipped over). These *bit skip instructions*, although being somewhat difficult to understand and apply when you are first learning to program the PIC MCU, are very powerful instructions that provide an amazing amount of programming flexibility and single-instruction capabilities rarely matched in other processors. And all this can happen in less than three instructions. In this experiment and the next, I will introduce you to the bit skip instructions and give you some pointers on how to implement them in traditional programming methodologies.

The two-bit skip instructions are:

```
btfsc  Register, Bit ;  Skip the Next Instruction
                     ;  if "Bit" of "Register"
                     ;  is Reset ("0")
btfss  Register, Bit ;  Skip the Next Instruction
                     ;  if "Bit" of "Register" is
                     ;  Set ("1")
```

In asmFlash.asm, I have changed the jumps based on adding the zero flag of the STATUS register to "btfss STATUS, Z" instructions, which will execute the next instruction if the zero flag is reset or the one following it if the zero flag is set. By doing this, I have reduced the number of instructions to change the execution from seven to two, based on the zero flag.

```
       title  "asmFlash - PIC16F684 Flashing LED"
;
;   This Program is the equivalent to "cFlash.c"
;   but written in assembly language, taking
;   advantage of ALL the features of PIC MCU
;   assembly language - bit instructions, bit
;   skip instructions and decrement and skip
;   instructions.
;
;   Hardware Notes:
;    PIC16F684 running at 4 MHz Using the Internal
;    Clock
;    Internal Reset is Used
;    RA4 - LED Positive
;    RA5 - LED Negative
;
;
;   Myke Predko
;   04.11.17
;
    LIST R=DEC
  INCLUDE "p16f684.inc"

  __CONFIG _FCMEN_OFF & _IESO_OFF & _BOD_OFF &
_CPD_OFF & _CP_OFF & _MCLRE_ON & _PWRTE_ON &
_WDT_OFF & _INTOSCIO

;  Variables
  CBLOCK 0x20
Dlay:2
  ENDC

  PAGE
;  Mainline

  org    0

    nop                        ; For ICD Debug

    clrf   PORTA               ; Initialize I/O Bits to
                               ; Off
    movlw  7                   ; Turn off Comparators
    movwf  CMCON0
    bsf    STATUS, RP0         ; Execute out of Bank 1
    clrf   ANSEL ^ 0x080       ; All Bits are Digital
    bcf    TRISA ^ 0x080, 4 ; Replaces:
                               ; movlw   b'001111'
    bcf    TRISA ^ 0x080, 5 ; movwf   TRISA ^ 0x080
    bcf    STATUS, RP0         ; Return Execution to
                               ; Bank 0

Loop:                          ; Return Here after D0
                               ; Toggle
    clrf   Dlay + 1            ; High 8 Bits for Delay
    clrf   Dlay                ; Low 8 Bits for Delay
DlayLoop:
    goto   $ + 1               ; Three Cycle Delay
    nop
    movlw  1                   ; Decrement the Inside
                               ; Loop
    subwf  Dlay, f
    btfss  STATUS, Z           ; Skip if Zero Flag is
                               ; Set
```

```
        goto    DlayLoop            ; Else, Loop Again
                                    ; Replaces:
                                    ; movf    STATUS, w
                                    ; andlw   1 << 2
                                    ; addwf   PCL, f
                                    ; goto    DlayLoop
                                    ; nop
                                    ; nop
                                    ; nop
        decf    Dlay + 1, f         ; Decrement the High
                                    ; Byte Until
        btfss   STATUS, Z           ;  It Equals Zero
          goto  DlayLoop            ; Replaces:
                                    ; movlw   1
                                    ; subwf   Dlay + 1, f
                                    ; movf    STATUS, w
                                    ; andlw   1 << 2
                                    ; addwf   PCL, f
                                    ; goto    DlayLoop
                                    ; nop
                                    ; nop
                                    ; nop

        btfsc   PORTA, 4            ; Toggle RA4
          goto  RA4Set
RA4Reset:                           ; RA4 is Off, Turn it On
        bsf     PORTA, 4
        goto    Loop
RA4Set:                             ; RA4 is On, Turn it Off
        bcf     PORTA, 4
        goto    Loop                ; Repeat


        end
```

At the end of the application, when I toggle RA4 to turn the LED on or off, I test its current state. I do so using the btfsc instruction to illustrate that the btfsc and btfss instructions can be used on *any* bit in any register and not just on the status bits. This ability to test individual bits allows you to implement code like this:

```
if (1 == RA4) then
    RA4 = 0;
```

in two instructions like this:

```
btfsc   PORTA, 4            ; Skip Over if RA4 is
                             Reset
  bcf   PORTA, 4            ; Else, Make RA4
                             Reset
```

instead of these four:

```
movf    PORTA, w            ; Read in PORTA
andlw   1 << 4             ; Test RA4 to see if
                            it is high or low
btfss   STATUS, Z          ; If zero set, then
                            skip next
  bcf   PORTA, 4
```

which perform the same task. But the btfsc solution is obviously more efficient (as it is implemented in fewer instructions and executes faster) and is much easier to read.

Experiment 68—Conditional Execution

PC/ Simulator

After working through the previous experiment, you may be wondering how to implement the first decision structure that you were ever introduced to—the if statement. If you have worked through the subtraction instructions and the bit skip instructions experiments, then you probably realize they can be used together to implement code that will jump on condition. You may

also be afraid of the amount of work it will take to implement the function. You will likely be surprised to discover that there is a four-instruction template that you can use to implement this function for comparing two eight-bit values.

The four-instruction template that implements the "if Condition then Label" decision structure are:

```
movf    Subtractor, w
subwf   Subtractend, w
btfs#   STATUS, Flag
  goto  Label
```

where Table 8-2 lists the different values for the optional Subtractor, Subtractend, #, and Flag. For the less-than and greater-than comparisons, the subtraction is arranged so the expected lower value is the subtractend, and if it is less than the subtractor, the carry

Table 8-2
Different Values for Comparison Template

"if" Statement	Subtractor	Subtractend	#	Flag
if A > B then goto Label	A	B	s	C
if A >= B then goto Label	B	A	c	C
if A < B then goto Label	B	A	s	C
if A <= B then goto Label	A	B	c	C
if A == B then goto Label	A	B	c	Z
if A != B then goto Label	A	B	s	Z

flag is reset (zero). This code template may seem like an advanced usage of the carry flag, but you should be able to figure out operations like this by reading through the PIC16F684 datasheet and experimenting with the subwf instruction. Do your experimenting with different values previously loaded into WREG and check the results (both the numeric result of the subtraction operation as well as how the STATUS bits are set).

When using this template, remember that the WREG's contents as well as the state of the C, Z, and DC STATUS flags have been changed.

The asmCondition.asm program demonstrates how these four statements are used to implement different decision structures using the information in Table 8-2 (and embedded in the source code):

```
title   "asmCondition - Demonstrate 'if'
Operations in Assembler"
;
; This experiment demonstrates a simple set of
; four instructions that simulates the operation
; of a "if (Condition) goto Label" high level
; language statement.
;
;   The four instructions are:
;
;   movf    Subtractor, w
;   subwf   Subtractend, w
;   btfs#   STATUS, Flag
;    goto   Label
;
;   Where the "Subtractor", "Subtractend", "#"
; and "Flag" are Defined as:
;
; "if Statement"   |Subtractor|Subtractend|#|Flag
; ----------------+----------+-----------+-+------
;if A>B then Label |    A     |    B      |s| C
;if A>=B then Label|    B     |    A      |c| C
;if A<B then Label |    B     |    A      |s| C
;if A<=B then Label|    A     |    B      |c| C
;if A==B then Label|    A     |    B      |c| Z
;if A!=B then Label|    A     |    B      |s| Z
;
;   Hardware Notes:
;   PIC16F684 running at 4 MHz in Simulator
;
```

```
;
;  Myke Predko
;  04.09.21
;
   LIST R=DEC
   INCLUDE "p16f684.inc"

   CBLOCK 0x20              ; Variable
Declaration
i, j
   ENDC

   PAGE

   org     0
   nop                     ; Required for MPLAB ICD2

   movlw   12              ; Initialize Test Variables
   movwf   i
   movlw   34
   movwf   j

   movf    i, w            ; if i > j then ErrorLoop
   subwf   j, w
   btfss   STATUS, C
    goto   ErrorLoop

   movf    j, w            ; if i >= j then ErrorLoop
   subwf   i, w
   btfsc   STATUS, C
    goto   ErrorLoop

   movf    i, w            ; if i == j then ErrorLoop
   subwf   j, w
   btfsc   STATUS, Z
    goto   ErrorLoop


   goto    $               ; Finished, Everything Okay

ErrorLoop:                 ; Compare Operation didn't
                           ; work correctly
   goto    ErrorLoop

   end
```

Looking at the four-instruction template and the code for this experiment, you can probably understand how to apply the four instructions, but, you may be unsure how to apply it when one of the condition parameters is a constant. In these cases, the literal instructions movlw and sublw replace movf and subwf, respectively. For example, to implement the high-level if statement:

```
if (i > 47) then goto Label
```

you would use the code:

```
   movf    i, w            ; if i > 47 then Label
   sublw   47
   btfss   STATUS, C
    goto   Label
```

Using the four-instruction template for the "if Condition then goto Label" decision structure, I suggest that you try to come up with ways of implementing high-level statements such as:

```
if ((A < B) && (A > C)) then goto RangeLabel
```

or:

```
while (A < 47)
```

in PIC MCU assembly language. It is actually easier than you might imagine, although before you start looking at different programming structures, you should assume they can all be implemented using the if statement (presented in this experiment) along with the goto instruction.

Experiment 69—decfsz Looping

After the previous experiments showing how conditional execution and decision structures are implemented in the PIC MCU, the simple if, while, and for statements of C probably sound good. Although these statements are not built into the PIC MCU assembly language, there is one instruction that makes repeated looping quite a bit easier. It is the basis of just about every PIC MCU assembly language program ever written.

The decfsz (or decrement file register and skip on zero result) can be used to execute a loop a set number of times as demonstrated in this experiment's program. This instruction decrements a file in a similar manner to the decf instruction, which decrements the contents of the register by one and stores the result either back into the file register or the WREG. But the decfsz does not change any of the STATUS register bits. Instead if the result of the decrement is zero, then the program counter is incremented before the next instruction is fetched. When this instruction is used in a loop, it will count down and cause execution to fall out of the loop after a set number of iterations.

```
    title    "asmDecfsz - Demonstrate decfsz loop"
;
;   Demonstrate the Operation of the "decfsz"
;   instruction for Looping.
;
;   "C" Equivalent Code:
;
;   i = 7;                  // Loop Around 7x
;   while (i != 0)
;   {
;       nop;               // Do nothing in the loop
;       i = i - 1;         // Decrement Loop Counter
```

```
;   }  // elihw
;
;   while (1 == 1);        // Finished, Loop Forever
;
;   Hardware Notes:
;    PIC16F684 running at 4 MHz in Simulator
;
;
;   Myke Predko
;   04.04.14
;
   LIST R=DEC
  INCLUDE "p16f684.inc"

;   Variables
  CBLOCK 0x20
i
  ENDC

  PAGE

  org      0
  nop

  movlw    7              ; i = 7
  movwf    i

Loop:                     ; Loop here 7x
  nop                     ; Do nothing in Loop

  decfsz   i, f           ; i = i - 1, if result == 0,
                          ; skip
    goto   Loop

  goto     $              ; Finished, just loop around

  end
```

Along with the "decfsz" instruction there is the "incfsz" (increment file register and skip on zero result), but this instruction does not have a "standard" use like decfsz. As I familiarize you with how the PIC MCU works and how you can optimize your applications in very surprising ways, you will find that both the decfsz and incfsz instructions can be used to make your applications very efficient. Until then, as you first start working with the PIC MCU and write your first applications, I recommend that you remember to use decfsz as just a way to execute a loop repeatedly.

PC/ Simulator

When you are learning to program the PIC MCU in assembler, it may seem like every instruction has some special little trick to be aware of. Fortunately, this is really not the case for adding subroutines to your program. They can be added quite easily, and the issues that you have to be aware of are quite standard for small processors. The operation and different aspects of subroutines are demonstrated in the asmSubr.asm program, which can be simulated in MPLAB IDE.

```
title    "asmSubr - Demonstrate Subroutine
Operation"
;
;   This program shows how Subroutines work in
;   the PIC MCU.
;
;   Hardware Notes:
;     PIC16F684 running at 4 MHz in Simulator
;
;
;   Myke Predko
;   04.11.20
;
  LIST R=DEC
  INCLUDE "p16f684.inc"

  PAGE

org      0
  nop                    ; Required for MPLAB ICD2

  movlw   0x55           ; Call Subroutine with "retlw"
  call    SubRetlw

  movlw   0x7B           ; Try again with a Subroutine
  call    SubReturn      ; With "return"

  call    Sub1           ; See if you can nest 10
                         ; Subroutines

  goto    $              ; When Execution Returns Here,
                         ; Loop Forever

; Subroutines
SubRetlw:                ; Return a Value in WREG
  retlw   0xAA

SubReturn:               ; Standard Subroutine Return
  return
```

```
Sub1:                    ; Depth Check, Call Next
  call    Sub2
  return
Sub2:                    ; Depth Check, Call Next
  call    Sub3
  return
Sub3:                    ; Depth Check, Call Next
  call    Sub4
  return
Sub4:                    ; Depth Check, Call Next
  call    Sub5
  return
Sub5:                    ; Depth Check, Call Next
  call    Sub6
  return
Sub6:                    ; Depth Check, Call Next
  call    Sub7
  return
Sub7:                    ; Depth Check, Call Next
  call    Sub8
  return
Sub8:                    ; Depth Check, Call Next
  call    Sub9
  return
Sub9:                    ; Finished, Return...
  return

  end
```

Data is typically passed to the subroutine either by using global variables or loading WREG (and, optionally, FSR) with the input data. Returning data can be passed using either one of these two methods, by setting STATUS flag (carry, digit carry, or zero) values, or by returning a constant in WREG. The first subroutine in asmSubr.asm returns data using the retlw, which loads WREG with a constant value before returning. This method probably seems somewhat redundant and even useless, but it is critical to implementing read-only arrays in the PIC MCU. It is also the method used to return from a subroutine in the low-range PIC MCUs, so it doesn't hurt to be aware of it.

The program counter stack in the PIC MCU is only eight-values deep. Calling more than eight nested subroutines (as I do in this experiment) will result in the code becoming lost in the subroutines and continually running through each of them. You may have discovered this when you simulated asmSubr.asm, but you will never approach this limit in normal programming. The only time I would expect you to have some problems with losing execution return for a subroutine is when implementing a *recursive subroutine* (or a subroutine that calls itself).

Experiment 71—Defining and Implementing Arrays

PC/ Simulator

PICkit™ 1

starter kit

```
cblock 0x20
   :
Array:6
   :
 endc
```

The last major assembly language programming concept to learn is declaring and accessing the data within arrays. Like other small processors, the PIC16F684 microcontroller processor provides a single *index register* (called FSR in the PIC MCU), which can access all the registers, both variable memory and the special function registers. The concept of indexed addressing is simply using an index register to access array elements. The PIC MCU implements indexed addressing a bit differently from other chips, as the registers pointed to by the index register are accessed via the use of a shadow register called INDF. In practical terms, indexed addressing is not any more difficult to implement in the PIC MCU than in other processors: The only issue is remembering to specify INDF when reading from or writing to the register pointed to by FSR.

Therefore, to read a byte from an array at position n, you would use the code:

```
movf    n, w     ; Get Offset of Array Element
addlw   Array    ; Add starting address of the
                 ; Array
movwf   FSR      ; Store in FSR (Index Register)
movf    INDF, w  ; Load WREG with "Array[n]"
```

Similarly, to write the contents of WREG into an array element n, you would use the code:

```
movwf   Temp     ; Temporarily store the value
movf    n, w     ; Get Offset of the  Array
                 ; Element
addlw   Array    ; Add starting address of the
                 ; Array
movwf   FSR      ; Store Array Element Address
movf    Temp, w  ; Retrieve saved Value
movwf   INDF     ; Save value.
```

Declaring array variables is very easy to do with the CBLOCK directive, as is the ability to specify the difference to the next label. For example, for the previous reading and writing example code, if the Array variable was six bytes in size and you wanted to make sure the next variable did not occupy the same space, you would put in the following statement:

The last operation that you should be aware of is that of initializing array elements. The simplest way of accomplishing this is to write directly to each address of the variable. For example, if Array was to have the first six prime numbers, it could be initialized as:

```
movlw    1              ; First Element
movwf    Array
movlw    2              ; Second Element
movwf    Array + 1
movlw    3              ; Third Element
movwf    Array + 2
movlw    5              ; Forth Element
movwf    Array + 3
movlw    7              ; Fifth Element
movwf    Array + 4
movlw    11             ; Sixth Element
movwf    Array + 5
```

When doing this type of initialization (or even directly accessing array elements), remember that the first element has an offset (and index) of zero, and each element has an index/offset that is one less than the element number.

This is really all there is to working with indexed addressing for byte arrays.

asmPCKLED 3.asm is a demonstration application in which the LEDs on the PICkit 1 starter kit flash from D0 to D7, and it was written using arrays for the PORTA and TRISA values.

```
title  "asmPKLED 3 - PICkit Running LED"
;
;  This program copies the operation of cPKLED.c
;  and Flashes the 8 LEDs on the PICkit 1 PCB in
;  Sequence by Storing the Values for the LEDs
;  and the TRIS registers in Arrays
;
;  Hardware Notes:
;   PIC16F684 running at 4 MHz Using the
;   Internal Clock
;   Circuit Runs on PICkit 1 PCB
;
;  Original Code:
;
;  int i, j, k;
;  char LEDValue[8]  = {0b010000, 0b100000,
;                       0b010000, 0b000100,
;                       0b100000, 0b000100,
;                       0b000100, 0b000010};
;  char TRISValue[8] = {0b001111, 0b001111,
;                       0b101011, 0b101011,
;                       0b011011, 0b011011,
;                       0b111001, 0b111001};
;
```

```
; main()
; {
;
;   PORTA = 0;
;   CMCON0 = 7;              // Turn off Comparators
;   ANSEL = 0;              // Turn off ADC
;
;   k = 0;                  // Start at LED 0
;
;   while(1 == 1)            // Loop Forever
;   {
;     for (i = 0; i < 255; i++) // Simple Delay Loop
;       for (j = 0; j < 129; j++);
;
;       PORTA = LEDValue[k];    // Specify LED Output
;     TRISA = TRISValue[k];
;
;     k = (k + 1) % 8;        // Increment k within
;                               range of 0-7
;
;   } // elihw
; } // End cPKLED
;
;
;
;  Myke Predko
;  04.11.07
;
  LIST R=DEC
  INCLUDE "p16f684.inc"
  __CONFIG _FCMEN_OFF & _IESO_OFF & _BOD_OFF &
_CPD_OFF & _CP_OFF & _MCLRE_ON & _PWRTE_ON &
_WDT_OFF & _INTOSCIO

; Variables
CBLOCK 0x20
LEDValue:8, TRISValue:8
Dlay
k
  ENDC

  PAGE
;  Mainline of asmPKLED

  org     0

  nop                       ; For ICD Debug

  clrf    PORTA             ; Initialize I/O Bits
                            ; to Off
  movlw   7                 ; Turn off Comparators
  movwf   CMCON0
  bsf     STATUS, RP0
  clrf    ANSEL ^ 0x80      ; All Bits are Digital
  bcf     STATUS, RP0

  clrf    k                 ; Start with D0

  movlw   b'010000'         ; Load LEDValue Explicitly
  movwf   LEDValue
  movlw   b'100000'
  movwf   LEDValue + 1
  movlw   b'010000'
  movwf   LEDValue + 2
  movlw   b'000100'
  movwf   LEDValue + 3
  movlw   b'100000'
  movwf   LEDValue + 4
  movlw   b'000100'
  movwf   LEDValue + 5
  movlw   b'000100'
  movwf   LEDValue + 6
  movlw   b'000010'
  movwf   LEDValue + 7
```

```
  movlw   b'001111'         ; Load TRISValue
Explicitly
  movwf   TRISValue
  movwf   TRISValue + 1 ; NOTE: LEDs are in Pairs
  movlw   b'101011'
  movwf   TRISValue + 2
  movwf   TRISValue + 3
  movlw   b'011011'
  movwf   TRISValue + 4
  movwf   TRISValue + 5
  movlw   b'111001'
  movwf   TRISValue + 6
  movwf   TRISValue + 7

Loop:                       ; Return after Delay and
                            ; RB0 Toggle
  clrf    Dlay              ; High 8 Bits for Delay
  clrw
  goto    $ + 1
  goto    $ + 1
  addlw   -1                ; Decrement the contents of
                            ; WREG
  btfss   STATUS, Z         ; 256x
   goto   $ - 4
  decfsz  Dlay, f           ; Repeat 255x
   goto   $ - 6

; PORTA = LEDValue[k]
  movf    k, w              ; Load the Offset for
                            ; LEDValue
  addlw   LEDValue          ; Add to Start of LEDValue
                            ; Array
  movwf   FSR               ; Store Address in Index
                            ; Register
  movf    INDF, w           ; Get LEDValue[k]
  movwf   PORTA             ; Save in PORTA Register

; TRISA = TRISValue[k]
  movf    k, w              ; Get the Offset for
                            ; TRISValue
  addlw   TRISValue         ; Add to the Start of Array
  movwf   FSR               ; Store Address in Index
                            ; Register
  movf    INDF, w           ; Get TRISValue[k]
  bsf     STATUS, RP0       ; Save in TRISA Register
  movwf   TRISA ^ 0x80
  bcf     STATUS, RP0

  incf    k, w              ; Increment the Active LED
  andlw   7                 ; Keep Within range of 0-7
  movwf   k

  goto    Loop              ; Repeat

  end
```

For Consideration

In this chapter, I have reviewed the basic assembly language instructions and given you a basic programming template to work from. Along with this, you have experience with the MPLAB IDE and with programming in C. You should be ready to start programming a PIC MCU in assembler, only you may not feel that confident in your abilities to do so from scratch. Chances are if you have a programming assignment, you will be able to copy asmTemplate.asm into a specific subdirectory for your application, rename it to the application name, load it into MPLAB IDE, and then just stare at it because you don't know where to begin. Before going on to more complex programming assignments, I would like to give you some guidance in structuring and writing PIC MCU assembly language programs.

The basic parts and sequence of the assembly language program are as follows:

1. Use nop as the first instruction so an MPLAB IDE incircuit debugger (MPLAB ICD) module can be used to help you debug your application in circuit. Although you may not have an ICD (or ICD 2) that you can use for debugging, it's a good idea to always put this instruction in so you are in the habit of being ready to debug your code with the ICD.

2. Perform the necessary bank 0 initializations. I like to put in the variable initializations first, followed by the hardware initializations. Remember that variables should always be initialized (even to zero) using a clrf instruction.

3. Perform the necessary bank 1 initializations. These should just be hardware initializations; you should not put any variables in bank 1 until you are very familiar with PIC MCU assembly language programming. Remember to XOR the register addresses with 0x080 to ensure you do not get a message indicating that you are accessing a register in the wrong bank. When you work with other PIC MCUs that have hardware (and variable) registers in banks 2 and 3, you will have to repeat this step for each one of the registers.

4. Virtually all microcontroller applications will execute forever, taking in input, processing it, and then passing it back out. After the initializations, the code should start executing the loop code.

5. Next, read your inputs. This can be simply polling the I/O pins or reading hardware register values.

6. The inputs are then processed. This processing can include comparisons to expected values or arithmetic computations.

7. Once you have processed the inputs and understand what should come out of the application, you can output them. Just as when you read your inputs, outputting could consist of writing directly to I/O pins or to hardware registers.

8. If some delay is required in the program, it should be done here. Creating delays will be explained in the next section.

9. The loop code is complete. Jump to the start of the loop and begin again.

10. Define the variables to be used in the program. I like to define them last because chances are I added one or two when I created my program, and this way I can go back and make sure everything is defined using MPASM assembler (when the program assembles cleanly, then all the variables are defined).

To illustrate how a program is created using this methodology, I created asmButton.asm. In this program, I have put in comment lines that start with #### to indicate which of the 10 steps is being implemented in that and the following lines of code.

```
  title   "asmButton - PIC16F684/PICKit Button
LED On"
;
;  This Program Turns on the PICKit's D0 LED
;  when the Button at RA3 is pressed.
;
;  Hardware Notes:
;   PIC16F684 running at 4 MHz Using the
;   Internal Clock
;   External Reset is Used
;   RA3 - PICKit Button
;   RA4 - LED Positive
;   RA5 - LED Negative
;
;
;  Myke Predko
;  04.11.07
;
  LIST R=DEC
  INCLUDE "p16f684.inc"

  __CONFIG _FCMEN_OFF & _IESO_OFF & _BOD_OFF &
_CPD_OFF & _CP_OFF & _MCLRE_OFF & _PWRTE_ON &
_WDT_OFF & _INTOSCIO

;  Variables
CBLOCK 0x20
;  #### - 10.  Put in the Variables (if required)
Temp
  ENDC

  PAGE
;  Mainline of cButton

  org    0

; #### - 1. Put in "nop" instruction required for
;          ICD
  nop                    ; For ICD Debug

; #### - 2. Put in "Bank 0 Initializations"
  movlw  0xFF            ; Make All PORTA Bits High
  movwf  PORTA
```

```
    movlw   7               ; Turn off Comparators          movwf   Temp            ; Store for RA3 Check
    movwf   CMCON0                                     ; #### - 6. Process Inputs
; #### - 3. Put in "Bank 1 Initializations"             movlw   B'011111'       ; D0 On?
    bsf     STATUS, RP0                                    btfsc   Temp, 3         ; If RA3 is Set then "Yes"
    clrf    ANSEL ^ 0x80 ; All Bits are Digital            movlw   B'111111'       ; No, Off
    movlw   0x0FF ^ ((1 << 4) | (1 << 5))          ; #### - 7. Output Processed Input
                    ; RA4/RA5 are Digital Outputs          movwf   PORTA
    movwf   TRISA ^ 0x80                           ; #### - 8. Delay (if Required)
    bcf     STATUS, RP0                            ; #### - 9. Return to the Start of the Loop
                                                          goto    Loop            ; Repeat
; #### - 4. Loop Entry Point
Loop:
; #### - 5. Read in Inputs                               end
    movf    PORTA, w    ; Want to check RA3
```

PIC® Microcontroller Assembly Language Resource Routines

PC/ Simulator

PICkit™ 1

starter kit

Required Parts

1 PIC16F684

2 Red LEDs

1 Yellow LED

1 0.01 μF capacitor (any type)

2 100Ω resistors

1 SPST switch

1 Breadboard

1 3-cell AAA battery pack

3 AA batteries

Tool Box

DMM

Needle-nose pliers

Wiring kit

If you were to ask me whether using C or assembler language is the most efficient method of creating complex PIC MCU applications, I would answer that the best applications are written using *both* programming languages. C is excellent for high-level program flow where complex conditional and arithmetic statements are normally used. Assembly language programming is necessary when you have a precisely timed interface requirement or logic/arithmetic operations that must be very efficiently implemented. Although in the rest of the book, I concentrate on assembly language programming, I want to point out that you have the opportunity to add assembly language code to your PICC Lite™ compiler applications. There are a number of things you will have to be aware of before you start adding assembly language code to your PICC Lite compiler application.

The first issue to be addressed is whether or not your application code will consist of a mix of C statements and assembly language instructions or if the application code will have only one of the two programming languages. Even without looking at how the PICC Lite compiler's built-in assembler works (which is required for an application that has both C state-

ments and assembly language instructions), I would tend to go towards having separate files for assembly language and C source files. The two code sources would be individually compiled into .obj files (often called object files), which would then be linked together. MPLAB® IDE handles the compiling, linking, and source-code-level debugging chores very efficiently, with each source code file displayable. An important advantage of using multiple source code files is that subroutine files can be shared very easily, especially as source code, which would eliminate the possibility of a developer changing code that is already running.

The problem with linking multiple files into an application is the additional learning and work that is required. For the applications in this book, I don't feel that it is appropriate to attempt linking multiple files together. Instead, the examples of mixed code sources will primarily use PICC Lite compiler C source code with in-line assembler statements. The assembly language instructions can be added individually using the directive:

```
asm("PIC MCU Instruction");
```

or they can be added as a predefined function statement. Earlier in the book, I used the NOP(); built-in function in a number of applications. This function inserts a single nop instruction in the code and allows me to set simulator breakpoints in the application where the statements or operations either do not seem to skip randomly to other instructions or to set breakpoints results in an MPLAB IDE simulator message. Along with the nop instruction, the *clear the Watchdog Timer* (clrwdt) instruction can also be added to a PICC Lite compiler application in this manner.

If you have a number of assembly language instructions, you should indicate to the compiler that assembly language instructions will follow by using the #asm directive, and that you are finished by using the #endasm directive. These directives and instructions are best placed in unique subroutines like the following one:

```
int AssemblyDemo(int SqValue)
{  // Calculate the Square of "Value"

 ai = SqValue;
    // Save Value to Square in Memory Location
    // assembler can access uniquely
#asm
 movf   _ai, w        // Get Multiplicand
 clrf   _Square
 clrf   _Square + 1
ASLoop:                ; Loop Back to Here until
                       "_ai" == 0
 addwf  _Square, f  ; Add Value to Square
 btfsc    status, 0 ; Overflow?
  incf   _Square + 1, f
 decfsz _ai, f      ; Decrement Repeated
                       Addition Counter
```

```
 goto   ASLoop        ; Return to Try again
#endasm

   return Square;

}  // End AssemblyDemo
```

There are five things you should note about the code in this function:

- The addition of the leading underscore to variable names in the in-line assembler code.
- The need to declare most variables as "*char unsigned*" variables. This isn't readily apparent in the code, but each byte variable should be declared as "char unsigned" to avoid negation issues when execution crosses between C and assembler.
- The lack of bit labels for status or any other registers.
- Declared registers are all lowercase.
- You do not have the $ operator to help you avoid coming up with your own labels. You will have to put in labels for every for loop.

For the most part, these differences do not make writing in-line assembler more difficult, just different. This is why I recommend that you add in-line assembler instructions only when you have to. The PICC Lite compiler is a remarkably efficient compiler, and there will be only limited situations where it is best to combine C statements and assembly language instructions together.

Experiment 72—Logic Simulation Using the PIC16F684

PC/ Simulator

PICkit™ 1

starter kit

Parts Bin

1 PIC16F684
2 Red LEDs
1 Yellow LED
1 0.01 μF capacitor (any type)
2 100Ω resistors
1 SPST switch
1 Breadboard
1 3-cell AAA battery pack
3 AAA batteries

Tool Box

DMM
Needle-nose pliers
Wiring kit

I hope that I have impressed you with the power and capabilities of the PIC MCU. Now I'm going to try to underwhelm you and show you the minimum that they can be programmed to do. This is a good time to go back and show you how truly good the PICC Lite compiler is and show you a little trick for creating your own PIC MCU applications. In this application, I am going we will create a two-input AND gate from a PIC16F684.

The circuit that I have come up with is very simple (see Figure 9-1) and can be wired on a breadboard (see Figure 9-2) in just a couple of minutes. I wanted the LEDs to be lit when the input was a 1 and off when an input is 0. To do this, I take advantage of the half Vdd voltage threshold on the PIC MCU pins. Red LEDs have a voltage drop of 2.0 volts, which is greater than the 1.5-volt threshold of the PIC MCU input pins when the PIC MCU is powered by 3 volts. With a 100Ω current-limiting resistor in series, the LED can either be shorted to ground (and turned off) for a 0 or left on in circuit to pass a 2.0-volt input to the PIC MCU. I'm pointing this out because this circuit will *not* work with a 5-volt power supply.

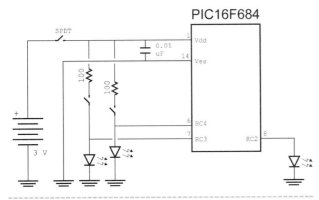

Figure 9-1   *PIC microcontroller AND gate*

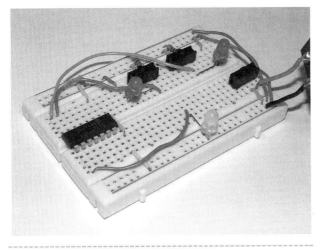

Figure 9-2   *Breadboard AND gate circuit*

Here is the first experiment for you to try out. Find the voltage level at which the PIC device transitions from a 0 to a 1, and when the red LEDs are wired as in this circuit. As part of this exercise, make a guess at what this *translation voltage* will be before you add the variable voltage supply.

The assembly language code for the application, asmLogicGate.asm, is very simple as you would expect. It makes all the input pins digital I/O and then polls the two input pins (RC3 and RC4). Then when the inputs are both high, the output of RC2 is driven high, and the LED connected to it is lit.

```
    title   "asmLogicGate - Demonstrate Logic
Simulation"
;
; This program causes the PIC16F684 to behave
; like a single 2 Input AND Gate.
;
; Hardware Notes:
;   PIC16F684 running at 4 MHz Using the Internal
;   Clock
;
;
; Hardware Notes
; RC4:RC3 - Input
; RC2     - Output
;
; Myke Predko
; 04.11.19
;
    LIST R=DEC
    INCLUDE "p16f684.inc"

    __CONFIG _FCMEN_OFF & _IESO_OFF & _BOD_OFF &
_CPD_OFF & _CP_OFF & _MCLRE_ON & _PWRTE_ON &
_WDT_OFF & _INTOSCIO

;   Variables
CBLOCK 0x20
ENDC

    PAGE
; Mainline

    org     0

    nop                     ; For ICD Debug

    clrf    PORTC           ; Initialize I/O Bits
                            ; to Off
    movlw   7               ; Turn off Comparators
    movwf   CMCON0
    bsf     STATUS, RP0
    clrf    ANSEL ^ 0x80    ; All Bits are Digital
    movlw   b'111011'       ; RC4:RC3 Inputs, RC2
                            ; Output
    movwf   TRISC ^ 0x80
    bcf     STATUS, RP0

Loop:                       ; Return Here after
                            ; Delay
    btfss   PORTC, 4        ; Test Input Bits
    goto    NoMatch4
    btfss   PORTC, 3
    goto    NoMatch3
    nop                     ; Match Cycles to Set
    bsf     PORTC, 2        ; Both are Set
    goto    Loop
```

```
NoMatch4:                    ; Two Cycles Past
                               NoMatch3
    goto     NoMatch3
NoMatch3:
    bcf      PORTC, 2       ; Reset the Output
    goto     Loop


    end
```

To compare the operation of my assembly language code to what the PICC Lite compiler produces, I created a similar application, cLogicGate.c. This program polls RC3 and RC4 and outputs a 1 on RC2 when both inputs are 1, just like in asmLogicGate.asm.

```
#include <pic.h>
/* cLogicGate.c - Simulate an AND Gate on a
PIC16F684

This Program outputs a "1" if Two Inputs are a
"1".

RC4 - Input
RC3 - Input
RC2 - Output

myke predko
04.11.19

*/

__CONFIG(INTIO & WDTDIS & PWRTEN & MCLRDIS &
UNPROTECT \
    & UNPROTECT & BORDIS & IESODIS & FCMDIS);


int i, j;

main()
{

    PORTC = 0;
    CMCON0 = 7;            // Turn off Comparators
    ANSEL = 0;             // Turn off ADC
    TRISC2 = 0;            // Make RC2 Output

    while(1 == 1)          // Loop Forever
    {
     if ((1 == RC4 ) && (1 == RC3))
       RC2 = 1;    // Both High, Output High
     else
       RC2 = 0;             // Else, Low Output

    }  // elihw
}  // End cLogicGate
```

As a follow-up to the introduction of this chapter, I want to show you the efficiency of the assembler instructions generated by the compiler. After I tested the operation of both applications, I took a look at the code generated by the PICC Lite compiler by clicking on "View" and then on "Disassembly Listing." The following code was displayed in a new window on the MPLAB IDE desktop:

```
---   C:\Writing\PIC Evil
Genius\Code\cLogicGate\cLogicGate.c ---------
1:  #include <pic.h>
```

```
2:  /* cLogicGate.c - Simulate an AND Gate on a
PIC16F6
3:
4:  This Program outputs a "1" if Two Inputs are
a "1".
5:
6:  RC4 - Input
7:  RC3 - Input
8:  RC2 - Output
9:
10: myke predko
11: 04.11.19
12:
13: */
14:
15: __CONFIG(INTIO & WDTDIS & PWRTEN & MCLRDIS &
    UNPROTE
16: & UNPROTECT & BORDIS & IESODIS & FCMDIS);
17:
18:
19: int i, j;
20:
21: main()
22: {
23:
24:   PORTC = 0;
0003F3  1283  BCF 0x3, 0x5
0003F4   187  CLRF 0x7
25:   CMCON0 = 7;              // Turn off Compara
0003F5  3007  MOVLW 0x7
0003F6   099  MOVWF 0x19
26:   ANSEL = 0;               // Turn off ADC
0003F7  1683  BSF 0x3, 0x5
0003F8   191  CLRF 0x11
27:   TRISC2 = 0;              // Make RC2 Output
0003F9  1107  BCF 0x7, 0x2
28:
29:   while(1 == 1)            // Loop Forever
30:   {
31:    if ((1 == RC4 ) && (1 == RC3))
0003FA  1283  BCF 0x3, 0x5
0003FB  1A07  BTFSC 0x7, 0x4
0003FC  1D87  BTFSS 0x7, 0x3
0003FD  2BF9  GOTO 0x3f9
32:    RC2 = 1;                // Both High, Output
0003FE  1507  BSF 0x7, 0x2
```

You can make a couple of conclusions from the resulting assembly language code. The first being, the *disassembly listing* is not complete. The closing brace for the "while(1 == 1)" statement *and* a "goto 0x3fa" instruction, which returns execution to the start of the comparison, are both missing. To simulate this application, I used the asynchronous "pin stimulus" feature of the debugger to change the states of the input pins.

The second conclusion you can make is that the PICC Lite compiler isn't stupid. It was able to convert the two-term if statement into just three assembly language instructions. And at the same time, it was able to note that resetting the TRIS bit RC2 in bank 1 is the same as resetting pin RC2 in bank 0. Later in this section, I will provide you with a set of macros that will allow you to virtually write your own high-level code in PIC MCU assembly language without having to understand how the different instructions operate. But the code produced by these macros will in no way be as efficient as the previous snippet of code.

The three-instruction code demonstrates how sophisticated operations can be created in the PIC MCU using just a few instructions. Two-bit-input AND statements and OR statements can be produced using the following templates:

```
; 3 Instruction AND Code
  btfsc   Register1, bit1  ; If First Bit Input
                             is 0, Jump to not
                             True
  btfss   Register2, bit2  ; If Second Bit Input
                             is 1, Jump to True
  goto    ANDNotTrue       ; One or Both Bit
                             Inputs is 0, "not"
                             True
ANDTrue:                   ; Code Executed if
                             Both Bits are 1

; 3 Instruction OR Code
  btfss   Register1, bit1  ; If First Bit is 1,
                             Jump to "True"
  btfsc   Register2, bit2  ; If Second Bit is 0,
                             Jump to "False"
  goto    ORTrue:          ; One or Both Bits are
                             1
ORNotTrue:                 ; Code Executed if
                             Both Bits are 0
```

These two snippets assume that True is high. They can be easily modified if True is low for one or both bits; simply change the btfss to btfsc, and visa versa.

If you are a student, I should remind you that it is not a good idea to write your application in C and then use the PICC Lite compiler to generate assembly language code for your application for a number of reasons. First, the display does not label registers; if you were to hand in the code from the disassembly window unchanged, I suspect that even the most dim-witted teacher would notice that something was amiss. Second, the code displayed in the Disassembly Listing window does not have any goto labels, and you will find it to be a significant challenge to correctly move the location of the code produced by the PICC Lite compiler to the locations required for your application. Chances are it will take more time and effort than if were to do it yourself. Finally, the worst thing that could happen is that you correctly label the registers and add correct goto labels, and then your teacher is so impressed by your work that you are asked to explain to the whole class how the code works. If you're having problems solving an application in assembler, create a C program and run it through the PICC Lite compiler to see how a professional would code the application. But remember, it's in your best interest to take what you learn from the Disassembly Listing and apply it to your application, rather than trying to use the code that's displayed.

Experiment 73—Implementing the C "Switch" Statement in Assembly Language

PC/ Simulator

If asked when I first learned to program, I like to reply that it has been going on for 25 years now. The implication is not that I am somewhat dim, but instead that there are so many different ways to solve programming problems. I bring this up because the code for this experiment, which was written after the code for the following experiment, is a lot more efficient (i.e., smaller) than the code in the following experiment. Logically, the programs should become more sophisticated and efficient as you progress through the book. I wanted to illustrate a particular point, so I placed this more efficient program before the less efficient one that follows it.

In this experiment, I would like to show how the C switch statement can be implemented in assembler:

```
switch (Variable) {    // Value to be tested
  case 2:
// Statements to Execute if "Variable == 2"
  break;              // Exit the "Switch" Statement
  case 23:
// Statements to Execute if "Variable == 23
  break;
  case 47:
// Statements to Execute if "Variable == 47
  break;
  default:            // "Variable" != 0 or 47
// Statements to Execute if "Variable != 0" and
  "Variable != 47"
  }  // hctiws
```

There are several ways in which the switch and case statements could be implemented in this block of code to execute specific code if the contents of a variable match a specific value, but for this experiment, I will focus on the general case code.

One way to implement this function is with multiple comparisons. If this were to be written in a language with an "if then goto" statement, the switch statement could be implemented as:

```
    if (2 == Variable) then goto Not2
// == First Switch Case
// Statements to Execute if "Variable == 2"
    goto SwitchDone
Not2:
    if (23 == Variable) then goto Not23
// == Second Switch Case
// Statements to Execute if "Variable == 23
    goto SwitchDone
Not23:
    if (47 == Variable) then goto Not47
// == Third Switch Case
// Statements to Execute if "Variable == 47
    goto SwitchDone
Not47:
// "Default" - Statements to Execute if
    "Variable != 2" and
//
    "Variable != 23" and
//
    "Variable != 47"
SwitchDone:
```

This isn't a terrible base to work from, but we can do better. The secret to doing better is to load the variable into the WREG and then compare its contents using the xorlw instruction repeatedly. Take advantage of the MPASM™ assembler built-in calculator and XOR the value to be tested with the previous value. By doing this, the previous XORed value in the WREG is returned to the original state and then XORed with the new value. To show you what I mean, I can consider the if equivalent to the switch statement as:

```
    movf    Variable, w
    xorlw   2               ; "if (2 == Variable)
                              then goto Not2"
    btfss   STATUS, Z
    goto    Not2
// Statements to Execute if "Variable ^ 2" equal
    zero
    goto    SwitchDone
Not2:
    xorlw   23 ^ 2          ; WREG = (WREG ^ 2) ^ (23
                              ^ 2)
                            ; = WREG ^ 23 ^ 2 ^ 2
                            ; = WREG ^ 23 (Because 2
                              ^ 2 = 0)
    btfss   STATUS, Z
    goto    Not23
// Statements to Execute if "Variable == 23
    goto SwitchDone
Not23:
    xorlw   47 ^ 23         ; WREG = (WREG ^ 23) ^
                              (47 ^ 23)
                            ; = WREG ^ 47 ^ 23 ^ 23
                            ; = WREG ^ 47 (Because 23
                              ^ 23 = 0)
    btfss   STATUS, Z
    goto    Not47
// Statements to Execute if "Variable == 47
    goto    SwitchDone
Not47:
// "Default" - Statements to Execute if
    "Variable != 2" and
//
    "Variable != 23" and
```

```
//
    "Variable != 47"
SwitchDone:
```

This method of implementing the switch statement in assembler is quite efficient and easy to code, as I show in the code for this experiment, asmPKLED 2.asm. In this code I have tried, as much as possible, to directly convert the high-level statements of cPKLEN.c to assembly language statements:

```
title  "asmPKLED 2 - PICkit 1 starter kit
Running LED"
;
; This program directly copies the operation of
cPKLED.c and Flashes
;  the 8 LEDs on the PICkit 1 starter kit in
Sequence.
;
; Hardware Notes:
;  PIC16F684 running at 4 MHz Using the Internal
Clock
;  External Reset is Used
;  Circuit Runs on PICkit 1 starter kit
;
; Original Code:
;
; int i, j, k;
;
; main()
; {
;
;    PORTA = 0;
;    CMCON0 = 7;          // Turn off Comparators
;    ANSEL = 0;           // Turn off ADC
;
;    k = 0;               // Start at LED 0
;
;    while(1 == 1)
// Loop Forever
;    {
;        for (i = 0; i < 255; i++)
// Simple Delay Loop
;            for (j = 0; j < 129; j++);
;
;        switch (k) {
// Select Which LED to Display
;    case 0:
;    PORTA = 0b010000;
;    TRISA = 0b001111;
;    break;
;    case 1:
;    PORTA = 0b100000;
;    TRISA = 0b001111;
;    break;
;    case 2:
;    PORTA = 0b010000;
;    TRISA = 0b101011;
;    break;
;    case 3:
;    PORTA = 0b000100;
;    TRISA = 0b101011;
;    break;
;    case 4:
;    PORTA = 0b100000;
;    TRISA = 0b011011;
;    break;
;    case 5:
;    PORTA = 0b000100;
;    TRISA = 0b011011;
;    break;
;    case 6:
```

```
;    PORTA = 0b000100;
;    TRISA = 0b111001;
;    break;
;    case 7:
;    PORTA = 0b000010;
;    TRISA = 0b111001;
;    break
;    }  // hctiws
;
;    k = (k + 1) % 8;
// Increment k within range of 0-7
;
;        }  // elihw
;    }  // End cPKLED
;
;
;  Myke Predko
;  04.11.28
;
  LIST R=DEC
  INCLUDE "p16f684.inc"

  __CONFIG _FCMEN_OFF & _IESO_OFF & _BOD_OFF &
_CPD_OFF & _CP_OFF & _MCLRE_ON & _PWRTE_ON &
_WDT_OFF & _INTOSCIO

;  Variables
  CBLOCK 0x20
Dlay
ActiveBit                ; Record the Active Bit
  ENDC

  PAGE
;  Mainline

  org     0

  nop                    ; For ICD Debug

  clrf    PORTA          ; Initialize I/O Bits to Off
  movlw   7              ; Turn off
Comparators
  movwf   CMCON0
  bsf     STATUS, RP0
  clrf    ANSEL ^ 0x80   ; All Bits are Digital
  bcf     STATUS, RP0

  clrf    ActiveBit      ; Start with D0

  movlw   TRISA          ; Point FSR to TRISA
                         ;  for Fast TRISA
  movwf   FSR            ;  Updating

Loop:                    ; Return Here after
                         ;  Delay

  clrf    Dlay           ; High 8 Bits for Delay
  clrw
  addlw   -1             ; Decrement the
                         ;  contents of WREG
  btfss   STATUS, Z      ;  256x
   goto   $ - 2
  decfsz  Dlay, f        ; Repeat 255x
   goto   $ - 4
  clrf    Dlay           ; Repeat Again for ~
                         ;  400 ms total Delay
  clrw
  addlw   -1             ; Decrement the
                         ;  contents of WREG
  btfss   STATUS, Z      ;  256x
   goto   $ - 2
  decfsz  Dlay, f        ; Repeat 255x
   goto   $ - 4

  movf    ActiveBit, w   ; Load the Bit Number
```

```
  xorlw   0              ; Start with D0
  btfss   STATUS, Z
   goto   TryD1
  movlw   B'010000'      ; Turn on D0
  movwf   PORTA
  movlw   B'001111'      ; Setup TRIS for the
                         ;  Next Display
  movwf   INDF
  goto    LEDDone        ; Finished, Repeat

TryD1:                   ; Display D1?
  xorlw   1 ^ 0
  btfss   STATUS, Z
   goto   TryD2
  movlw   B'100000'      ; Turn on D1
  movwf   PORTA
  movlw   B'001111'      ; Setup TRIS for the
                         ;  Next Display
  movwf   INDF
  goto    LEDDone        ; Finished, Repeat

TryD2:                   ; Display D2?
  xorlw   2 ^ 1
  btfss   STATUS, Z
   goto   TryD3
  movlw   B'010000'      ; Turn on D2
  movwf   PORTA
  movlw   B'101011'      ; Setup TRIS for the
                         ;  Next Display
  movwf   INDF
  goto    LEDDone        ; Finished, Repeat

TryD3:                   ; Display D3?
  xorlw   3 ^ 2
  btfss   STATUS, Z
   goto   TryD4
  movlw   B'000100'      ; Turn on D3
  movwf   PORTA
  movlw   B'101011'      ; Setup TRIS for the
                         ;  Next Display
  movwf   INDF
  goto    LEDDone        ; Finished, Repeat

TryD4:                   ; Display D4?
  xorlw   4 ^ 3
  btfss   STATUS, Z
   goto   TryD5
  movlw   B'100000'      ; Turn on D4
  movwf   PORTA
  movlw   B'011011'      ; Setup TRIS for the
                         ;  Next Display
  movwf   INDF
  goto    LEDDone        ; Finished, Repeat

TryD5:                   ; Display D5?
  xorlw   5 ^ 4
  btfss   STATUS, Z
   goto   TryD6
  movlw   B'000100'      ; Turn on D5
  movwf   PORTA
  movlw   B'011011'      ; Setup TRIS for the
                         ;  Next Display
  movwf   INDF
  goto    LEDDone        ; Finished, Repeat

TryD6:                   ; Display D6?
  xorlw   6 ^ 5
  btfss   STATUS, Z
   goto   TryD7
  movlw   B'000100'      ; Turn on D6
  movwf   PORTA
  movlw   B'111001'      ; Setup TRIS for the
                         ;  Next Display
  movwf   INDF
  goto    LEDDone        ; Finished, Repeat
```

```
TryD7:                          ; Display D7?
  xorlw    7 ^ 6
  btfss    STATUS, Z
  goto     LEDDone
  movlw    B'000010'            ; Turn on D7
  movwf    PORTA
  movlw    B'111001'            ; Setup TRIS for the
                                  Next Display
  movwf    INDF
  goto     LEDDone              ; Finished, Repeat

LEDDone:                        ; Finished Displaying
                                  LED
  incf     ActiveBit, w         ; Increment the Active
                                  LED
  andlw    7                    ; Keep Within range of
                                  0-7
  movwf    ActiveBit

  goto     Loop                 ; Repeat

  end
```

There is one last point I would like to make about asmPKLED 2.asm, and that is how I was able to quickly change the values of the TRISA register. In the code, you see that I loaded the FSR register with the address of TRISA, and when I had to change the value of TRISA, I simply write to the register pointed to by the FSR. Another way of implementing this code would be to change the bank and write directly to TRISA (without involving the FSR register), using the following code:

```
  movlw    NewPORTAValue        ; Write the new PORTA
                                  Value
  movwf    PORTA
  movlw    NewTRISAValue        ; Setup TRISA for the
                                  Next Display
  bsf      STATUS, RP0          ; Execute in Page 1
  movwf    TRISA ^ 0x080        ; Store the New TRISA
                                  Value
  bcf      STATUS, RP0          ; Return Execution to
                                  Page 2
```

Experiment 74—Defines

If you have worked through all the experiments in the previous section, you will have noticed that accessing I/O pins in assembly language can be somewhat awkward due to the need to specify both the port and the bit number of the pin in the bit instructions. As you start programming your own applications, you will discover cases where you mix up which port is used with which pin; the program will not work and the problem will not be easily found by reading through the program. You will be able to find the problem by carefully simulating the application and watching its operation, but there is an easier way to avoid the problem all together. Instead of associating a pin with an I/O port register for the bit operations, you can declare them together using the #define directive.

The format of the #define directive is:

```
#define Label[(argument[,...])] String
```

and it behaves similarly to the C language define. That is, when "Label" is encountered, "String" is substituted in with and any arguments put into the string. The string substitution provided by the #define directive differs from the operation of a macro in the scope of the substitution. The #define substitutes a string on a line of assembly language code, whereas a macro substitutes the entire line (and maybe adds additional lines).

The most common use of the #define directive in PIC microcontroller applications is for the combining the port and bit number for a specific pin. For example, if you wanted to declare RA3 as an LCD's clock (E) pin, you could use the statement:

```
#define LCDE PORTA, 3
```

Now, when the label LCDE is encountered, as in the bit statement:

```
  bsf      LCDE
```

the string "PORTA, 3" will be substituted into the instruction.

To demonstrate the use of the #define for port pins, I created the asmPKLED.asm, which is a "port" (or language translation) of cPKLED.c and cycles through the eight LEDs of the PICkit 1 starter kit.

```
title   "asmPKLED - PICkit 1 starter kit Running
LED"
;
;  This program copies the function of cPKLED.c
;  and Flashes the 8 LEDs
;  on the PICkit 1 starter kit in Sequence.
;
;  Hardware Notes:
;   PIC16F684 running at 4 MHz Using the
;   Internal Clock
;   External Reset is Used
;   Circuit Runs on PICkit 1 starter kit
;
;  LED Defines:
 #define D0Anode   PORTA, 4
 #define D0Cathode PORTA, 5
 #define D1Anode   PORTA, 5
 #define D1Cathode PORTA, 4
 #define D2Anode   PORTA, 4
 #define D2Cathode PORTA, 2
 #define D3Anode   PORTA, 2
 #define D3Cathode PORTA, 4
 #define D4Anode   PORTA, 5
 #define D4Cathode PORTA, 2
 #define D5Anode   PORTA, 2
 #define D5Cathode PORTA, 5
 #define D6Anode   PORTA, 2
 #define D6Cathode PORTA, 1
 #define D7Anode   PORTA, 1
 #define D7Cathode PORTA, 2
;
;
;  Myke Predko
;  04.10.04
;
  LIST R=DEC
  INCLUDE "p16f684.inc"

  __CONFIG _FCMEN_OFF & _IESO_OFF & _BOD_OFF &
_CPD_OFF & _CP_OFF & _MCLRE_ON & _PWRTE_ON &
_WDT_OFF & _INTOSCIO

; Variables
CBLOCK 0x20
Dlay
ActiveBit                 ; Record the Active Bit
 ENDC

  PAGE
; Mainline

 org    0

  nop                     ; For ICD Debug

  clrf   PORTA            ; Initialize I/O Bits
                            to Off
  movlw  7                ; Turn off Comparators
  movwf  CMCON0
  bsf    STATUS, RP0
  clrf   ANSEL ^ 0x80     ; All Bits are Digital
  bcf    STATUS, RP0

  clrf   ActiveBit        ; Start with D0

Loop:                     ; Return Here after
                            Delay
  incf   ActiveBit, w     ; Load the Bit Number

  addlw  -1               ; Start with D0
  btfss  STATUS, Z
   goto  TryD1
  bsf    STATUS, RP0
  bsf    D7Anode          ; Turn OFF Previously
                            Displayed LED
  bsf    D7Cathode
  bcf    D0Anode          ; Enable D0 LEDs
  bcf    D0Cathode
  bcf    STATUS, RP0
  bsf    D0Anode
  bcf    D0Cathode
  goto   LEDDlay          ; Finished, Repeat

TryD1:                    ; Display D1?
  addlw  -1
  btfss  STATUS, Z
   goto  TryD2
  bsf    STATUS, RP0
  bsf    D0Anode          ; Turn OFF Previously
                            Displayed LED
  bsf    D0Cathode
  bcf    D1Anode          ; Enable D0 LEDs
  bcf    D1Cathode
  bcf    STATUS, RP0
  bsf    D1Anode
  bcf    D1Cathode
  goto   LEDDlay          ; Finished, Repeat

TryD2:                    ; Display D2?
  addlw  -1
  btfss  STATUS, Z
   goto  TryD3
  bsf    STATUS, RP0
  bsf    D1Anode          ; Turn OFF Previously
                            Displayed LED
  bsf    D1Cathode
  bcf    D2Anode          ; Enable D0 LEDs
  bcf    D2Cathode
  bcf    STATUS, RP0
  bsf    D2Anode
  bcf    D2Cathode
  goto   LEDDlay          ; Finished, Repeat

TryD3:                    ; Display D3?
  addlw  -1
  btfss  STATUS, Z
   goto  TryD4
  bsf    STATUS, RP0
  bsf    D2Anode          ; Turn OFF Previously
                            Displayed LED
  bsf    D2Cathode
  bcf    D3Anode          ; Enable D0 LEDs
  bcf    D3Cathode
  bcf    STATUS, RP0
  bsf    D3Anode
  bcf    D3Cathode
  goto   LEDDlay          ; Finished, Repeat

TryD4:                    ; Display D4?
  addlw  -1
  btfss  STATUS, Z
   goto  TryD5
  bsf    STATUS, RP0
  bsf    D3Anode          ; Turn OFF Previously
                            Displayed LED
  bsf    D3Cathode
  bcf    D4Anode          ; Enable D0 LEDs
  bcf    D4Cathode
  bcf    STATUS, RP0
  bsf    D4Anode
  bcf    D4Cathode
  goto   LEDDlay          ; Finished, Repeat

TryD5:                    ; Display D5?
  addlw  -1
  btfss  STATUS, Z
   goto  TryD6
  bsf    STATUS, RP0
```

```
    bsf     D4Anode         ; Turn OFF Previously       incf    ActiveBit, w   ; Increment the Active
                              Displayed LED                                        LED
    bsf     D4Cathode                                   andlw   7              ; Keep Within range of
    bcf     D5Anode         ; Enable D0 LEDs                                      0-7
    bcf     D5Cathode                                   movwf   ActiveBit
    bcf     STATUS, RP0
    bsf     D5Anode                                     goto    Loop           ; Repeat
    bcf     D5Cathode
    goto    LEDDlay         ; Finished, Repeat          end

TryD6:                      ; Display D6?
    addlw   -1
    btfss   STATUS, Z
     goto   TryD7
    bsf     STATUS, RP0
    bsf     D5Anode         ; Turn OFF Previously
                              Displayed LED
    bsf     D5Cathode
    bcf     D6Anode         ; Enable D0 LEDs
    bcf     D6Cathode
    bcf     STATUS, RP0
    bsf     D6Anode
    bcf     D6Cathode
    goto    LEDDlay         ; Finished, Repeat

TryD7:                      ; Display D7?
    addlw   -1
    btfss   STATUS, Z
     goto   LEDDlay
    bsf     STATUS, RP0
    bsf     D6Anode         ; Turn OFF Previously
                              Displayed LED
    bsf     D6Cathode
    bcf     D7Anode         ; Enable D0 LEDs
    bcf     D7Cathode
    bcf     STATUS, RP0
    bsf     D7Anode
    bcf     D7Cathode
    goto    LEDDlay         ; Finished, Repeat

LEDDlay:                    ;
    clrf    Dlay            ; High 8 Bits for Delay
    clrw
    addlw   -1              ; Decrement the
                              contents of WREG
    btfss   STATUS, Z       ;  256x
     goto   $ - 2
    decfsz  Dlay, f         ; Repeat 255x
     goto   $ - 4
    clrf    Dlay            ; Repeat Again for ~
                              400 ms Delay
    clrw
    addlw   -1              ; Decrement the
                              contents of WREG
    btfss   STATUS, Z       ;  256x
     goto   $ - 2
    decfsz  Dlay, f         ; Repeat 255x
     goto   $ - 4
```

Chances are that if you were to code this application (with six I/O port register writes for each of the eight LEDs) without the use of the #define directive for the different port bits, you would make several mistakes. The #define directive simplifies tedious code operations such as the ones required for this application.

I don't want to leave you with the impression that the #define directive is only for simplifying I/O pin accesses; like the C language define, it can be used as a shorthand way of adding constant arithmetic operations to the application without having to type them repeatedly. Additionally, bit-flag variables can be declared using #define directives, in the same way the I/O port bits are declared in this application. Remember that like macros, #define substitutions take place *before* the assembly step. As far as the assembler is concerned, the strings are constant values when it executes. #defines can also be used for defining strings that are used to control the operation of the code build (This will be discussed in more detail in the next experiment). But the most common define is created by the MPLAB IDE to specify which PIC MCU part number information is to be used by the assembler.

As a final point, you'll see that I implemented the C switch equivalent code in a different manner than in the previous experiment. In this experiment, I knew that the values to turn on each LED are arranged sequentially. So rather than implementing a general-case solution to the problem, I took advantage of being able to decrement the value by one for each LED and finally stopping and displaying the LED when the result is zero. Remember that there is always more than one way to skin a cat.

The example application shown in the previous experiment was a pretty clumsy piece of code. The task of lighting each of the eight LEDs on the PICkit 1 starter kit is not something that lends itself to algorithmic programming. Although the code for each LED is the same, but there is no consistency between pairs, which necessitates writing code for each LED. The final result is an application that is quite long and cumbersome. What is needed is a way to minimize the development effort. The task of lighting each LED in sequence is not a problem that is easily addressed by using a single data table. Turning off the previous LED and turning on the next one, as I have done here, is not a terribly inefficient way of implementing the application. The problem rather is how to efficiently code the application.

A potential solution to the problem is to use the conditional assembly directives built into MPASM assembler. These directives allow you to select which code is to be assembled or even to create code algorithmically that can be assembled into the application. In the interest of clarity, I must point out that these directives execute before the code is assembled and are used to select which code is and isn't part of the assembly. Although conditional assembly is not a critical concept, it is something that can make your life quite a bit easier and allow you to produce higher quality code.

The basic conditional assembly directives are ifdef, ifndef, else, and endif. The ifdef and ifndef directives pass any statements that follow them if the specified #defined label is present. The else directive provides alternative code, and the endif directive indicates that the ifdef/ifndef test is complete. I commonly use these directives with the Debug #define to avoid code that will not operate properly in the MPLAB IDE simulator. For example:

```
#define Debug
  :
  ifndef Debug
  movf    ADRESH, w    ; Load in ADC Result
  else                 ; In Simulator
  movlw   0x088        ; Simulated Read
  endif
```

will return 0x088 in WREG if Debug has been #defined, skipping over an invalid register access. When using ifdef and "ifndef" to add and take away instructions to avoid problems simulating the application, you should make sure that both paths have the same number of instructions and execute in the same number of instruction cycles. This will ensure the application will execute the same way in the simulator and once it is burned into a chip.

As I indicated in the previous experiment a common declared #define by MPLAB IDE is the PIC MCU part number, which consists of two underscore characters followed by the part number. For example, the PIC16F684 used in this book would have the #define __16F684. In your application you can test for this define using the ifdef or ifndef directives. This capability is useful in cases where different PIC microcontrollers might be used in the same application. For example, when you are designing a product, you may want to use a part that has Flash and can be easily reprogrammed (like the PIC16F684), but there may be a cheaper *one-time programmable* (OTP) part that can perform the same functions at a fraction of the costs. A conditional assembly statement will allow you to use the same source code for both PIC MCU part numbers, with the only difference being the numbers specific to the different chips.

A single-comparison expression can be evaluated by the if directive, which also uses the else and endif directives. You also can define numeric variables in your application that execute *before* application assembly. For example, a variable i could be declared using the variable directive, and later in the application, it could be tested against different values:

```
variable i
  :
if i > 7
  bsf    PORTA, 5
else
  bsf    PORTA, 0
i = i + 2
  endif
```

Labels declared using the variable directive are just that—labels. To change the value of a variable, it has to be placed in the first column of the application source code as I have done in the previous snippet.

The final conditional-assembly directives are the while and wend directives, which allow you to place the same code repeatedly in an application. The format of these instructions should not be surprising; if you wanted to put in four nop instructions, you could use the directives:

```
variable i
  :
i = 0                       ; Initialize Counter
 while i < 4                ; Repeat 4x
  nop                       ; Put in nop
i = i + 1                   ; Increment Counter
 endw                       ; Loop End
```

When using the while directive to implement multiple instances of application code, you can use a little-known directive, which is in the format #v(expr) and inserts the numeric value of expr into the application code. For example, if the variable i had the value of 7 and the statement:

```
    movf    Var#v(i)Value
```

the assembler would process the instruction:

```
    movf    Var7Value
```

This last directive probably seems somewhat esoteric, but it allows you access individual labels as if they were array elements. I have done exactly that in asmCond.asm, which is a rewrite of the sequencing LED application and uses the while, if, and #v directives discussed here to produce the identical code to the previous experiment's asmPKLED.asm), but it avoids the need to put in the same code repeatedly.

```
title   "asmCond - Conditional Assembly"
;
;  This program takes advantage of the
;    conditional assembly
;    features of MPASM and instead of repeating a
;    block of code
;    seven times that cycles the LEDs on the
;    PICkit 1 starter kit, the
;    code is put in algorithmically.
;
;  Hardware Notes:
;    PIC16F684 running at 4 MHz Using the
;    Internal Clock
;    External Reset is Used
;    Circuit Runs on PICkit 1 starter kit
;
;  LED Defines:
CD0Anode    EQU 4
CD0Cathode  EQU 5
CD1Anode    EQU 5
CD1Cathode  EQU 4
CD2Anode    EQU 4
```

```
CD2Cathode  EQU 2
CD3Anode    EQU 2
CD3Cathode  EQU 4
CD4Anode    EQU 5
CD4Cathode  EQU 2
CD5Anode    EQU 2
CD5Cathode  EQU 5
CD6Anode    EQU 2
CD6Cathode  EQU 1
CD7Anode    EQU 1
CD7Cathode  EQU 2
;
;
;  Myke Predko
;  04.10.04
;
  LIST R=DEC
  INCLUDE "p16f684.inc"

  __CONFIG _FCMEN_OFF & _IESO_OFF & _BOD_OFF &
_CPD_OFF & _CP_OFF & _MCLRE_ON & _PWRTE_ON &
_WDT_OFF & _INTOSCIO

;  Variables
 variable LEDCount = 0  ; Only 7 Bits to Display
 CBLOCK 0x20
Dlay
ActiveBit                   ; Record the Active Bit
 ENDC

    PAGE
;  Mainline of asmPKLED

  org     0

  nop                       ; For ICD Debug

  clrf    PORTA             ; Initialize I/O Bits to
                            ;   Off
  movlw   7                 ; Turn off Comparators
  movwf   CMCON0
  bsf     STATUS, RP0
  clrf    ANSEL ^ 0x80      ; All Bits are Digital
  bcf     STATUS, RP0

  clrf    ActiveBit         ; Start with D0

Loop:                       ; Return Here after
                            ;   Delay
  incf    ActiveBit, w      ; Load the Bit Number

 while LEDCount < 8         ; Repeat the Code 8x
  addlw   -1                ; Decrement Bit Count
  btfss   STATUS, Z
   goto   $ + 10
  bsf     STATUS, RP0
 if 0 == LEDCount           ; Turn off D7 if D0
                            ;   Being Turned on
  bsf     PORTA, CD7Anode   ; Turn OFF Previously
                            ;     Displayed LED
  bsf     PORTA, CD7Cathode
 else
  bsf     TRISA ^ 0x80, CD#v(LEDCount-1)Anode
  bsf     TRISA ^ 0x80, CD#v(LEDCount-1)Cathode
 endif
  bcf     TRISA ^ 0x80, CD#v(LEDCount)Anode
                            ; Enable LED's I/O Pins
  bcf     TRISA ^ 0x80, CD#v(LEDCount)Cathode
  bcf     STATUS, RP0
  bsf     PORTA, CD#v(LEDCount)Anode
  bcf     PORTA, CD#v(LEDCount)Cathode
  goto    LEDDlay           ; Finished, Repeat
LEDCount += 1               ; Do the Next LED
 endw
```

```
LEDDlay:              ;
    clrf    Dlay      ; High 8 Bits for Delay
    clrw
    addlw   -1        ; Decrement the contents
                        of WREG
    btfss   STATUS, Z ;  256x
    goto    $ - 2
    decfsz  Dlay, f   ; Repeat 255x
    goto    $ - 4
    clrf    Dlay      ; Repeat Again for ~ 400
                        ms Delay
    clrw
    addlw   -1        ; Decrement the contents
                        of WREG
    btfss   STATUS, Z ; 256x
    goto    $ - 2
    decfsz  Dlay, f   ; Repeat 255x
    goto    $ - 4

    incf    ActiveBit, w ; Increment the Active
                           LED
    andlw   7         ; Keep Within range of
                        0-7
    movwf   ActiveBit

    goto    Loop      ; Repeat

    end
```

When you look through the code, you'll see that I have replaced the pin defines with equates. This was done because the #v(expr) directive cannot be used with other directives. As I have emphasized, the previous and the next experiments' directives execute *before* the application's instructions are assembled. Only one pass is made to the preprocessor that executes the directives, and if any directives are changed and require additional execution, the application will fail.

When I created asmPKLED.asm, I did so with an eye toward repeating it with conditional assembly statements, but it is still not that inefficient in terms of execution and program size. The problem with asmP-KLED.asm is in its creation; although the instructions for each LED can be cut and pasted, I found it a chore to change which pin defines are used for each LED (and I made two mistakes). By using the while directive and other conditional assembly directives, I eliminated the task of repeating the same instructions over and over and minimized the opportunity for errors in the application. This is really the major advantage of using conditional assembly directives: They help you minimize opportunities for making mistakes in your application.

Experiment 76—Macros

PC/ Simulator

The next level of code substitution after the #define, in MPASM assembly language, is the macro. Macros consist of instruction statements that are inserted into an application along with the other directives when the preprocessor is active. Although there is not one all-purpose macro that can be used to illustrate the concept, macros can greatly simplify your application programming. This is especially true if you take advantage of the #define and conditional assembly directives presented in the previous experiments.

The format for declaring a macro is quite simple:

```
MacroName macro MacroParameter[,...]
; Macro Code
    endm
```

The MacroParameter is a single string or multiple strings that are substituted into the macro at assembly time. For example, if a macro was required to load WREG with a numeric value divided by two, the macro could be defined as:

```
movlwDiv2 macro Div2Value
    movlw   Div2Value / 2
    endm
```

If the macro was invoked with a Div2Value of 47, the following code would be inserted into the program:

```
    movlw   47 / 2
```

Macros can take advantage of the conditional assembly directives of MPASM assembler, and surprisingly sophisticated functions can be created from them as I will show in this experiment.

The reasons for using macros in your application include the following:

- Minimizing program keying
- Reducing program execution time
- Minimizing program .hex file size
- Simplifying the source code

- Providing opportunity for optionally compiled functions
- Providing advanced build-time options
- Simplifying the effort required by new programmers to create their own assembly language programs

Functions and subroutines can be written as macros, and, depending on their size and the number of times they are accessed, they can decrease the program execution time, complexity, and size. Macros can be added to the program, in the same way a subroutine or function call is added. The macro avoids the overhead of the call/return instructions, and placing parameters in subroutine- or function-specific variables can reduce both the total space and variable requirements of an application. They also provide the assembly language code with variably added functions, and they do so without having to save the functions in a library and hope that they will not be linked into the final application.

One important note to newcomers in the programming world about the use of macros: Although you may not have the abilities (yet) to implement macros for simplifying your programming workload, I'm sure you can identify situations where assembly language programming seems particularly difficult and, therefore, would benefit from a macro. One such example, which I address in asmMacro.asm, is the if decision structure in PIC MCU assembler. In this experiment, I have created a macro that uses conditional assembly directives to insert the appropriate four-instruction template code (presented in Section 8) for a conditional jump based on two variable values.

```
  title   "asmMacro - Demonstrate the 'ifgoto'
Macro"
;
;  This application demonstrates the "ifgoto"
;  Macro which compares
;  two values and executes according to the
;  specified condition.
;
;
;  Hardware Notes:
;  PIC16F684 running at 4 MHz in Simulator
;  Reset is tied directly to Vcc via
;  Pullup/Programming Hardware
;
;
;  Myke Predko
;  04.09.17
;
  LIST R=DEC
 INCLUDE "p16f684.inc"

; Variables
CBLOCK 0x20
i, j
 ENDC
```

```
;  Macro
ifgoto Macro Var1, Comp, Var2, Dest
  if (1 Comp 0)            ; Check for "!="/">"/">="
   if (0 Comp 1)           ; "!="
    movf   Var1, w
    subwf  Var2, w
    btfss  STATUS, Z
   else                    ; Else ">"/">="
    if (1 Comp 1)          ; ">="
     movf   Var2, w
     subwf  Var1, w
     btfsc  STATUS, C
    else                   ; ">"
     movf   Var1, w
     subwf  Var2, w
     btfss  STATUS, C
    endif
   endif
  else
   if (0 Comp 1)           ; Check for "<"/"<="
    if (1 Comp 1 )         ; "<="
     movf   Var1, w
     subwf  Var2, w
     btfsc  STATUS, C
    else                   ; "<"
     movf   Var2, w
     subwf  Var1, w
     btfss  STATUS, C
    endif
   else
    if (1 Comp 1)          ; "=="
     movf   Var2, w
     subwf  Var1, w
     btfsc  STATUS, Z
    else
     error Unknown "if" Condition
    endif
   endif
  endif
   goto  Dest
endm

  org     0
  nop                      ; Required for MPLAB ICD2

  movlw  15                ; Get Test Values
  movwf  i
  movlw  20
  movwf  j

ifgoto i, >, j, ErrorDone

  nop
ifgoto i, >=, j, ErrorDone

  nop
ifgoto j, <, i, ErrorDone

  nop
ifgoto j, <=, i, ErrorDone

  nop
ifgoto i, ==, j, ErrorDone

  movf   i, w              ; Set Equals for Not
                           Equals Test
  movwf  j

ifgoto i, !=, j, ErrorDone

  nop
```

```
ifgoto i, ==, j, GoodDone ; Should be match,
                            Jump to Proper End

ErrorDone:                 ; Code Should NEVER
                             be Here
  goto    ErrorDone

GoodDone:                  ; Could Should End up
                             Here
  goto    GoodDone


 end            .
```

The ifgoto macro in asmMacro is useful to people other than new PIC MCU assembly language programmers. I should point out that the preprocessor in MPASM assembler will not recognize the difference between variable labels and constant values. This is unfortunate because you will have to modify this macro if you want to perform a conditional jump based on the contents of a variable and a constant. The background for the changes required to modify this macro to support constant values can be found in the "Conditional Execution" experiment in Section 8.

Experiment 77—16-Bit Values/Variables with Addition, Subtraction, and Comparison

I find that when I am doing actual PIC microcontroller applications, eight-bit values are simply not enough. Sixteen-bit variables provide you with a significant amount of additional programming flexibility in a variety of different areas, including the creation and processing of data. Programming for 16-bit variables is not beyond your capabilities if you have been following along. In fact, you will find that 16-bit variables are not that hard to work with, especially if you follow the rules that I set out in this section.

When I first started PIC MCU programming, the directives that I will introduce to you in this experiment were not available in the Microchip assembler products. For example, to declare a 16-bit variable, you would use the equ directive:

```
iLo equ 0x20   ; Declare a 16 bit variable
                 starting
iHi equ 0x21   ;  at address 0x20
```

to declare a 16-bit variable as two eight-bit variables. The declaration format shown in this snippet is known as *little endian* because the least significant bytes is declared at the lower-file register. This is the format used in many 16-bit processors (e.g., the Intel 8086), and I feel that it makes intuitive sense. The problem with this format comes when you are using a debugger to display variables; in the debugger a number like

0x01234 would be displayed in memory as "34 12," which requires you mentally to rearrange the bytes. This mental rearrangement is not required in the MPLAB IDE simulator. You can display 16-bit data in MPLAB IDE by setting the properties of the Watch window to 16-bit data, in little endian or big endian formats.

With the availability of the cblock directive, you can declare the same 16-bit variable as:

```
cblock 0x20
i:2           ; "i" is a 16 bit (2 Byte) Variable
 endc
```

and the low address byte (in little endian format) is accessed as the variable, and the high address byte as the variable + 1. To show how they are accessed, here is an initialization statement:

```
movlw  0x1234 - ((0x1234 / 0x0100) * 0x0100)
movwf  i             ; i = 0x1234
movlw  0x1234 / 0x0100
movwf  i + 1
```

The first instruction of these four lines finds the low eight bits of the constant value by doing an integer divide by eight bits (0x0100) and then multiplying it again by eight bits. After the multiplication, the values lower than eight bits will all be reset. Subtracting this value from the original will return the value of the lower eight bits. The upper eight bits are simply calculated by eight bits, or shifted to the right by eight bits. These two calculations could be built into defines, but Microchip has provided the HIGH and LOW directives, which perform the same calculations. The four previous statements then become:

```
movlw  LOW 0x1234    ; i = 0x1234
movwf  i
```

```
movlw   HIGH 0x1234
movwf   i + 1
```

Understanding how to declare and access 16-bit values and variables is a good fraction of the work required to use 16-bit values in your assembly language application. The remaining part is to understand how arithmetic operations pass, carry, and borrow information from the low byte to the high byte of a variable. In this experiment's program (asm16Bit.asm), I demonstrate how declarations, initializations, increments, decrements, addition, subtraction, and conditional jumps are implemented.

```
title   "asm16Bit - Demonstrate 16 Bit
Operations"
;
;  Demonstrate 16 Bit Variable Declaration,
;    Initialization,
;    addition, subtraction, incrementing
;    decrementing and comparison
;    operations.
;
;  Hardware Notes:
;    PIC16F684 running at 4 MHz in Simulator
;    Reset is tied directly to Vcc via
;    Pullup/Programming Hardware
;
;
;  Myke Predko
;  04.09.21
;
   LIST R=DEC
   INCLUDE "p16f684.inc"

   CBLOCK 0x20            ; Variable Declaration
i:2                       ; 16 Bit (2 Byte)
                          ; Counter

j:2, Destination:2
   ENDC

   PAGE

   org     0
   nop                    ; Required for MPLAB
                          ; ICD2

                          ; Initialize "i" with
                          ; 12345 decimal
   movlw   HIGH 1234      ; High Byte First
   movwf   i + 1
   movlw   LOW 1234
   movwf   i

   movlw   HIGH 56789     ; Initialize "j"
   movwf   j + 1
   movlw   LOW 56789
   movwf   j

   incfsz  i, f           ; Increment "i"
    decf   i + 1, f       ; Increment High Byte
                          ; if Result of
   incf    i + 1, f       ; Incrementing Low Byte
                          ; == 0

   movf    i, f           ; Decrement "i"
   btfsc   STATUS, Z      ; Decrement High Byte
                          ; if Low Byte is
    decf   i + 1, f       ; initially equal to
                          ; zero
   decf    i, f
```

```
   movf    i, w           ; Destination = i + j
   addwf   j, w
   movwf   Destination
   movf    i + 1, w       ; Increment High Byte
                          ; if Result of
   btfsc   STATUS, C      ; Low Byte Add is >
                          ; 0x0FF (Carry Set)
    incf   i + 1, w       ; Note: Written for Low
                          ; and Mid-Range
   addwf   j + 1, w       ; PIC MCUs so no "addlw
                          ; 1"
   movwf   Destination + 1

   movf    j, w           ; Destination =
                          ; Destination - j
   subwf   Destination, f
   movf    j + 1, w
   btfss   STATUS, C      ; If Carry Reset (Low
                          ; Byte Negative)
    incf   j + 1, w       ; Take One More Away
                          ; from the High Byte
   subwf   Destination + 1, f

   movf    i, w           ; if (i > j) goto Error
   subwf   j, w
   movf    i + 1, w
   btfss   STATUS, C
    incf   i + 1, w
   subwf   j + 1, w
   btfss   STATUS, C
    goto   ErrorLoop      ; i is Less than j, if
                          ; Jump Taken, Error

   goto    $              ; Finished, Everything
                          ; Okay

ErrorLoop:                ; Compare Operation
                          ; didn't work correctly
   goto    ErrorLoop

   end
```

Note that the conditional jump has a bit-skip instruction that follows the same rules as the eight-bit conditional jump code presented in Section 8. This means that the tables used to define which value will be subtracted (and from where), and the tables used to define the flag state for the bit-skip instruction are exactly the same. The conditional jump code for a value being equal or not equal to zero is a special case that can be tested by simply ORing each of the two bytes of a 16-bit variable together and checking the zero flag:

```
   movf    Label, w       ; OR Low Byte with
                          ; High Byte
   iorwf   Label + 1, w   ; Z Set if result (both
                          ; bytes) is zero
   btfsc   STATUS, Z
    goto   SixteenBitZero ; Jump is sixteen bit
                          ; value is zero
```

This example illustrates that although the different operations presented in asm16Bit are useful and can usually be used without modification, some cases exist where better code can be produced without the same number of instructions. For example, think about how to implement adding an eight-bit value to a 16-bit variable and what would be changed from the previous code to do so.

When you have worked through the different operations listed in asm16Bit, you should have a good idea how to perform bitwise operations. You should also recognize that the bitwise operations are somewhat simpler than addition and subtraction operations because you do not have to carry or borrow. Multiplication and division are somewhat different than bitwise operations or addition and subtraction, but you should be able to come up with some ideas on how they can be implemented. (Section 12 will examine different methods of multiplication and division that can be used as a basis of these 16-bit operations.)

PC/ Simulator

When I discussed delays in C programming, I probably seemed pretty nonchalant about it; basic delays could be implemented using for loops, and more precise delays using built-in timers and critical signals would be output from custom hardware built into the PIC MCU. For C programming and relatively feature-rich devices like the PIC16F684, this is a pretty good attitude to take. But when you are working with simpler PIC microcontrollers, which do not have multiple timers or CCP modules, you are going to have to program a delay yourself. Programming delays in PIC MCU can be done surprisingly efficiently as the macro in this experiment demonstrates.

The PIC's clock is divided into four parts, each of which provides a different function in the instruction cycle. By knowing the clock speed, you can easily derive the instruction-cycle period as the reciprocal of the PIC MCU's clock speed divided by four. Every instruction, except the ones that change the program counter execute in one instruction cycle, and this knowledge can be used to create simple delays. As you might expect the nop instruction can be used to delay one instruction cycle and the following instruction:

```
goto    $ + 1       ; Delay two instruction
                      cycles
```

will delay two instruction cycles using only one instruction.

If you need to delay more instruction cycles, you can use a small loop like this one:

```
movlw    Cycles       ; Load Counter
movwf    Dlay
decfsz   Dlay, f      ; Decrement the Counter
 goto    $ - 1        ; Loop Back to Counter
                        Decrement
```

Each loop executes in three cycles, and a maximum of 769 instruction cycles can be delayed using this method. Note that I am using the relative address for looping: Multiple delay loops can quickly run out of your list of stock labels like BitDelayLoop, and you'll end up with labels like BitDelayLoop7. These are not recommended because the loops are probably copied and pasted from other parts of the program, and you may not always remember to update all the label references. A jump to an incorrect label can be very difficult to see in a program. By avoiding their use in delay loops, you will eliminate a possible source for programming errors.

For longer delays, you will probably see the need for a 16-bit counter and might want to use a loop like this one:

```
movlw    HIGH Count    ; Load 16 Bit Counter
movwf    Dlay + 1
movlw    LOW Count
movwf    Dlay
decfsz   Dlay, f       ; Decrement Low Byte
 goto    $ - 1
decfsz   Dlay + 1, f   ; Decrement High Byte
 goto    $ - 3         ; Loop Back to Low
                         Byte
```

This code works well, but it is extremely difficult to come up with a 16-bit value that will delay a specified number of cycles. I use this code to give me a quick-and-dirty approximate 200 ms delay by resetting the 16-bit Dlay variable before entering the loop: It's fast, easy to code, and doesn't require any thinking on my part.

The following code is the recommended way to use a 16-bit delay counter:

```
movlw    HIGH (Count + 256) ; Load 16 Bit
                              Counter
movwf    Dlay
movlw    LOW (Count + 256)
```

```
       addlw   -1              ; Decrement Low
                                   Byte
   btfsc  STATUS, Z
    decfsz Dlay, f             ; Decrement High
                                   Byte
        goto  $ - 3            ; Loop Back
```

This code has a constant loop delay of five instruction cycles and requires one fewer instruction and file register than the more obvious solution. The 256 is added to the cycle count to ensure the correct number of loops (otherwise, if the high byte of cycles is zero, the code will loop 255 extra times, and when it is not zero, it will loop one time less than you want it to). I should point out that the last pass through the loop executes in four cycles, and with the three overhead cycles before the loop, the cycle delay can be written out as the following simple formula:

```
CyclesDelay = (Count * 5) + 2
```

In the program asmDlay, I have taken advantage of this delay and the one- and two-instruction-cycle delays shown previously to create a macro that provides a precise delay to a specified cycle count. In the program, I have listed the macro and put in a number of tests to make sure that it executes for the specified number of cycles. The nops after each macro invocation are used to set breakpoints before and after in order to time the code's execution using MPLAB IDE's Stopwatch to test the operation of the macro.

```
   title  "asmDlay - Create a Universal Delay
Macro"
;
;   The "Dlay" macro in this routine will provide
delays ranging
;   from 1 to 1,000,000 cycles.
;
;   Hardware Notes:
;   PIC16F684 running at 4 MHz in Simulator
;   Reset is tied directly to Vcc via
Pullup/Programming Hardware
;
;
;   Myke Predko
;   04.09.22
;
   LIST R=DEC
   INCLUDE "p16f684.inc"


   __CONFIG _FCMEN_OFF & _IESO_OFF & _BOD_OFF &
_CPD_OFF & _CP_OFF & _MCLRE_ON & _PWRTE_ON &
_WDT_OFF & _INTOSCIO

CBLOCK 0x20                    ; Variable
                                   Declaration
DlayValue:2                    ; Requires 24 Bit
                                   Counter (w/ WREG)
   ENDC

Dlay Macro Cycles
   variable CyclesLeft         ; Keep Track of
                                   Remaining Cycles
   variable LargeNum
CyclesLeft = Cycles
   local LongLoop
```

```
   if Cycles > 0x04FFFF00      ; Can't Handle > 83
                                   Seconds (@ 4 MHz)
   error "Required Delay is longer than 'Dlay'
Macro can support"
   endif
   if Cycles > 327681          ; Need Large Loop?
LargeNum = CyclesLeft / 327681
     movlw    LargeNum
     movwf    DlayValue + 2     ; Calculate Number of
                                    Loops
LongLoop:                       ; Repeat for Each
                                    Loop
     clrf     DlayValue + 1     ; Do Maximum Possible
                                    Loop Count
     clrf     DlayValue
     decf     DlayValue, f
     btfsc    STATUS, Z
      decfsz DlayValue + 1, f
       goto   $ - 3
     decfsz DlayValue + 2, f ; Repeat Loop
     goto    LongLoop
CyclesLeft = CyclesLeft - ((LargeNum * 327681) +
1 + (LargeNum * 3))
   endif ; Need Large Loop
   if Cycles > 14              ; Will a Loop be
                                   required?
     movlw    HIGH (((CyclesLeft - 3) / 5) + 256)
     movwf    DlayValue + 1
     movlw    LOW (((CyclesLeft - 3)/ 5) + 256)
     movwf    DlayValue
     decf     DlayValue, f ; 5 Cycle Constant Delay
                                   Loop
     btfsc    STATUS, Z
      decfsz DlayValue + 1, f
       goto   $ - 3
CyclesLeft = CyclesLeft - (3 + (5 * ((CyclesLeft
- 3)/ 5)))
   endif                       ; Finished with Loop Code
   while CyclesLeft >= 2 ; Put in 2 Instruction
                                   Cycle Delays
     goto     $ + 1
CyclesLeft = CyclesLeft - 2
   endw
   if CyclesLeft == 1          ; Put in the Last
                                   Required Cycle
    nop
   endif
   endm

   PAGE

org       0
   nop                         ; Required for MPLAB ICD2

Dlay 3                         ; Should be goto & nop
   nop

Dlay 44                        ; Delay 44 Cycles
   nop

Dlay 1000                      ; 1 ms Delay
   nop

Dlay 100000                    ; Delay 0.1s
   nop

Dlay 5000000                   ; Delay 5s
   nop

Dlay 60000000                  ; Delay 60s

   goto     $                  ; Finished, Everything
                                   Okay

   end
```

I would consider this macro to be a combination of showboating and overachieving. If you were to read through the application, you would see that it is limited to 83 million cycles (83 seconds in a PIC MCU running at 4 MHz), and it provides accurate delays to the instruction cycle. This accuracy is usually required for long delays, but for short delays, the accuracy this macro provides is not required. You might want to cut the macro down for your own use and avoid the initial loop, which is required for delays longer than 327,681 cycles (a third of a second when running at 4 MHz).

If you compare this application to others in the book that are designed to run only in the MPLAB IDE simulator, you might think that my inclusion of the __CONFIG directive is an oversight. Generally it's not required in the simulator. However, in this case the __CONFIG directive is required. In the testing of delays that are longer than 2.9 seconds, the simulated PIC MCU would reset itself due to the Watchdog Timer resetting the processor. To avoid this problem, I added the standard __CONFIG statement to make sure the Watchdog Timer is disabled.

Experiment 79—High-Level Programming in Assembler

starter kit

When I first started working after university, I was part of a team developing a high-performance functional tester for electronic circuits. One of the tasks we had was to come up with a programming language for the tester, and I have always been intrigued by the solution that was chosen. Rather than come up with a custom compiler for the programming language, a standard compiler was selected and its macro processor would be used to convert programming statements into the data required by the tester. To some extent, this effort was successful as it did produce data, unfortunately the compiler's macro processor could not produce *enough* of it. Much of the data was in the form of tables, and the maximum table size the supported by the macro processor was 32 KB. So far in this section, I have given you a macro for conditional execution (or if) statements and a macro for precision delays. And, I wanted to see if I could come up with a macro that would support assignment statements, because with such a macro, I felt I could claim that I had developed a high-level language that ran under the MPLAB IDE assembler.

Unfortunately, I couldn't do it in *one* macro. It required two. The first macro, ASSIGN, was used to assign the contents of a variable or a constant (literal) value into a variable. Its format was:

```
ASSIGN Dest, EqualSign, dType, Source
```

and would produce the following statements:

```
[bsf    STATUS, RP0]
movf    i, w
[bcf    STATUS, RP0]
[bsf    STATUS, RP0]
movwf   j
[bcf    STATUS, RP0]
```

for the macro invocation:

```
ASSIGN j, =, " ", i
```

or

```
movlw   47
[bsf    STATUS, RP0]
movwf   i
[bcf    STATUS, RP0]
```

for the macro invocation:

```
ASSIGN i, =, ".", 47
```

The setting and resetting of the RP0 bit of STATUS is dependant on the position of the variables. I could have optimized the macro to eliminate one set of bcf/bsf statements if both variables were in bank 1, but I felt this was good enough. The blank ( ) or period (.) dType parameter before the source macro parameter indicates whether the source is a variable or a literal. There is no way the macro processor can determine which is which.

I originally wanted the ASSIGN macro to be able to handle not only straight assignment statements to variables, but also two-parameter operations (like add or subtract). Unfortunately, I wasn't able to discover a method to determine whether or not a parameter was present and, in any case, if any parameters were missing the macro processor would flag an error. So, I added the second macro:

```
OPERATE Dest, EqualSign, dType1, Source1, Oper,
dType2, Source2
```

and would produce statements like:

```
movlw  47
[bsf    STATUS, RP0]
subwf  j, w
[bcf    STATUS, RP0]
[bsf    STATUS, RP0]
movwf  i
[bcf    STATUS, RP0]
```

for the macro invocation:

```
OPERATE i, =, " ", j, -, ".", 47
```

The OPERATE macro puts in the second parameter was first to load WREG with the subtractor if the subtraction operation is selected.

To test the ASSIGN and OPERATE macros, I created asmAssign.asm:

```
title  "asmAssign - Experiment with an
Assignment Macro"
;
;  This program demonstrates how an
;  assignment macro for
;  the PIC MCU could be created.
;
;  Hardware Notes:
;   PIC16F684 running at 4 MHz in Simulator
;
;
;  Myke Predko
;  04.09.28
;
  LIST R=DEC
  INCLUDE "p16f684.inc"

CBLOCK 0x20               ; Variable Declaration
i, j
  ENDC

ASSIGN macro Dest, EqualSign, dType, Source
 variable _aai = 7, _aaj
_aaj EqualSign _aai
 if (_aai != _aaj)
  error "No Equals ('=') in Assignment Statement
 else                     ; Equals Present
  if (dType != ".")       ; Variable Direct
   if (Source > 0x7F)
     bsf STATUS, RP0       ; Bank 1 Variable
   endif
     movf Source & 0x7F, w
   if (Source > 0x7F)
     bcf STATUS, RP0       ; Bank 1 Variable
   endif
  else                    ; Literal
     movlw Source
  endif
  if (Dest > 0x7F)
    bsf STATUS, RP0        ; Bank 1 Variable
  endif
     movwf Dest & 0x7F
  if (Dest > 0x7F)
    bcf STATUS, RP0        ; Bank 1 Variable
  endif
 endif
 endm
```

```
OPERATE macro Dest, EqualSign, dType1, Source1,
Oper, dType2, Source2
 variable _aai = 7, _aaj
_aaj EqualSign _aai
 if (_aai != _aaj)
  error "No Equals ('=') in Assignment Statement
 else                     ; Equals Present
  if (dType2 != ".")      ; Variable Direct
   if (Source2 > 0x7F)
    bsf STATUS, RP0       ; Bank 1 Variable
   endif
    movf Source2 & 0x7F, w
   if (Source2 > 0x7F)
    bcf STATUS, RP0       ; Bank 1 Variable
   endif
  else                    ; Literal
    movlw Source2
  endif
 if (dType1 != ".")       ; Variable Direct
  if (Source1 > 0x7F)
    bsf STATUS, RP0        ; Bank 1 Variable
  endif
  if ((1 Oper 1) == 2)     ; Addition Operator
    addwf Source1 & 0x7F, w
  else
   if ((47 Oper 25) == 22) ; Subtraction
                                  Operator
    subwf Source1 & 0x7F, w
   else
    if ((3 Oper 2) == 3)   ; OR Operator
     iorwf Source1 & 0x7F, w
    else
     if ((3 Oper 2) == 2)  ; AND Operator
      andwf Source1 & 0x7F, w
     else
      if ((3 Oper 2) == 1) ; XOR Operator
       xorwf Source1 & 0x7F, w
      else
       error "Unknown Operator"
      endif
     endif
    endif
   endif
  endif
 if (Source1 > 0x7F)
   bcf STATUS, RP0         ; Bank 1 Variable
 endif
 else                     ; Literal
  if ((1 Oper 1) == 2)     ; Addition Operator
    addlw Source1
  else
   if ((47 Oper 25) == 22) ; Subtraction
                                  Operator
    sublw Source1
   else
    if ((3 Oper 2) == 3)   ; OR Operator
     iorlw Source1
    else
     if ((3 Oper 2) == 2)  ; AND Operator
      andlw Source1
     else
      if ((3 Oper 2) == 1) ; XOR Operator
       xorlw Source1
      else
       error "Unknown Operator"
      endif
     endif
    endif
   endif
  endif
 endif
 if (Dest > 0x7F)
   bsf STATUS, RP0         ; Bank 1 Variable
 endif
```

```
      movwf Dest & 0x7F
    if (Dest > 0x7F)
      bcf STATUS, RP0        ; Bank 1 Variable
    endif
  endif
endm

  PAGE

  org      0
  nop                        ; Required for MPLAB
                               ICD2

  ASSIGN i, =, " ", j
  ASSIGN i, =, ".", 47
  ASSIGN j, =, " ", TRISC

  OPERATE i, =, " ", j, +, ".", 47
  OPERATE i, =, " ", j, &, ".", 47
  OPERATE i, =, " ", j, -, ".", 47
  OPERATE OPTION_REG, =, " ", OPTION_REG, &, ".",
0x7F

  goto     $               ; Finished, Everything Okay

  end
```

When you test this application in the simulator, you are going to discover it is extremely difficult to follow the progress of the code; execution will bounce between the macro invocations and the lines inside the code. You might want to look at the Disassembly Listing, but common statements are listed one after the other in the macro definitions, making the code even more difficult to follow. Your best resource for seeing how the macros produce the necessary code is in the source code listing, asmAssign.lst, which provides the inserted macros and indicators for active statements. Unfortunately, the MPLAB IDE simulator does not step through the listing file.

From my perspective, this inability to simulate effectively the application is the biggest drawback to writing code with these macros. A secondary concern is the possible difficulty in reading what the source code is doing from the macro invocations. To illustrate what I mean, I created asmMFlash.asm, which flashes D0 once per second and is written in just the ASSIGN, OPERATE, Dlay, and ifgoto macros (the body of which I have deleted from the following listing):

```
  title    "asmMFlash - Flash D0 Using High Level
Macros"
;
;  This program demonstrates how the High Level
Operations
;   Macros can be used to Flash the PICkit's D0
LED.
;
;  Hardware Notes:
```

```
;   PIC16F684 running in PICkit 1 starter kit
;
;
;  Myke Predko
;  04.09.28
;
  LIST R=DEC
  INCLUDE "p16f684.inc"

  __CONFIG _FCMEN_OFF & _IESO_OFF & _BOD_OFF &
_CPD_OFF & _CP_OFF & _MCLRE_ON & _PWRTE_ON &
_WDT_OFF & _INTOSCIO

  CBLOCK 0x20              ; Variable Declaration
i, j
DlayValue:2               ; Dlay Requires 24 Bit
                            Counter (w/ WREG)
  ENDC

ASSIGN macro Dest, EqualSign, dType, Source
;;  ASSIGN macro source not shown

OPERATE macro Dest, EqualSign, dType1, Source1,
Oper, dType2, Source2
;;  OPERATE macro source not shown

Dlay Macro Cycles
;;  Dlay macro source not shown

ifgoto Macro Var1, Comp, Var2, Dest
;;  ifgoto macro source not shown

  PAGE

  org      0
  nop                       ; Required for MPLAB
                              ICD2

  ASSIGN PORTA, =, ".", 0  ; Initialize Variables
  ASSIGN PORTA, =, ".", 0xff
  ASSIGN CMCON0, =, ".", 7
  ASSIGN ANSEL, =, ".", 0
  ASSIGN TRISA, =, ".", b'001111'
  ASSIGN j, =, ".", 0

Loop:                       ; Return Here when
                              Done
  Dlay 500000               ; Delay a 1/2 Second
  OPERATE i, =, " ", PORTA, &, ".", (1 << 4)
  ifgoto i, ==, j, SetD0
  ASSIGN PORTA, =, ".", 0  ; Turn off the LED
  goto   Loop
SetD0:
  ASSIGN PORTA, =, ".", (1 << 4)  ; Turn ON the
                                    LED
  goto   Loop               ; Repeat

  end
```

After a few moments of study, you should be able to understand what is happening in the application code. However, if the application code did not execute as expected when it is programmed into a PIC MCU, I would think it would be a major effort to find the problem in the program logic or the macro code.

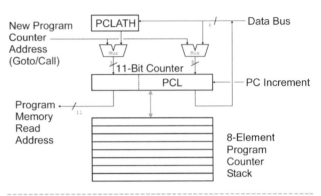

Figure 9-3 *Full program counter*

Read-only arrays are useful and effective solutions to a number of different programming problems. Typical applications range from storing ASCIIZ strings employed as user messages in a format that is reasonably space efficient, to providing a translation table for different applications, and using a jump table to different locations in an application. In a von Neumann processor architecture, a read-only array is normally implemented in program memory as a series of bytes that can be accessed by indirect addressing. The Harvard architecture of the PIC MCU does not lend itself to this simple solution, and more complex software is required to return read-only array data.

To access table data in the PIC MCU, you will have to perform what is known as a *computed goto* wherein the processor will calculate the address where the data is located and write this address into the program counter. If you remember that the PIC16F684 has 2,048 instructions (11 bits wide) and the data bus width is eight bits, you may be wondering how this is accomplished.

In the PIC16F684 (and all the other mid-range PIC microcontrollers), two registers can be written to that allow program access to the processor's program counter. The first register is called PCL and is actually the lower eight bits of the program counter (see Figure 9-3). The upper three bits in the PIC16F684 (and up to five bits in PIC MCUs with a full 8k of program memory) of the program counter are loaded from the five-bit PCLATH register, the contents of which are written into the program counter any time PCL is updated. This means that for computed gotos, the address space is effectively broken up into 256-instruction address blocks. Unfortunately Microchip has not give this block of instruction addresses a label, so you will see awkward terms like "256-instruction address block." To implement a computed goto, you must first load PCLATH with the appropriate 256-instruction address block value and then write the lower eight bits of the address to PCL.

Although it seems that I've wandered off topic a bit, I want to note that performing the computed gotos is half the problem of implementing a read-only array. The second half of the problem is to load the appropriate array value. Fortunately, the designers of the PIC microcontroller have provided you with the retlw instruction, which provides a simple method of returning a read-only array element in WREG. To take advantage of the retlw instruction, a subroutine with a computed goto is called. If you look at different PIC MCU applications, you may see a read-only array implemented in the same manner as the following subroutine, which returns the ASCII character for the nybble in WREG:

```
HexTable:            ; Convert Nybble in WREG to Hex
                       Char
    addwf    PCL, f
    retlw    '0'
    retlw    '1'
    retlw    '2'
    retlw    '3'
    retlw    '4'
    retlw    '5'
    retlw    '6'
    retlw    '7'
    retlw    '8'
    retlw    '9'
    retlw    'A'
    retlw    'B'
    retlw    'C'
    retlw    'D'
    retlw    'E'
    retlw    'F'
```

There are two issues with this function. The first is the use of 16 retlw instructions. To ease the workload of creating read-only arrays, Microchip added the dt directive to the language to pass its parameters to an appropriate number of retlw instructions. The dt directive's parameters can be ASCII strings set in double quotes, ASCII characters set in single quotes, or numeric values. Different data types can be concatenated, or linked, together using commas. The nybble to ASCII conversion subroutine could be rewritten as follows:

```
HexTable:       ; Convert Nybble in WREG to Hex
                  Char
  addwf    PCL, f
  dt       "012345678", 0x039, 'A', 'B', 'C',
'D', "EF"
```

The other problem with the read-only array is that it will work only if three conditions are met: (1) the table data must follow the "addwf PCL, f" instruction (which adds the contents of WREG to PCL and stores the result back in PCL); (2) the table must be totally within the first 256-instruction address block (within the address range of 0x0000 to 0x00FF); and (3) the PCLATH register (which is loaded when PCL has the contents of WREG added to it) must be zero and not changed anywhere else in the program. For many simple applications written by beginners, these conditions are not unreasonable.

For more complex applications, these conditions can be very difficult to meet, especially if multiple read-only arrays are built into the application. For these cases, a more general read-only array subroutine is required. In this experiment's application, asmTable.asm, I have put in what I feel is the best read-only array subroutine. This subroutine saves the index value in PCLATH, loads PCLATH with the 256-instruction address block of the start of the table, and swaps this value (in WREG) with the index value in PCLATH. I must point out that PCLATH is only five bits, so the maximum number of table entries that can use this code is 32. If you want more, you will have to use code like that presented in the following scheme. The index value is then added to the lower eight bits of the address of the table start (and PCLATH is incremented if this value is greater than 0x0FF) and finally loaded into PCL.

```
 title   "asmTable - A Read Only Array with a
Message"
;
;   This application copies the ASCIIZ string
;     from a read only table and puts it into
;     a variable array.
;
;
;   Hardware Notes:
;     PIC16F684 running at 4 MHz in Simulator
;     Reset is tied directly to Vcc via
;     Pullup/Programming Hardware
;
;
;   Myke Predko
;   04.09.17
;
  LIST R=DEC
```

```
 INCLUDE "p16f684.inc"

;   Variables
 CBLOCK 0x20
String:18
i
 ENDC

  org      0

  nop

  movlw    String            ; Point to the
                               Destination Array
  movwf    FSR
  clrf     i
Loop:                         ; Loop For Each
                               Character to Null
  movf     i, w              ; Get Next Array
                               Element
  call     ReadTable
  movwf    INDF              ; Store Character in
                               Variable Array
  incf     i, f              ; Point to Next Read
                               Array Element
  incf     FSR, f            ; Point to Next
                               Variable Array
                               Element

  iorlw    0                 ; Character == 0?
  btfss    STATUS, Z
  goto     Loop

  goto     $                 ; Finished, Loop
                               Forever

;   Subroutines
ReadTable:
  movwf    PCLATH            ; Save Offset
  movlw    high _ReadTable   ; Calculate PCLATH
                               Value
  xorwf    PCLATH, f         ; Swap Contents of
                               PCLATH & WREG
                             ; PCLATH = PCLATH ^
                               WREG
  xorwf    PCLATH, w         ; WREG = WREG ^
                               (PCLATH ^ WREG)
                             ; WREG = PCLATH
  xorwf    PCLATH, f         ; PCLATH = PCLATH ^
                               (PCLATH ^ WREG)
                             ; PCLATH = WREG
  addlw    low _ReadTable    ; Calculate Table
                               Offset
  btfsc    STATUS, C         ;  > 256?
   incf    PCLATH, f         ; Yes, Point to Next
                               256 Address Block
  movwf    PCL               ; Jump to Table
                               Element
_ReadTable:                   ; Read Only Table
                               Sentence
  dt       "Myke "           ; Subject
  dt       'a' + 'E' - 'A'   ; Cryptic Verb
  dt       'A' - 'W' + 'w'
  dt       29 * 4
  dt       'w' - 4
  dt       16 * 2
  retlw 0x77                 ; Noun
  retlw 0x6F
  retlw 0x72
  retlw 0x6D
  retlw 0x73
  retlw 0                    ; Zero At End of
                               String

  end
```

When you read through and test asmTable, you will see that it loads a file register array with the contents of the read-only array. And, if you display the file registers, you will see a message that has been (simplistically) encrypted in the read-only array.

As an exercise, I would suggest that you try to come up with your own subroutine code that implements the general read-only array (i.e., it can be anywhere in the PIC16F684's memory; it is greater than 32 bytes, and its contents can straddle a 256-instruction address block). When you try to come up with your own routine, remember that you have a number of directives, such as the if directive that can be used as I have in the following routine:

```
HexTable:                       ; Return Hex ASCII Char
                                   for LSNybble
   clrf    PCLATH               ;  in WREG
if ((_HexTable & 0x0100) != 0)
   bsf     PCLATH, 0
endif
if ((_HexTable & 0x0200) != 0)
   bsf     PCLATH, 1
endif
if ((_HexTable & 0x0400) != 0)
   bsf     PCLATH, 2
endif
   andlw   0x00F                ; Just want lower 4 Bits
   addlw   LOW _HexTable        ; Calculate the Correct
                                   Offset
   btfsc   STATUS, C
   incf    PCLATH, f
   movwf   PCL
_HexTable:
   dt      "0123456789ABCDEF"
```

Experiment 81—Data Stacks

One of the first applications I designed for microcontrollers was a Multi-PC printer switch and data spooler. This device would monitor the printer ports of multiple PCs, and when one started sending print commands to a printer, the device would pass them along to the printer. And if there were other PCs sending data to the printer later, their data would be held in a memory buffer until the first PC had finished sending data to the printer. When the first PC had finished (I waited 15 seconds without a new character coming in), the circuit would then assign one and then pass the data of the other PC's data to the printer. Along with saving data from other PCs, the spooling function allowed the PCs to pass data to the device, which "appeared" to be a printer, at full speed. Real direct-connect printers hold the PC from sending more data until it has finished the operation it has been given. These functions allowed all the PCs connected to the printer to work at full speed and have a minimal program delay when sending data to the printer. Today, modern multitasking operating systems such as Windows and Linux provide separate processes for applications and printer control, and network-enabled printers allow data to be sent directly without the need of the printer switch and data spooler. But the data storage methods used in this application are important to know, as they will be required from time to time in your applications.

There are two basic methods of storing data temporarily. The first is known as the *stack* and can be modeled like the stack of papers in an executive's in-basket. When data is stored on the stack, other pieces can be stored on top of it (see Figure 9-4) with the interesting result that the first piece of data on the stack will always be the last piece removed from the stack. For this reason, stacks are known as *last in first out* (LIFO) memory.

To demonstrate the operation of the stack, I created asmStack.asm, in which I have set up a 16-byte data array in bank 1 for the storage of data. To store and retrieve data on the stack, I created the Push subroutine:

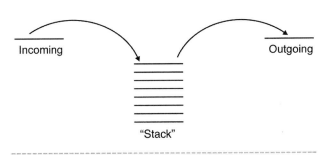

Figure 9-4   *Stack*

```
int Push(Data)         // Put Data in Array "Stack"
{                      // At Element "StackPtr"

  if (StackPtr == StackTop)
    return OverFlowError;
// The Stack Pointer at end of stack
  else                          // Still Space Left
  {
    Stack[StackPtr] = Data;  // Save Data
    StackPtr = StackPtr + 1; // Point to Next
                                     Element
    return PushGood;
    }  // fi
}  // end Push
```

And the Pop subroutine, which operates as follows:

```
int Pop()          // Return Data at the Top of
                              the Stack
{

  if ((StackPtr - 1) = StackBottom)
    return UnderFlowError;
// Nothing in Stack
  else              // "Pop" data off Stack
    {
      StackPtr = StackPtr - 1;  // Move to Last
                                    Stack Element
    return Stack[StackPtr];
    }
}  // end Pop
```

I feel that limit checking is important, even if it is ignored by the calling code; it ensures that no memory outside the allocated stack array area is accessed. To simplify the limit testing in asmStack.asm, I took advantage of the bit arrangement of a 16-bit buffer with bit 4 of the address always being zero. If it is ever a one, then I know I have gone outside the stack array and can stop trying to access invalid memory locations. It also means that data can be lost, as you will observe when you simulate asmStack.asm.

```
  title  "asmStack - Demonstrate Simple Stack"
;
;  This program demonstrates how a 16-Element
;   Stack could be
;   Implemented in the PIC MCU.
;
;  Hardware Notes:
;   PIC16F684 running at 4 MHz in Simulator
;
;
;  Myke Predko
;  04.12.29
;
  LIST R=DEC
  INCLUDE "p16f684.inc"

  CBLOCK 0x20     ; Variable Declaration
i, j
_Temp
  ENDC

  PAGE

  org     0
```

```
  nop              ; Required for MPLAB ICD2

  movlw   0xA0     ; Start Stack at 0x20 in Bank 1
  movwf   FSR

  movlw   20
  movwf   i
  clrf    j

PushLoop:                   ; Push Data Onto Stack
  movf    j, w
  call    TableRead      ; Get Random Value
  call    Push
  incf    j, f           ; Goto Next Value
  decfsz  i, f
  goto    PushLoop

  movlw   20              ; See What's on Stack
  movwf   i
PopLoop:
  call    Pop            ; Pop Top Stack Element
  decfsz  i, f
  goto    PopLoop

  goto    $               ; Finished, Loop Forever

;  Subroutines
Push:                       ; Push Value onto Stack
  bsf     STATUS, C
  btfsc   FSR, 4          ; Already have 16
                             Elements?
  return                   ; 16 Elements, Return
                             Error
  bcf     STATUS, C
  movwf   INDF
  incf    FSR, f
  return

Pop:                        ; Pop Top of Stack
  movlw   0xA0
  xorwf   FSR, w          ; Nothing There?
  bsf     STATUS, C
  btfsc   STATUS, Z
  return                   ; Yes, Return Error
  decf    FSR, f          ; Else, Return top Byte
  movf    INDF, w
  bcf     STATUS, C
  return

TableRead:                  ; Random Table
  movwf   _Temp
  movlw   HIGH _Table
  movwf   PCLATH
  movf    _Temp, w
  addlw   LOW _Table
  btfsc   STATUS, C
  incf    PCLATH, f
  movwf   PCL
_Table:                     ; PORTA and TRISA Bit
                              Table Values
  dt      47, 23, 33, 2, 190, 44, 31, 83, 42, 21,
98, 74, 79
  dt      8, 68, 54, 29, 37, 91, 100

  end
```

When you run asmStack.asm and execute the PopLoop, you will see that the data is retrieved in opposite order in which it was input. This makes a stack a poor choice for spooling data in the application mentioned at the start of this experiment. However, it makes a stack an excellent choice for saving data that may be temporarily overwritten and in need of retrieval once the operation that overwrote the data has finished executing. Stacks are used in the PIC MCU for saving the program counter when calling subroutines. The original program counter is saved before jumping to the subroutine. It is then retrieved when the subroutine has completed and the original program counter value is required.

Experiment 82—Circular Buffers

In this experiment, I will introduce you to the circular buffer, which retrieves data in the same order as it was written and is ideally suited for the printer switch and buffer application I mentioned in the previous experiment.

The circular buffer consists of a circular, singly linked list of data elements (see Figure 9-5). The buffer has two pointers, one pointing to the next location where data will be stored and one pointing to the next location to be read from. If both the next store pointer and next read pointer are pointing to the same element, the buffer is empty. Conversely, if the next store pointer is pointing to the element just prior to the next read element, the buffer is full. Just a few instructions are required to implement these basic rules.

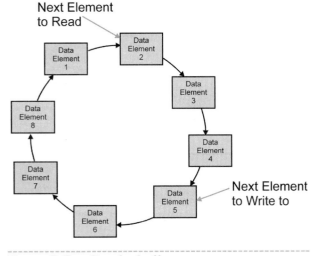

Figure 9-5   *Circular buffer*

Although I've drawn the circular buffer as a linked list, it can be easily implemented as a simple array. For example, the Put function can be modeled as:

```
int Put(Data)           // Put Data Into the
                               Circular Buffer
{

  if (((PutElement + 1) % BufferSize) ==
GetElement)
    return BufferFull;  // PutNext + 1 ==
                            ReadNext (Full)
  else                  // Can Store the Data Element
    {
    Buffer[PutElement] = Data;
    PutElement = (PutElement + 1) % BufferSize;
    return PutGood;
    }  // fi
}   // End Put
```

And the Get function is:

```
int Get()               // Get the Next Value in
                            the Buffer
{
int Temp;

  if (GetElement == PutElement)  // Nothing to
                                     Get
    return BufferEmpty;
  else                  // Element to Retrieve
    {
    Temp = Buffer[GetElement];  // Get the
                                    Element
    GetElement = (GetElement + 1) % BufferSize;
    return Temp; // Return the Element
    }  // fi
}   // End Get
```

To demonstrate the circular buffer function, I created asmCBuffer.asm, which uses the same 16-byte data area that was used by asmStack.asm (0xA0 to 0xAF). This allowed me to use the same trick to make sure that data outside the allocated memory is not allocated (if bit 4 is set, the pointer is outside of the circular buffer).

```
  title   "asmCBuffer - Demonstrate Circular
Buffer"
;
;   This program demonstrates how a 16-Element
    Circular
;   Buffer could be Implemented in the PIC MCU.
;
;   Hardware Notes:
```

```
;    PIC16F684 running at 4 MHz in Simulator
;
;
;  Myke Predko
;  04.12.29
;
  LIST R=DEC
  INCLUDE "p16f684.inc"

  CBLOCK 0x20              ; Variable Declaration
i, j
_Temp
_CBufferPut, _CBufferGet
  ENDC

  PAGE

  org      0
  nop                      ; Required for MPLAB
                             ICD2

  movlw    0xA0            ; Start Buffer at 0x20
                             in Bank 1
  movwf    _CBufferPut
  movwf    _CBufferGet

  movlw    20
  movwf    i
  clrf     j

PutLoop:                   ; Put Data Into Buffer
  movf     j, w
  call     TableRead       ; Get Random Value
  call     Put
  incf     j, f            ; Goto Next Value
  decfsz   i, f
   goto    PutLoop

  movlw    20              ; See What's on Stack
  movwf    i
GetLoop:
  call     Get             ; Get Next Buffer
                             Element
  decfsz   i, f
   goto    GetLoop

  goto     $               ; Finished, Loop Forever

;  Subroutines
Put:                       ; Put Value into
                             Circular Buffer
  movwf    _Temp
  movf     _CBufferPut, w ; Can't have Put == Get
  movwf    FSR            ; If Unread Data in
                             Buffer
  addlw    1
  andlw    b'11101111'    ; WREG has Final Get
                             Value
  xorwf    _CBufferGet, w
  bsf      STATUS, C
  btfsc    STATUS, Z
   return                  ; No More Space
  movf     _Temp, w        ; Store Value
  movwf    INDF
  incf     _CBufferPut, f ; Increment Pointer
  bcf      _CBufferPut, 4 ; Keep Within Range
  bcf      STATUS, C       ; No Error
  return

Get:                       ; Get Next Value in C.
                             Buffer
  movf     _CBufferGet, w ; IF Put == Get, then
                             Nothing
  movwf    FSR
  xorwf    _CBufferPut, w
  bsf      STATUS, C
  btfsc    STATUS, Z
   return                  ; Nothing to Get
  movf     INDF, w         ; Get the Value
  incf     _CBufferGet, f
  bcf      _CBufferGet, 4
  bcf      STATUS, C
  return

TableRead:                 ; Random Table
  movwf    _Temp
  movlw    HIGH _Table
  movwf    PCLATH
  movf     _Temp, w
  addlw    LOW _Table
  btfsc    STATUS, C
   incf    PCLATH, f
  movwf    PCL
_Table:                    ; PORTA and TRISA Bit
                             Table Values
  dt       47, 23, 33, 2, 190, 44, 31, 83, 42, 21,
98, 74, 79
  dt       8, 68, 54, 29, 37, 91, 100

  end
```

When you have simulated asmStack.asm, you should have noticed that data is retrieved in exactly the same order it was placed in the stack. You should also have noticed that the maximum storage of data in the circular buffer is one byte less than its total size. This can be a concern to new programmers who expect the data structures to store everything available to the device, but in practical terms, it really isn't a problem.

The important consideration when using a circular buffer is that the average rate of data removal must be greater than the average rate of data insertion. For the printer switch and spooler circuit, I found I had to increase the size of the buffer to 256 kilobytes of DRAM before I could reasonably expect to avoid a condition where the incoming data would fill the entire circular buffer. At 256 KB, the printer switch buffer could store about 100 pages of text, which was usually more than three people working on PCs could generate before the printer finished printing. In most cases, it is not practical to keep increasing the size of a circular buffer to meet the expanding needs of an application; you will probably have to put in some kind of "handshaking" to tell the sending device the circular buffer is full and enable data sending only when space is available for more data.

Experiment 83—Reading and Writing the EEPROM Data Memory

PC/ Simulator

PICkit™ 1

starter kit

Earlier in the book, I introduced you to the PIC MCU's built-in EEPROM data memory, or in other words a number of bytes built from electrically erasable programmable read-only memory and can be used for data storage between power on or reset cycles of the PIC MCU. Accessing the EEPROM in assembler is a bit trickier than in C where HT-Soft has provided you with macros that will read and write the EEPROM for you. In this experiment, I will present you with the information and code you need to access the EEPROM as well as demonstrate its operation using the same application, but I'll do so in assembler, as I used earlier.

The EEPROM control hardware is accessed by four registers located in bank 1 of the PIC16F684. In other PIC MCUs, you will find these registers located in multiple banks, which makes the coding of the accesses more difficult. Two of the registers are used for passing a data byte at a specific EEPROM address and are given the labels EEDAT and EEADR, respectively. There are two control registers: EECON1 (described in Table 9-1) and EECON2, which is a key sequence register used to ensure that data is written correctly. These four registers are used to read and write the EEPROM data memory.

To perform an *EEPROM read*, the following instruction sequence is used:

Table 9-1
EECON1 Register Bits

Bits	Label	Function
7:4		Unused (Return 0 when Read)
3	WRERR	Set when write operation did not complete correctly
2	WREN	Write enable bit, set before start of write operation
1	WR	Write start bit, set to initiate write operation
0	RD	Read start bit

```
movf     Address, w         ; Get the Address
                                 to be Read
bsf      STATUS, RP0
movwf    EEADR ^ 0x80       ; Set the Address
                                 to Read
bsf      EECON1 ^ 0x80, RD
movf     EEDAT ^ 0x80, w    ; Load in Byte at
                                 "Address"
bcf      STATUS, RP0
```

Performing an *EEPROM write* is a bit more involved as it requires the two-byte write to EECON2 to allow the write to take place. In the sequence below, notice that I store the address in a temporary register at an address that is shadowed between the two banks before changing execution to bank 1.

```
movf     Address, w         ; Get the Address
                                 to be Read
movwf    0x70               ; Store in
                                 shadowed
                                 register
movf     Data, w
bsf      STATUS, RP0
movwf    EEDAT ^ 0x80
movwf    0x070, w           ; 0x070 is same
                                 register in Bank
                                 0/1
movwf    EEADR ^ 0x80       ; Set the Address
                                 to Write
bsf      EECON1 ^ 0x80, WREN ; Start Write
                                 Operation
movf     EEDAT ^ 0x80, w    ; Load in Byte at
                                 "Address"
bcf      STATUS, RP0
nop
movlw    0x55               ; Start of
                                 Required
                                 Sequence
movwf    EECON2 ^ 0x80
movlw    0xAA
movwf    EECON2 ^ 0x80
bsf      EECON1 ^ 0x80, WR  ; End of Required
                                 Sequence
btfsc    EECON1 ^ 0x80, WR  ; Wait for Write
                                 to Complete
goto     $ - 1
bcf      STATUS, RP0
```

This write sequence is different from what you would see in a datasheet. I wrote it as a general case, and it will not return until the write has completed. In most applications (like the code used for this experiment), I will not poll the WR bit for completion at the end of the write; instead I will check it before starting another EEPROM access, as shown in asmEEPROM.asm:

```
        title   "asmEEPROM - Display/Save a Value using
the EEPROM"
;
;   This program Displays a Saved EEPROM
;   value Unless the Button on
;   RA3 is pressed and then the ADC is Displayed.
;   When the Button is
;   Released, the ADC Value is Saved in EEPROM.
;    This code is based on
;    "cADC.c".
;
;   Hardware Notes:
;    PIC16F684 running in PICkit 1 starter kit
;
;
;   Myke Predko
;   04.12.28
;
  LIST R=DEC
  INCLUDE "p16f684.inc"

;  Registers

  __CONFIG _FCMEN_OFF & _IESO_OFF & _BOD_OFF &
_CPD_OFF & _CP_OFF & _MCLRE_OFF & _PWRTE_ON &
_WDT_OFF & _INTOSCIO

;  Variables
CBLOCK 0x20
i, j
Display, OldDisplay
ADCState, ButtonState
Dlay, Temp
  ENDC

  PAGE
;  Mainline

  org     0
  clrf    PORTA
  movlw   7                   ; Turn Off Comparators
  movwf   CMCON0
  movlw   b'00000001'         ; Turn on the ADC with:
                              ; Left Justified
                              ; Vdd instead of Vref
                              ;   Channel 0
  movwf   ADCON0
  bsf     STATUS, RP0         ; Do Bank 1 Inits
  movlw   1
  movwf   ANSEL ^ 0x80        ; Just RA0 as Analog
                                 Input
  movlw   b'00010000'         ; Select the Clock as
                              ; Fosc/8
  movwf   ADCON1 ^ 0x80
  bcf     STATUS, RP0
  clrf    ADCState            ; Init State Variables
  clrf    ButtonState

  clrf    FSR                 ; EEPROM(0) == 0?
  call    EEPROMRead
  xorlw   0
  btfss   STATUS, Z
   goto   InitEEPROM
  incf    FSR, f
  call    EEPROMRead          ; EEPROM(1) == 0xFF?
  xorlw   0xFF
  btfss   STATUS, Z
   goto   InitEEPROM
  incf    FSR, f
  call    EEPROMRead          ; EEPROM(2) == 0x55?
  xorlw   0x55
  btfss   STATUS, Z
   goto   InitEEPROM
```

```
  incf    FSR, f
  call    EEPROMRead          ; EEPROM(3) == 0xAA?
  xorlw   0xAA
  btfsc   STATUS, Z
  goto    HaveEEPROM          ; Yes, Pattern Good
InitEEPROM:                   ; Initialize the EEPROM
  clrf    FSR
  movlw   0
  call    EEPROMWrite
  incf    FSR, f
  movlw   0xFF
  call    EEPROMWrite
  incf    FSR, f
  movlw   0x55
  call    EEPROMWrite
  incf    FSR, f
  movlw   0xAA
  call    EEPROMWrite
  incf    FSR, f
  movlw   0
  call    EEPROMWrite         ; Start with Nothing

HaveEEPROM:
  movlw   4                   ; Initial Display Value
                              = EEPROM(4)
  movwf   FSR
  call    EEPROMRead
  movwf   Display

Loop:                         ; Display Loop
  movlw   7                   ; Loop through Each of
                              the 8 LEDs
  movwf   i
  movf    Display, w
  movwf   Temp
DisplayLoop:
  movf    i, w                ; Get the PORTA Value
  movwf   FSR
  call    PORTATableRead
  rlf     Temp, f             ; Shift up Temp to Test
  btfss   STATUS, C
   movlw  0                   ; Bit Reset, No Write
  movwf   PORTA
  movlw   8                   ; Get TRIS Pattern
  addwf   i, w
  call    PORTATableRead
  bsf     STATUS, RP0
  movwf   TRISA ^ 0x80
  bcf     STATUS, RP0
  movlw   HIGH ((2000 / 5) + 256)
  movwf   Dlay
  movlw   LOW ((2000 / 5) + 256)
  addlw   -1                  ; Display LED for 2 ms
  btfsc   STATUS, Z
   decfsz Dlay, f
   goto   $ - 3
  clrf    PORTA               ; Turn Off LED
  movlw   1                   ; Take 1 Away from the
                              Display
  subwf   i, f
  btfsc   STATUS, C
   goto   DisplayLoop         ; Loop Until i == 0xFF

;  Handle Button Press or Ignore
  btfss   PORTA, 3            ; RA3 Pressed?
   goto   ButtonPress

  clrf    ADCState            ; Button Released
  btfss   ButtonState, 0      ; Have New ADC Value?
   goto   Loop                ; No, Just Return
  clrf    ButtonState
  movlw   4                   ; Write New Value
  movwf   FSR
```

```
        movf    Display, w
        call    EEPROMWrite
        movf    OldDisplay, w  ; Restore Display Value
        movwf   Display
        goto    Loop

ButtonPress:
        movf    ADCState, w    ; What is the ADC
                                 State?

        btfss   STATUS, Z
         goto   BPADState1
        bsf     ADCON0, GO     ; Start ADC Operation
        incf    ADCState, f    ; Increment to the Next
                                 State

        goto    Loop
BPADState1:                     ; Read the ADC Value?
        xorlw   1
        btfss   STATUS, Z
         goto   BPADState2
        btfss   ADCON0, GO     ; Wait for ADC Read to
                                 Finish
         incf   ADCState, f
        goto    Loop
BPADState2:                     ; Save the Sample Value
        movf    Display, w     ; Save Display Value?
        btfss   ButtonState, 0
         movwf  OldDisplay
        movlw   1              ; Mark Display Value
                                 Saved

        movwf   ButtonState
        movf    ADRESH, w      ; Get ADC Value
        movwf   Display
        clrf    ADCState       ; Reset State Machine
        goto    Loop


;  Subroutines
PORTATableRead:                 ; Get Table Values for
                                  Light Display

        movwf   PCLATH
        movlw   HIGH _PORTATable
        xorwf   PCLATH, f
        xorwf   PCLATH, w
        xorwf   PCLATH, f
        addlw   LOW _PORTATable
        btfsc   STATUS, C
         incf   PCLATH, f
        movwf   PCL
_PORTATable:                    ; PORTA and TRISA Bit
                                   Table Values
        dt      b'010000', b'100000', b'010000',
b'000100'
        dt      b'100000', b'000100', b'000100',
b'000010'
        dt      b'001111', b'001111', b'101011',
b'101011'
        dt      b'011011', b'011011', b'111001',
b'111001'


EEPROMWrite:                    ; Write Contents of
                                  WREG at Address in
                                  FSR
        bsf     STATUS, RP0
        btfsc   EECON1 ^ 0x80, WR ; Wait for Previous
                                     Write to Complete
         goto   $ - 1
        movwf   EEDAT ^ 0x80     ; Store Word to Write
        movf    FSR, w
        movwf   EEADR ^ 0x80
        bsf     EECON1 ^ 0x80, WREN
        nop
        movlw   0x55             ; Start of Required
                                   Sequence
        movwf   EECON2 ^ 0x80
```

```
        movlw   0xAA
        movwf   EECON2 ^ 0x80
        bsf     EECON1 ^ 0x80, WR
        bcf     STATUS, RP0
        return

EEPROMRead:                     ; Read EEPROM at
                                  Address in FSR
        bsf     STATUS, RP0
        btfsc   EECON1 ^ 0x80, WR ; Wait for Previous
                                     Write to Complete
         goto   $ - 1
        movf    FSR, w
        movwf   EEADR ^ 0x80     ; Store Address
        bsf     EECON1 ^ 0x80, RD ; Read Byte at
                                     Address
        movf    EEDAT ^ 0x80, w
        bcf     STATUS, RP0
        return


        end
```

The asmEEPROM.asm code is a direct translation of cEEPROM.c and performs identically to it (the value saved in EEPROM is shown unless the PICkit 1 starter kit's button (on RA3) is pressed. When the PICkit 1 starter kit's button is pressed, the value of the potentiometer on RA0 is read by an ADC and then displayed on the LEDs. When the button is released, the original value is displayed and the last ADC read value is stored in EEPROM, to be displayed the next time the PIC MCU is reset or powered up.

For Consideration

Table 9-2 lists the different parameter options for the 37 instructions that the PIC16F684's (and all mid-range PIC MCU's) processor recognizes. The instructions are listed in Table 9-3. I recommend that you try to memorize these instructions, a brief explanation of what they do, and the registers they affect). You will be able to program more efficiently and be able to think about what instruction and the algorithm to be used to solve a programming problem.

The MPASM assembler directives are listed in Table 9-4. Many of the directives are listed for completeness and will not be required during normal programming; these cases are marked in the table. Also note that the list directive has a number of arguments; these are listed in Table 9-5.

In Table 9-4, the message "Use of this directive is not recommended due to unknown operation" indicates the directive is carrying out its function by inserting instructions. The instructions used are unknown, may affect the contents of other registers, and may change with different versions of MPLAB IDE/MPASM assembler.

Table 9-2

Data and Destination Symbols Used in the PIC
MCU Assembler Instructions Listed in Table 9-3

Symbol	Function
D	Destination instruction: can be either f for register or w for WREG
F	Register address: seven bits for direct addressing within a bank
k	Bit specification for instructions, only 0 to 7 valid
kk	Eight-bit constant value: literal instruction argument
kkk	Eleven-bit constant value: goto and call instruction address
p	PORT register: for PIC16F684, only 5 (PORTA) and 7 (PORTC) are valid.
Z	STATUS, Z: zero bit
C	STATUS, C: carry bit
DC	STATUS, DC: digit carry bit

Table 9-3

PIC16F684 (and Mid-Range PIC) Microcontroller Instruction Set

Instruction	Flags Affected	Instruction Operation
addlw kk	Z, C, DC	Add constant kk to contents of WREG and store result in WREG.
addwf f, d	Z, C, DC	Add contents of register f to contents of WREG and store sum either in register f or in WREG.
andlw kk	Z	AND constant kk with contents of WREG and store result in WREG.
andwf f, d	Z	AND contents of register f with contents of WREG and store sum either in register f or in WREG.
bcf f, k	None	Clear bit k in register f. Instruction operation is: f = f & (0x0FF ^ (1 << k)).
bsf f, k	None	Set bit k in register f. Instruction operation is f = f	(1 << k).
btfsc f, k	None	Skip following instruction if bit k in register f is reset (0).
btfss f, k	None	Skip following instruction if bit k in register f is set (1).
call kkk	None	Save address of next instruction and load PC with new address kkk.
clrf f	Z = 1	Clear (load with 0s) register f.
Clrw	Z = 1	Clear WREG.
clrwdt	None	Clear the PIC MCU's Watchdog Timer (WDT).
comf f, d	Z	The bits in register f are inverted and the result can be stored either in register f or WREG.
decf f, d	Z	Decrement the contents of register f by one and store the result in either register f or WREG.
decfsz f, d	None	Decrement the contents of register f by one and store the result in either register f or WREG. If the result is equal to zero, then skip the next instruction.
goto kkk	None	Load PC with new address kkk.
incf f, d	Z	Increment the contents of register f by one and store the result in either register f or WREG.
incfsz f, d	None	Increment the contents of register f by one and store the result in either register f or WREG. If the result is equal to zero, then skip the next instruction.
iorlw kk	Z	Inclusive OR constant kk with contents of WREG and store result in WREG.
iorwf f, d	Z	Inclusive OR contents of register f with contents of WREG and store sum either in register f or in WREG.
movf f, d	Z	Pass the contents of register f through the PIC MCU processor's Algorithm/Logic Unit's (ALU) zero check and save them either back into register f or in WREG.
movlw kk	None	Load WREG with constant kk.
movwf f	None	Store the contents of WREG in register f.
nop	None	No-operation for one instruction cycle.

Table 9-3 *(continued)*
PIC16F684 (and Mid-Range PIC) Microcontroller Instruction Set

Instruction	Flags Affected	Instruction Operation
option	None	Store the contents of WREG in OPTION_REG. **Note:** use of this instruction is not recommended due to device compatibility issues.
retfie	N/A	Restore processor interrupt request hardware to waiting state and return to instruction interrupted.
retlw kk	None	Load WREG with kk before loading PC with address saved by call instruction.
return	None	Load PC with address saved by call instruction.
rlf f, d	C	Rotate register f to the left with bit 0 being loaded with the current carry flag. Bit 7 of register f is stored in the carry flag. The result is stored in either in register f or in WREG.
rrf f, d	C	Rotate register f to the right with bit 7 being loaded with the current carry flag. Bit 0 of register f is stored in the carry flag. The result is stored either in register f or in WREG.
sleep	N/A	Put the PIC MCU into a low-power mode.
sublw kk	Z, C, DC	Subtract the contents of WREG from constant kk and store the difference back into WREG.
subwf f, d	Z, C, DC	Subtract the contents of WREG from the contents of register f and store the difference back into register f or WREG.
swapf f, d	None	Swap the least significant nibble with the most significant nibble of register f. The swapped value can be stored back in either register f or in WREG.
tris p	None	Store the contents in WREG in the specified PORT register. Note: Use of this instruction is not recommended due to device compatibility issues.
xorlw kk	Z	Exclusive OR constant kk with contents of WREG and store result in WREG.
xorwf f, d	Z	Exclusive OR contents of register f with contents of WREG and store sum either in register f or in WREG.

Table 9-4
MPASM Assembler Directives

Directive	Function
__badram *expr*[-*expr*][, *expr*[-*expr*]]	Specify registers that should not be accessed. When a direct access to these registers is made, MPASM assembler flags the instruction with a warning. This directive is normally used only in the Microchip written part number .inc file to define the PIC MCU part number register ranges.
__badrom *expr*[-*expr*][, *expr*[-*expr*]]	Specify program memory address that should not be accessed. When instructions are found in these regions, MPASM assembler produces a warning or error. This directive is normally used only in the Microchip written part number .inc file to define the PIC MCU part number address ranges.
__config expr	Specify PIC MCU configuration word value. Option specification equates are found in the Microchip written PIC MCU part number .inc file.
__config expr	Specify 16 bits of data to be stored in non-program-accessible memory. Generally used for serial numbers or program revision information.
__maxram *expr*	Directive used to specify maximum file register address. Like __badram, this directive is normally used only in the Microchip written part number .inc file to define the PIC MCU part number register ranges.
__maxrom *expr*	Directive used to specify maximum program memory address. Like __badrom, this directive is normally used only in the Microchip written part number .inc file to define the PIC MCU part number register ranges.
#define L*abel* [*string*]	Declare a text substitution string. Defines do NOT replace complete lines like macros.
#include *include_file*	Load the contents of the specified file into the assembly language source file.
#undefine *label*	Delete label previously defined. Previous (above) instances are not affected.
#v(expr)	Insert ASCII value for expr in the string during macro processing. Used to differentiate labels algorithmically.
bankisel *label*	This directive will set the IRP bit of the STATUS register according to the value of label. Use of this directive is not recommended due to unknown operation.
banksel *label*	This directive will set RP1/RP0 bits of the STATUS register according to the value of label. Use of this directive is not recommended due to unknown operation.

Table 9-4 *(continued)*

cblock [*expr*]	Start variable, structure, and equates at specified value. Labels between cblock and endc are given incrementing values. Larger value numbers are specified by a colon (:) after the label.
[*label*] code [*ROM_address*]	Used when object file is created for linked applications. Not required for single source file applications.
constant *label=expr* [. . . ,*label=expr*]	Specify a constant value for a label. See also equ.
[*label*] da *expr* [, *expr2*, . . . , *exprn*]	Created a packed 14-bit ASCII string (i.e., seven bits per character) in program memory. The PIC16F684 cannot access this data.
[*label*] data *expr*,[,*expr*, . . . ,*expr*]	Store data as full instructions in program memory. The PIC16F684 cannot access this data.
[*label*] db *expr*[,*expr*, . . . ,*expr*]	Reserve program memory with packed eight-bit data. The PIC16F684 cannot access this data.
[*label*] de *expr* [, *expr*, . . . , *expr*]	Define data to be burned in data EEPROM during PIC MCU programming.
[*label*] dt *expr* [, *expr*, . . . , *expr*]	Define table bytes (retlw expr).
[*label*] dw *expr*[,*expr*, . . . ,*expr*]	Reserve program memory words. The PIC16F684 cannot access this data.
else	Used with if directive to indicate end of primary options and start of secondary.
end	Indicate that the end of the source file has been reached.
endc	End cblock label assignments.
endif	Used to end if conditional assembly. May follow else.
endm	End macro definition.
endw	End while loop.
label equ expr	Assign the value of expr to the label. See set.
error String	Force an error with a message string.
errorlevel {0|1|2|+*msgnum*|-*msgnum*}	Inhibit printing or displaying specific error and warning message numbers.
exitm	Exit from macro without processing any following macro statements.
expand	Expand all macro invocations in the listing file. Used with list and noexpand.
extern Label [, Label . . .]	Declare that the specified label is located in a linked file.
[*label*] fill *instruct*|*expr, count*	Fill program memory, starting at the current address, with the specified instruction or value.
global *label* [, *label* . . .]	Define a label that can be accessed by other linked files.
[*label*] idata [*RAM_address*]	Declare the start of initialization code in a linked application.
if *expr*	Begin block of conditionally assembled code. Completed with either else . . . endif or just with endif. The if code and other conditional assembly code can be nested.
ifdef *label*	Begin block of conditionally assembled code if label is present. The ifndef is a complementary directive. Completed with either else . . . endif or just with endif. Can be nested with other conditional assembly code.
ifndef *label*	Begin block of conditionally assembled code if label is *not* present. The ifdef is a complementary directive. Completed with either else . . . endif or just with endif. Can be nested with other conditional assembly code.
list [*list_option*, . . . , *list_option*]	If just the list directive is on a line, listing is printed. Uses parameters listed in Table 9-5.
local *label* [,*label* . . .]	Define conditional assembly labels local to the current macro.
label macro [*arg*, . . . , *arg*]	Declare a macro with optional parameters.
messg "*message_text*"	Generate a message in the listing file.
noexpand	Turn off macro expansion in the listing file.
nolist	Turn off code listing.
[*label*] org *expr*	Specify where address code is to start.
page	Start a new page in the listing file.
pagesel *label*	Update PCLATH page bits for the specified label. Use of this directive is not recommended due to unknown operation.
processor *processor_type*	Specify processor type. Use of this directive is not recommended due to the processor type being specified by MPLAB IDE.
radix *default_radix*	Specify number radix type. Use of this directive is not recommended. Instead the r= option of the list directive should be used.

Table 9-4 *(continued)*
MPASM Assembler Directives

Directive	Function
[*label*] res *mem_units*	Reserve specified program memory for other linked files.
label set *expr*	Assign numeric value to label. Similar to equ directive, but label value can be changed with additional set directives.
space *expr*	Insert expr blank lines into the listing file.
subtitle "*sub_text*"	Specify second line of program title.
title "*title_text*"	Specify string for top line of each listing file page.
[*label*] udata [*RAM_address*]	Declares uninitialized data area in linked object file.
[*label*] udata_acs [*RAM_address*]	Declares uninitialized data area in linked object file for PIC18 series line PIC microcontrollers.
[*label*] udata_ovr [*RAM_address*]	Declares a section of overlayed uninitialized data in linked object file.
[*label*] udata_shr [*RAM_address*]	Declares a section of shared uninitialized data in linked object file.
Variable Label[=expr][, . . .]	Declare variable with optional initial value for assembly.
while *expr*	Loop code within while . . . endw until the expr is not true (or zero).

Table 9-5
Optional Parameters Available to the List Directive

Parameter	Function
b=#	Set tab spaces. Eight is the default.
c=#	Set column width. Default is 132.
f=hexfileFormat	Specify hex file output format. Default is INHX8M and can be INHX32 or INHX8S.
Free	Allow free-format parser. Default is "fixed."
mm=OFF	Turn off memory map display in listing file.
n=#	Specify the number of lines per page. Default is 60.
p=type	Specify processor type. Default is defined by MPLAB IDE processor option. Use of this list parameter is not recommended.
pe=type	Specify processor type and enable extended instruction set. Available *only* in PIC18 series PIC microcontrollers.
r=radix	Specify oct or dec number radix. Default is hex.
st=OFF	Turn off symbol table print in listing file.
t=ON	Truncate lines. Default is OFF and will cause lines to wrap to the next one.
w=0 1 2	Set the reporting message level. See errorlevel directive.
x=OFF	Turn macro expansion off. Default is ON.

Sensors

PC/ Simulator

PICkit™ 1

starter kit

Tool Box

1 DMM

1 Wire strippers

1 Needle-nose pliers

2 Breadboards

1 Wiring kit

Required Parts

1 PIC16F684

1 PIC12F675

2 TLC251

1 Two-line by 16-character LCD display

1 74LS174

1 Sharp GP2D120 IR ranging module

2 38 KHz IR TV remote control receiver

1 2N3904 NPN transistor

1 U-shaped IR opto-interrupter (see text)

1 10-LED bargraph display

3 High-intensity white LEDs

4 LEDs (any color)

1 IR LED

1 1N914 (1N4148) silicon diode

2 10k breadboard-mountable poten-tiometers

Either

1 10k 10-pin SIP

Or

8 10k resistors

2 3.3M resistors

2 2.3M resistors

1 4.7k resistor

1 1k resistor

4 10k resistors

4 470Ω resistors

2 330Ω resistors

2 220Ω resistors

2 100Ω resistors

1 1k breadboard-mountable poten-tiometer

2 10k light-dependant resistors (R decreases with light)

2 47 µF electrolytic capacitor

4 0.1 µF capacitors

4 0.01 µF capacitors

1 16-button keypad interface

1 electret microphone

1 SPST breadboard-mountable switch

4 breadboard-mountable momentary on pushbuttons

1 Four-cell AA battery clip

2 Three-cell AA battery clips

10 AA batteries

1 6V lantern battery

1 55mm black heat-shrink tubing, 1 to 1.25inches (2.5 to 3 cm) long

When I passed around a proposed table of contents for this book, this section caught many eyes, as an introduction to sensors was felt to be useful for robot developers. At first, I was a bit disappointed by this response because the topics I cover in this section are actually required in many different microcontroller applications; it provides you with many of the basics needed to change environmental parameters into values that can be processed by the PIC MCU. This information is useful for both microcontroller application and robot developers. The topic of proper sensor interfacing and data processing would take up a book at least equal in size to this one, but in this section, I will introduce you to some of the different sensors that are typically used with microcontrollers as well as provide you with a simple interface that will allow you to create efficient multi-MCU applications.

An important aspect of implementing sensor interfaces is characterizing the parameter being monitored and the appropriate time interval between sensing operations. Instead of using *sensing operations*, I will use the more commonly used term *polling*. The polling interval can vary widely and according to the characteristics of what is being polled. In Table 10-1, I have listed some different things that are typically sensed in microcontroller applications along with some ideas about the appropriate polling interval.

The applications presented in this book are actually very simple sensor applications. There is very little processing done on the data other than magnitude comparison. And, in many commercial applications, the rate of change and projected final values are calcu-

lated. Beyond these calculations, when you are trying to identify an object or a total environment, the processing requirements can be formidable.

To simplify the experiments presented in this section as well as eliminate any integration issues you may have using the sensors with some kind of user interface, I am going to break from tradition and use a simple interface as a base for complex, multiprocessor applications. When most people think of a microcontroller interface, they think of an asynchronous serial interface using the *non-return-to-zero* (NRZ) protocol of RS-232 (the serial interface used in PC Modems). The problem with this protocol is the need for dedicated hardware to wait for and decode the incoming signal. The PIC16F684 and many lower-end PIC® MCUs do not have the built-in serial interface (called a UART), so different communications methods are necessary.

The method I choose for this section is a synchronous serial interface that I originally came up with for the *TAB Electronics Build Your Own Robot Kit*. This interface takes advantage of the Parallax *BASIC Stamp 2* (or BS2) built-in SHIFTIN and SHIFTOUT synchronous serial data transfers. The PIC MCU

Table 10-1
Environmental Parameters with Appropriate Sensors and Polling Intervals

Parameter	Sensor	Polling Interval	Comments
Ambient Light Level	Light Dependent Resistor (LDR/CDS Cell)	200 to 500 ms	LDR response is generally 100 ms.
Sector Light Level	Light Dependent Resistor (LDR/CDS cell)	200 to 500 ms	Multiple LDRs or a single LDR on moving turret used to characterize environment.
Object Location	Digitized Video	15 to 30 ms	Object identification; location determination can change with sensor movement.
Object Distance	Ultrasonic Ranging	100 to 300 ms	Sensor movement requires continual polling.
Object Collision	IR Reflection	50 to 100 ms	Usually used in safety applications (robot or operator); continuous polling required.
Object Collision	"Whisker" or Mechanical Sensors	20 to 100 ms	Usually used in safety applications (robot or operator), continuous polling required.
Sound Event	Filtered Sound Input	1 to 10 ms	Short, loud events can be easily missed.
Sound Characteristic	Filtered Sound Input	50 to 200 ms	Waiting for tone or other easily recognized characteristic that is present for up to several seconds.
Verbal Command	Digital Signal Processor	500 to 2,000 ms	Speech processing requires a significant amount of time to complete.
Ambient Temperature	Thermistor-Calibrated Temperature Sensor	1 to 10 s	The assumption is that the ambient temperature cannot change quickly.
Fire, High-Heat Sources	Digitized Video, Pyrometers, LDRs	50 to 250 ms	Typically a safety sensor with possible requirement for source location, general requirement to respond quickly.
Humans and Animals	Pyrometers or IR Reflection	250 to 1,000 ms	Typical applications include burglar alarms and automatic lights.
(Air) Pressure	Barometers or Pressure Sensors	5 to 60 s	Pressure changes generally take place very slowly except in safety applications.
Humidity	Humidity Sensors	15 to 60 s	Slow changes in environmental humidity.

peripheral clocks data in and out using the clock signal produced by the BS2. The PICSend (sending a seven-bit command to a PIC MCU from a BS2) and the PIC-SendReceive (sending a seven-bit command to a PIC MCU from a BS2 and the PIC MCU responding with an eight-bit response) are demonstrated here in BS2 IF Base.bs2:

```
' BS2 IF Base
'
' Basic BS2 to PIC MCU Interface Template
'
' "PICSend" and "PICSendReceive" are only
' supported Operations
'
'
' Myke Predko
'
' 04.11.26
'
'
'{$STAMP BS2}
'{$PBASIC 2.5}


' Variables


' Mainline
  HIGH PC           ' High PIC MCU I/O Pins
  HIGH PD
  PAUSE 100         ' Wait for PIC MCU to Set Up

' #### - Put in program Statements as Shown
'        Below

  PICData = 1       ' Send Basic Command to PIC
                      MCU
  GOSUB PICSend

  PAUSE 100         ' Wait for PIC to Finish
                      Previous
  PICData = $81     ' Send 0x81 and Receive
                      Response
  GOSUB PICSendReceive
  DEBUG "PIC Response", DEC(PICData), CR

  END
```

```
' PIC MCU Interface Code Follows
'
' Myke Predko
'
' PICSend           Commands are 1 - 8
' PICSendReceive Commands $81 - $88
' PIC MCU Interface Pins
PICData VAR Byte    ' Data Byte to Send to/Receive
PC   PIN 0          ' IO Pins
PD   PIN 1
PICSend:            ' Send the Byte in "PICData"
  LOW PC            ' Hold Low for 1 msec before
  PAUSE 1           ' Shifting in DATA
  SHIFTOUT PD, PC, LSBFIRST, [PICData]
  HIGH PC
  RETURN

PICSendReceive:     ' Send the Byte in "PICData"
  LOW PC            ' Hold Low for 1 msec before
  PAUSE 1           '  Shifting in Data
  SHIFTOUT PD, PC, LSBFIRST, [PICData]
  PAUSE 1           ' Wait for Operation to
                    ' Complete
  SHIFTIN PD, PC, LSBPOST, [PICData]
  HIGH PC
  RETURN
```

A PIC MCU application that interfaces to this code can either wait for a command from the BS2 or continually poll and process the sensor, polling the clock pin (PC in BS2 IF Base.bs2) at least once per ms in order to not miss the beginning of any commands. If the same requirement was given for an application that interfaced to another device using the NRZ protocol, the maximum data rate would be approximately 300 bps, due to the need to also poll the serial input line and still be able to perform some useful applications. In contrast, the BS2 takes about 2 milliseconds to send a byte, one-fifteenth of the time required for the 300 bps NRZ send time.

I realize that few people have BS2s available for working through the projects in this book, so in the first experiment, I will show how you can simulate the BS2 using a PIC MCU. With the addition of a simple LCD user interface, you can experiment with different sensors.

Experiment 84—PIC MCU BS2 User Interface

PC/Simulator

PICkit™ 1

starter kit

Parts Bin

1 PIC12F675

1 Two-line by 16-character LCD display

1 74LS174

1 1N914 (1N4148) silicon diode

2 10k breadboard-mountable potentiometers

1 10k resistor

2 0.01 µF capacitors

1 Breadboard-mountable momentary on push-buttons

1 Three-cell AA battery clip

3 AA batteries

Tool Box

DMM

Needle-nose pliers

Breadboard

Wiring kit

To create the PIC MCU-based, simulated, BS2 user interface, I used the two-wire LCD display first presented in Section 6. The application shown in this section simulates the waveform clock low/pause 1ms/shiftout/clock high of the PICSend BS2 subroutine and the waveform of PICSendReceive. In Figures 10-1 and 10-2, I have correlated the different features of these two waveforms to the BS2 statements that produced them (listed earlier in this section). The data transfer is based on a 14 µs clock pulse over a 46 µs period for each bit (see Figure 10-3). The work to create the application code was fairly substantial, and the need for precise timing gave me a good feeling for

writing assembly language code for the PICC Lite™ compiler. The final result is quite easy to work with, although I discovered a potential pitfall that you should be aware of.

The circuit used for the simulated BS2 interface (see Figure 10-4) uses all the available pins of a PIC12F675. In addition to the two LCD interface pins, the circuit assigns two pins to a potentiometer/button user interface and two pins to the simulated BS2 clock/data lines. The circuit was built on a breadboard (see Figure 10-5) and is about as complex as I would like a PIC MCU breadboard application to be. Included in Figure 10-5 is the application circuit from

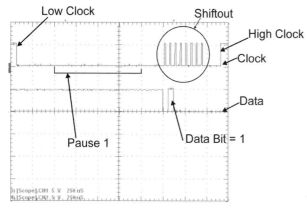

Figure 10-1 *BS2Send*

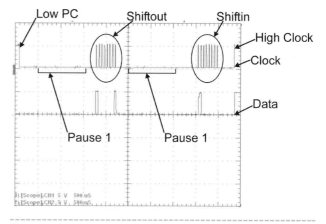

Figure 10-2 *BS2SendReceive*

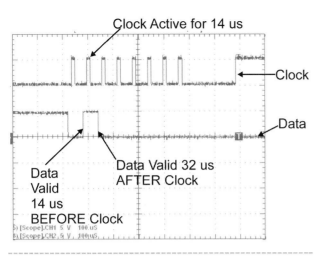

Figure 10-3 *BS2 Shiftout data*

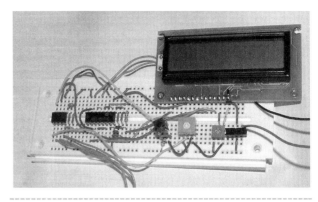

Figure 10-5 *BS2 simulator circuit on a breadboard and connected to test sensor application*

the next experiment, to test out the operation of the interface and the PIC MCU code.

The user interface consists of a potentiometer that selects which command is to be sent to the PIC MCU instrument circuit. The user interface is capable of sending the commands 0x01 through 0x08 as well as the response commands 0x81 to 0x88 by turning the potentiometer. To avoid the need of precisely setting the potentiometer, the first half turn selects 0x01 to 0x08 and the second half turn selects 0x81 to 0x88. Once the command is selected, the button is pressed to send the data. If a response command is sent, the response is displayed on the lower line of the LCD. This method of specifying and sending the command and then displaying the response to the instrument PIC MCU is surprisingly easy and fast to work with. If you are looking for a simple-to-use interface, you might want to consider using the method I provided here.

The application code cBS2 Sim.c provides the LCD, potentiometer, and button user interface as well as the two-wire BS2 communications simulation code. The

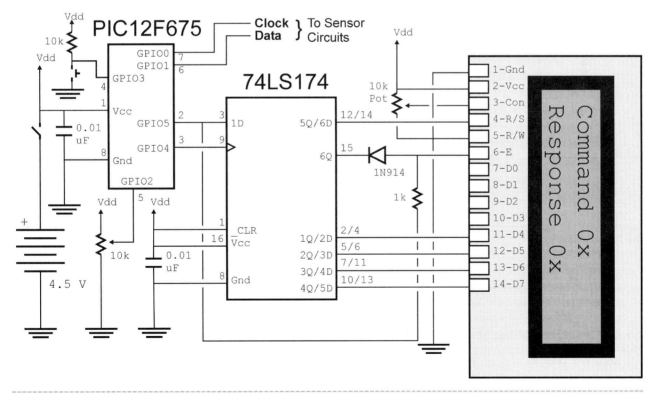

Figure 10-4 *BS2 simulator circuit*

basic code was written in C to take advantage of the code already written for the BS2 two-wire interface and the ADC (for the potentiometer), although the two BS2 simulated subroutines are written in assembler to ensure the timings are as close to the BS2's as possible. As a reference for the simulated BS2, an oscilloscope was used with the Parallax BS2 datasheets to make sure the signals were as precise as possible.

```c
#include <pic.h>
/* cBS2 Sim.c - Simulated BS2 Sensor Interface

This Program Initializes the Hitachi 44780
  Based LCD in 4 Bit Mode and then continually
  polls the Potentiometer on GPIO2 as well as the
  Button on GPIO3.  When the Button is pressed, the
  ADC Value (processed as below) is Sent to another
  PIC MCU as if it were a
  BS2 Command.

The ADC Value is processed as:

ADCValue = (ADRES >> 6) & 0x0F;
ADCValue = ((ADCValue & 0x08) << 4) = (ADCValue &
0x07);
ADCValue = ADCValue + 1;

This will manufacture Hex Values of either 0x01-
0x08 or 0x81-0x88

GPIO0 - BS2 Interface Clock Pin
GPIO1 - BS2 Interface Data Pin

LCD Write Information can be found at
http://www.myke.com

This application uses a combination of PICC Lite C
and PIC Assembler.

NOTE:  THIS APPLICATION USES THE PIC16F675!!!

GPIO5 - LCD Data Pin
CGIO4 - LCD Clock Pin

myke predko
04.11.23

*/

__CONFIG(UNPROTECT & UNPROTECT & BORDIS & MCLRDIS &
PWRTEN & WDTDIS & \
  INTIO);

int LastADC = -1;      // Last ADC Value Read In
int ButtonFlag = 0;    // Mark When Pressed

unsigned char aLastADC;  // assembler LastADC
unsigned char ai;        // assembler Counter
unsigned char aDlay;     // assembler Delay

//    0123456789012345
const char TopMessage[] = "  Command 0x00  ";
const char BotMessage[] = " Response 0x00  ";

#define Clk  GPIO4
// Define the LCD Serial Control Pins
#define Data GPIO5
// If these Change, Change assembler

const int Twentyms = 1250;
```

```c
// Declare a Constant for 20 ms Delay
const int Fivems = 300;
const int TwoHundredus = 10;

PICSend(int DataValue)
{

  aLastADC = DataValue;
// Save ADC Value to Send

#asm
 bcf        porta, 0         ; Clock Line is Low
 movlw      high ((1688 / 5) + 256)
                             ; Delay by 1,688 us
 movwf      _aDlay
 movlw      low ((1688 / 5) + 256)
PSStartLowDlay:
 addlw      -1
 btfsc      status, 2
  decfsz    _aDlay, f
 goto       PSStartLowDlay
 movlw      8                ; Shift Out the Data
 movwf      _ai
PSOutLoop:
 movf       porta, w         ; Set GPIO1
                             ; Appropriately
 andlw      3Dh
 rrf        _aLastADC, f
 btfsc      status, 0
  iorlw     02h
 movwf      porta            ; This Point 14 cycles
                             ; before Clock Hi
 iorlw      01h              ; Set Clock Bit
 bsf        _aDlay, 2
PSPreLoop:
 decfsz     _aDlay, f
  goto      PSPreLoop
 movwf      porta            ; Save Clock High
 andlw      3Eh              ; Turn Off Clock High
 bsf        _aDlay, 2
PSClockLoop:
 decfsz     _aDlay, f
  goto      PSClockLoop
 movwf      porta
 movlw      7
 movwf      _aDlay
 nop
PSPostLoop:                  ; Data Valid for 32
                             ; cycles
 decfsz     _aDlay, f
  goto      PSPostLoop
 decfsz     _ai              ; Repeat for 8 Cycles
  goto      PSOutLoop

 movlw      62               ; Return Clock High
                             ; after 188 cycles
 movwf      _aDlay
PSEndLowLoop:
 decfsz     _aDlay, f
  goto      PSEndLowLoop
 bsf        porta, 0         ; Finally, Clock is
                             ; High, Command Sent

#endasm
}  // End PICSend

int PICSendReceive(int DataValue)
{

  aLastADC = DataValue;
// Save ADC Value to Send

#asm
 bcf        porta, 0         ; Clock Line is Low
```

```
        movlw    high ((1688 / 5) + 256)                bcf      status, 0        ; After 26 Cycles, Poll
                          ; Delay by 1,688 us                                     ; Data
        movwf    _aDlay                                 btfsc    porta, 1
        movlw    low ((1688 / 5) + 256)                 bsf      status, 0
PSRStartLowDlay:                                        rrf      _aLastADC        ; Save in "aLastADC"
        addlw    -1                                     movlw    4
        btfsc    status, 2                              movwf    _aDlay
         decfsz  _aDlay, f                     PSRPDoneLoop:                      ; Delay 46 cycles to
          goto   PSRStartLowDlay                                                  ; Next Clock
        movlw    8                   ; Shift Out the Data
        movwf    _ai                                    decfsz   _aDlay, f
PSROutLoop:                                             goto     PSRPDoneLoop
        movf     porta, w           ; Set GPIO1
                          ; Appropriately                decfsz   _ai, f
        andlw    3Dh                                     goto     PSRPollLoop
        rrf      _aLastADC, f
        btfsc    status, 0                              movlw    high ((350 / 5) + 256)
         iorlw   02h                                                     ; Final Clock Low Delay
        movwf    porta              ; This Point 14 cycles  movwf    _aDlay, f
                          ; before Clock Hi              movlw    low ((350 / 5) + 256)
        iorlw    01h                ; Set Clock Bit     PSRFinalLoop:
        bsf      _aDlay, 2                              addlw    -1
PSRPreLoop:                                             btfsc    status, 2
         decfsz  _aDlay, f                              decfsz   _aDlay, f
          goto   PSRPreLoop                              goto     PSRFinalLoop
        movwf    porta              ; Save Clock High
        andlw    3Eh                ; Turn Off Clock High  bsf      status, 5        ; Make Data Pin Output
        bsf      _aDlay, 2                              bcf      porta, 1
PSRClockLoop:                                           bcf      status, 5
         decfsz  _aDlay, f
          goto   PSRClockLoop                           bsf      porta, 0         ; Finally, Clock is
        movwf    porta                                                           ; High, Command Sent
        movlw    7
        movwf    _aDlay                        #endasm
        nop
PSRPostLoop:                        ; Data Valid for 32        return   aLastADC;
                          ; cycles                       // Return the Response
         decfsz  _aDlay, f
          goto   PSRPostLoop                   }   // End PICSendReceive
         decfsz  _ai                ; Repeat for 8 Cycles
          goto   PSROutLoop                    NybbleShift(int LCDOut, int RS)
                                               // Shift Out the Nybble
        movlw    high ((1840 / 5) + 256)       {
                          ; Poll Response after
                          ; 1,850 us           int i;
        movwf    _aDlay
        movlw    low ((1840 / 5) + 256)        Data = 0;
PSRPollDlayLoop:                               for (i = 0; i < 6; i++) // Clear the Shift Register
        addlw    -1                            {
        btfsc    status, 2                       Clk = 1;              // Just Toggle the Clock
         decfsz  _aDlay, f                        Clk = 0;
          goto   PSRPollDlayLoop               }   // rof

        bsf      status, 5          ; Make Data Pin Input  LCDOut = LCDOut | (1 << 5) | ((RS & 1) * (1 << 4));
        bsf      porta, 1                      for (i = 0; i < 6; i++) // Shift Data Out
        bcf      status, 5                     {
        movlw    8                  ; Start Polling Response  if (0 != (LCDOut & (1 << 5)))
        movwf    _ai                              Data = 1;          // Shift Out the Highest
PSRPollLoop:                                                        //   Bit
        bsf      porta, 0           ; Clock High to Poll    else
        nop                                              Data = 0;
        movlw    3                             LCDOut = LCDOut << 1; // Shift Up the Next Bit
PSRPHighLoop:                                   Clk = 1;             // Clock the Bit into
        addlw    -1                                                 //   the S/R
        btfss    status, 2                       Clk = 0;
         goto    PSRPHighLoop                  }   // rof
        bcf      porta, 0           ; After 14 Cycles, Clock
                          ; Low                 NOP();
        movlw    5                             Data = 1;            // Clock the Nybble int
PSRPWaitLoop:                                                       //   the LCD
        nop                                    NOP();
        addlw    -1                            Data = 0;
        btfss    status, 2
         goto    PSRPWaitLoop                  }   // End NybbleShift
```

```
LCDWrite(int LCDData, int RSValue)
{                          // Send Byte to LCD

int i, j;

  NybbleShift((LCDData >> 4) & 0x0F, RSValue);
  NybbleShift(LCDData & 0x0F, RSValue);
                           // Shift out Byte

  if ((0 == (LCDData & 0xFC)) && (0 == RSValue))
   i = Fivems;             // Set Delay Interval
  else
   i = TwoHundredus;

  for (j = 0; j < i; j++); // Delay for Character

}  // End LCDWrite

main()
{

int i, j;

  GPIO = 0b110011;
// Start with Interface Bits High
  ADCON0 = 0b00001001; // ADC Turned On:
                       //   ADFM - Left Justified
                       //   VCFG - Vdd Reference
                       //   CHS - AN2
                       //   GO - Off
                       //   ADON - On
  ANSEL = 0b00010100;  // ANSEL Specified as
                       //   Tosc = Tosc*8
                       //   AN2 - Analog Input
  CMCON = 0b00000111;  // Disable Comparator
                       //   Module
  OPTION = 0b01111111; // Enable Pull Ups on RA0
                       //   & RA1
  WPU = 0b000011;
  TRISIO= 0b001100;
// Everything But GPIO3/AN2 Outputs

// Initialize LCD according to the Web Page
  j = Twentyms;
  for (i = 0; i < j; i++); // Wait for LCD to
                           //        Power Up

  NybbleShift(3, 0);   // Start Initialization
                       //        Process
  j = Fivems;          // Send Reset Command
   for (i = 0; i < j; i++);

  NybbleShift(3, 0);   // Repeat Reset Command
  j = TwoHundredus;
  for (i = 0; i < j; i++);

  NybbleShift(3, 0);   // Repeat Reset Command
                       //        Third Time
  j = TwoHundredus;
  for (i = 0; i < j; i++);

  NybbleShift(2, 0);   // Initialize LCD 4 Bit
                       //        Mode
  j = TwoHundredus;
  for (i = 0; i < j; i++);

  LCDWrite(0b00101000,  0);   // LCD is 4 Bit I/F,
                              //     2 Line

  LCDWrite(0b00000001,  0);   // Clear LCD

  LCDWrite(0b00000110,  0);   // Move Cursor After
                              //     Each Character

  LCDWrite(0b00001110, 0);    // Turn On LCD and
                              //     Enable Cursor

  for (i = 0; TopMessage[i] != 0; i++)
   LCDWrite(TopMessage[i], 1);

  LCDWrite(0b11000000, 0);    // Move Cursor to
                              //     the Second Line

  for (i = 0; BotMessage[i] != 0; i++)
   LCDWrite(BotMessage[i], 1);

  while(1 == 1)              // Loop Through Polling
                            //     GPIO3 and AN2
  {
    for (i = 0; i < Twentyms; i++);
                            //     Basic Delay for ADC

    GODONE = 1;             // Start ADC Operation
    while(1 == GODONE);
    j = (ADRESH >> 4) & 0x0F; // Process Using
                              //     Algorithm Above
  j = ((j & 0x08) << 4) + (j & 0x07);
  j = j + 1;
  if (j != LastADC)     // If Different, Save and
                        //     Display
  {
    LastADC = j;          // Save the New ADC Value
    LCDWrite(0b10001100, 0);  // Move to 10s of
                              //     Output
    j = (j >> 4) & 0x0F;      // Write High Digit
    LCDWrite(j + '0', 1);
    LCDWrite((LastADC & 0x0F) + '0', 1);
  }  // fi

  if ((0 == GPIO3) && (0 == ButtonFlag))
  {                          // Debounce Button Press
    i = 0;
    while (i < Twentyms)
    for (i = 0; (i < Twentyms) && (0 == GPIO3);
        i++);
    ButtonFlag = 1;     // Button Pressed
    if (LastADC < 0x80)
     PICSend(LastADC);
    else                 // Get Remote Status/Value
    {
      LastADC = PICSendReceive(LastADC);
      LCDWrite(0b11001100, 0);
      j = (LastADC >> 4) & 0x0F;
                           // Write High Digit
      if (j > 9)
       LCDWrite(j + 'A' - 10, 1);
      else
       LCDWrite(j + '0', 1);
      j = LastADC & 0x0F;      // Write Low Digit
      if (j > 9)
       LCDWrite(j + 'A' - 10, 1);
      else
       LCDWrite(j + '0', 1);
    }  // fi
  } else if ((1 == GPIO3) && (1 == ButtonFlag))
  {                          // Look for Button Released
    i = 0;
    while (i < Twentyms)
    for (i = 0; (i < Twentyms) && (1 == GPIO3);
        i++);
    ButtonFlag = 0;  // Button Released
  }  // fi
}  // elihw
}  // End cBS2 Sim
```

The assembly language used in cBS2 Sim.c took longer than I would have expected, this was due to two factors. The first was a mistake on my part: I did not disable the comparator on the PIC12F675's GPIO0 and GPIO1 pins (used for the BS2 interface), which I discovered often results in unexpected resetting of a pin when writing to a single but different pin. In Section 8, I went to great pains to explain how the bsf and bcf instructions work, and, in this program, I ended up with a great example of how they can unintentionally change a pin value.

In my application code, I set the data pin (GPIO1) and then I then set the clock pin (GPIO0), using the standard instructions:

```
bsf    GPIO, 1
bsf    GPIO, 0
```

but when I looked at the signals on an oscilloscope, I saw that the data pin (GPIO1) would go low when the clock pin (GPIO0) was set high. When I wrote the application, I was under the impression that the PIC12F675 had an ADC but not a comparator, but I was wrong. Like the PIC16F684, the PIC12F675 has a comparator with its input pins being GPIO0 and GPIO1. And like the PIC16F684, if the CMCON register is not written to turn off the comparator at power up, these two pins are assumed to be analog inputs (which always return zero when the port register is read or used by the bsf instruction). Once I turned off the comparator, the pins behaved as I expected them to.

I'm explaining this because if you had an application that didn't work or if it behaved as mine did, you might be inclined to assume that the PIC MCU chip was bad. Remember that you should always assume first that the chip is good and the problem is software related, as was the case here: I didn't properly initialize the chip. Assume second that your circuit/wiring is bad.

Over the 10 years I've been working with PIC MCU chips, I've encountered only two chips that were bad. The first time was when I was writing *Programming and Customizing the PICmicro® Microcontroller*, in which I seemed to have worn out the Flash program memory of a PIC16F84; several instructions could no longer be programmed. The second encounter happened when I was debugging this application and I miswired a PIC16F684 instrument application to this BS2 simulator; I accidentally connected the BS2 simulator's ground to the instrument application's positive power. After correcting the error, the instrument application would work periodically, but required a special sequence of actions to get it working. (I had to power up the instrument application first, next power up the BS2 simulator, then disconnect the two applications, and finally reconnect the applications.) Programming another PIC16F684 with the same application did not require any of these actions and would, in fact, work perfectly regardless of which PIC MCU application was powered up first.

The root cause of this problem was the use of two breadboard circuits, each powered separately and connected by three wires (two for clock/data and one for ground) as shown in Figure 10-5. This connection reflects how I wanted the BS2 simulator and the instrument interface applications—each as separate as possible, having only the signal and ground reference in common. If I were using this application for anything but prototyping, I would create polarized connections that could not be plugged in error.

Experiment 85—PIC MCU BS2 Keypad Interface

PC/Simulator

PICkit™ 1

starter kit

Parts Bin

1 PIC16F684

1 16-button keypad interface

1 Two-line by 16-character LCD display

1 10k breadboard-mountable potentiometer

Either 1 10k 10-pin SIP

Or 8 10k resistors

1 0.01 µF capacitor

1 SPST breadboard-mountable switch

1 Three-cell AA battery clip

3 AA batteries

Tool Box

DMM

Needle-nose pliers

Breadboard

Wiring kit

Although the potentiometer-controlled PIC12F675 interface works well, I wanted to see how easy it would be to create a keypad-controlled interface while at the same time eliminating the need for the 74LS174 used as a shift register for the LCD display. Pin allocation can be difficult in this situation. A four-by-four (16-button) keypad requires eight I/O pins; an LCD requires a minimum of six if the two-wire shift register is not used; and to top it off, two wires are required to provide the BS2 interface. The total number of pins required for this application appears to be 16. However, as I will show, the keypad column and the LCD data pins can be shared, allowing this application to be wired into a single PIC16F684 with only 12 I/O pins.

The trick to sharing pins in an application like this is to look for output pins that can be shared between interface devices. It should be obvious that clocking or control pins cannot be shared between interfaces, as the device being accessed will be confusing. The LCD data pins and R/S pin can be shared with the four column drivers used in a four-by-four switch matrix keypad. The combination of these two components on the same data bits worked very well with just a couple of complications.

For this application, I wanted to keep the LCD data pins on RC3:RC0 to simplify the programming of the LCD writes (keeping the data bits in the lower nybble of a byte). And I also wanted to keep the wiring as simple as possible on the breadboard. These two condi-

tions were met (see Figure 10-6), although it was difficult to show the simple wiring to the keypad. The schematic might be a bit confusing; rather than draw the rows and columns of the keypad as four individual wires, I used a bus, which is a thicker line. These four lines are passed to the rows and columns of the keypad as shown: The keypad I used (bought from a surplus store in Toronto) had the rows and columns in two groups of four pins, which simplified the breadboard wiring for my prototype.

Note in Figure 10-7, the four column bits (the bus lines that intersect with the LCD bits) do not use all four LCD data bits. Instead three data bits and the R/S bit are used, as this worked best with the breadboard. This made the software a bit more complex but was worth it, as I did not have to wire each keypad pin individually to the breadboard/PIC16F684. Instead I could simply push it into the same column as the pin-1 side of the PIC MCU.

In the application, I pulled up all eight pins of the keypad. Like when I started the application, I didn't know which pins were which, and I had to decode them. When you look at the source files for this application, you will see that I created asmSCtrl1.asm through asmSCtrl4.asm to decode the keypad and make sure the operation was (reasonably) intuitive. The final result, asmSCtrl5.asm, takes in a maximum two-hex number command (taking advantage of my keypad's 2ND key, which I used to shift the key inputs

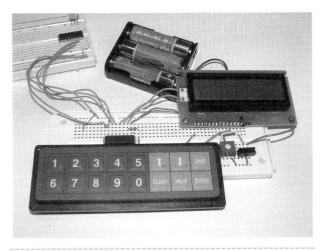

Figure 10-6   *Breadboard-wired keypad-based BS2 simulator wired to sensor breadboard*

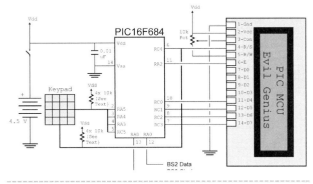

Figure 10-7   *Keypad BS2*

of 1 through 6 to A through F). The CLEAR key was used to erase the two-character command, and ENTER sends the command (and, if the value is 0x80 or greater, it also looks for a reply).

Depending on the keypad you use for this application, you may not be able to wire it as simply as I have

for my prototype or the keys on your keypad may require you to work out a different user interface. When you create your own keypad interface, follow the steps that I have: First, get the LCD working; next, work at decoding the keypad, and model the user interface; and, finally, add the BS2 interface for the peripheral functions. You will probably find that working through these steps will take you a few days, but they should be fairly uneventful, and when you have to perform this task again later, you'll find that it will take only a day or so.

Experiment 86—PIC MCU Instrument Interface

PC/ Simulator

PICkit™ 1

starter kit

Parts Bin

1 PIC16F684
1 LED
1 0.01 µF capacitor
1 Three-cell AA battery clip
3 AA batteries

Tool Box

PIC12F675 or PIC16F684-based BS2 command simulator interface

DMM

Needle-nose pliers

Wire strippers

Breadboard

Wiring kit

With the BS2 and PIC MCU equivalent interfaces available, we can create a simple application that turns on and off the LED connected between RA4 and RA5 of a PIC16F684 remotely (see Figure 10-8). This experiment has only three commands: (1) LED On, (2) LED Off, and (3) 0x81 — Return LED State. And its code is based on asmBS2 Template.asm:

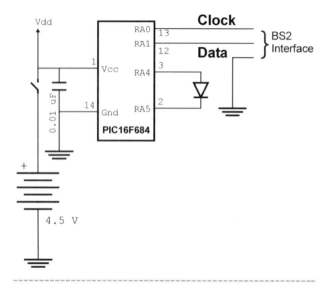

Figure 10-8 *BS2 Demo*

```
    title  "asmBS2 Template - BASIC Stamp II PIC
MCU Interface"
;
;   This Program provides the basis for
;   an Intelligent
;   BS2 Interface.
;
;  BS2 Commands (to be defined by application)
;    1 - Command 1
;    2 - Command 2
;    3 - Command 3
;    4 - Command 4
;    5 - Command 5
;    6 - Command 6
;    7 - Command 7
;    8 - Command 8
;  Data Respond Commands (to be defined by
;  application)
;   128 - Respond Command 1
;   129 - Respond Command 2
;   130 - Respond Command 3
;   131 - Respond Command 4
;   132 - Respond Command 5
;   133 - Respond Command 6
;   135 - Respond Command 7
;   136 - Respond Command 8
;
;   Hardware Notes:
;   PIC16F684 running at 4 MHz Using the
;   Internal Clock
;   Internal Reset is Used
;   RA5 - Clock Input
;   RA4 - Data I/O
;   Remaining Pins are for Various I/O and
Functions

 #define ComData    PORTA, 1
 #define ComClock   PORTA, 0
;
;
;  Myke Predko
;  04.11.25
```

```
;
   list R=DEC
   INCLUDE "p16f684.inc"


; Register Usage
 CBLOCK 0x20    ; Start of GP Registers
i, j, k
BS2Byte          ; BS2 Send/Receive Value
DlayValue:3      ; Three Bytes Needed for Delay
                 ; Macro
 ENDC



 PAGE

 __CONFIG _FCMEN_OFF & _IESO_OFF & _BOD_OFF &
_CPD_OFF & _CP_OFF & _MCLRE_OFF & _PWRTE_ON &
_WDT_OFF & _INTOSCIO

 PAGE
; Mainline

 org      0
  nop                           ; For ICD2

  movlw    0x0FF                ; Make All PORTA
                                ; Bits High
  movwf    PORTA
  movlw    7                    ; Turn off
                                ; Comparators
  movwf    CMCON0
  bsf      STATUS, RP0
  clrf     ANSEL ^ 0x080        ; All Bits are
                                ; Digital
  bcf      OPTION_REG ^ 0x80, 7 ; Enable PORTA
                                ; Pull Ups
  movlw    b'000011'            ; Enable RA0/RA1
                                ; Pull Ups
  movwf    WPUA ^ 0x80
; #### - Modify/Put in Additional Port
Bit/Peripheral Initializations
  bcf      STATUS, RP0

;  Loop Waiting for BS2 Command
Loop:
  btfss    ComClock             ; Read BS2 For
                                ; Command
  call     BS2Read

; #### - Perform any Repeating Sub-ms
operations here
  goto     Loop


;  Read/Respond to BS2 Commands
BS2Read:                        ; Decode the BS2
                                ; Command
  movlw    8                    ; Number of Bits
                                ; to Read
  movwf    i

  movlw    HIGH (3800 + 256)
  movwf    j
  movlw    LOW (3800 + 256)
BS2RLoop1:                      ; Wait for Data to
                                ; become Active
```

```
        btfsc   ComClock                                        nop     ; #### - Command 2 Code
        goto    BS2RSkip3                                       xorlw   3 ^ 2                   ; Command 3?
        addlw   -1              ; Count Down                    btfsc   STATUS, Z
        btfsc   STATUS, Z                                       nop     ; #### - Command 3 Code
        decfsz  j, f                                            xorlw   4 ^ 3                   ; Command 4?
        goto    BS2RLoop1                                       btfsc   STATUS, Z
                                                                nop     ; #### - Command 4 Code
BS2RError:              ; Nothing Valid                         xorlw   5 ^ 4                   ; Command 5?
                       ; Received                               btfsc   STATUS, Z
        bsf     STATUS, RP0                                     nop     ; #### - Command 5 Code
        bsf     ComData         ; Make Sure Data                xorlw   6 ^ 5                   ; Command 6?
                               ; is Input                       btfsc   STATUS, Z
        bcf     STATUS, RP0                                     nop     ; #### - Command 6 Code
        return                  ;  Return                       xorlw   7 ^ 6                   ; Command 7?
                                                                btfsc   STATUS, Z
BS2RSkip3:             ; Clock High, Wait                       nop     ; #### - Command 7 Code
                       ; for it to go low                       xorlw   8 ^ 7                   ; Command 8?
        movlw   4                                               btfsc   STATUS, Z
        movwf   j                                               nop     ; #### - Command 8 Code
BS2RLoop3:             ; Wait 15 µs for                         goto    BS2RError
                       ; Clock Pulse                                    ; Otherwise, Unknown/Completed Command
                       ; Active
        btfss   ComClock                                ;   #### - Put Command Functions Here with Final
        goto    BS2RSkip4                               ;   "goto BS2RError" to
        decfsz  j, f                                    ;   #### -  avoid having funny jumps over
        goto    BS2RLoop3                               ;   assembler "switch" comparisons
        goto    BS2RError

BS2RSkip4:             ; Clock Low/Read
                       ; the Data Bit                   BS2RRespondCommand:
        bcf     STATUS, C                               ;   Return the Requested Information within
        btfsc   ComData                                 ;   1 ms to "BS2Byte"
        bsf     STATUS, C                               ;    and "goto BS2RRespond:
        rrf     BS2Byte, f      ; Shift in the Bit              bsf     STATUS, RP0     ; Make ComData an
                                                                                        Output
        decfsz  i, w            ; At the Last Bit?               bcf     ComData
        goto    BS2RSkip2       ;  - No...                       bcf     STATUS, RP0
                                                                movf    BS2Byte, w      ; Get the Command
        btfsc   BS2Byte, 7      ; BS2 Respond                    xorlw   0x81            ; Compare to Respond
                               ; Command?                                               ; Command 1
        goto    BS2RRespondCommand ; if No, Wait for             btfsc   STATUS, Z
                               ;  Final Clock High               nop     ; #### - Respond Command 1 Code
                                                                xorlw   0x82 ^ 0x81     ; Compare to Respond
BS2RSkip2:             ; Clock Still                                                     ; Command 2
                       ; Low/Wait for High                      btfsc   STATUS, Z
        movlw   100                                             nop     ; #### - Respond Command 2 Code
        movwf   j                                               xorlw   0x83 ^ 0x82     ; Compare to Respond
BS2RLoop2:             ; Wait 500 µs for                                                 ; Command 3
                       ; Clock High                             btfsc   STATUS, Z
        btfsc   ComClock                                        nop     ; #### - Respond Command 3 Code
        goto    BS2RSkip5                                       xorlw   0x84 ^ 0x83     ; Compare to Respond
        decfsz  j, f                                                                     ; Command 4
        goto    BS2RLoop2                                       btfsc   STATUS, Z
        goto    BS2RError                                       nop     ; #### - Respond Command 4 Code
                                                                xorlw   0x85 ^ 0x84     ; Compare to Respond
BS2RSkip5:             ; Clock High                                                      ; Command 5
                       ; Again/Finished                         btfsc   STATUS, Z
                       ; with Bit                               nop     ; #### - Respond Command 5 Code
                       ; Read Another                           xorlw   0x86 ^ 0x85     ; Compare to Respond
        decfsz  i, f            ; Bit?                                                    ; Command 6
                                                                btfsc   STATUS, Z
        goto    BS2RSkip3                                       nop     ; #### - Respond Command 6 Code
                                                                xorlw   0x87 ^ 0x86     ; Compare to Respond
BS2PrimaryCommand:     ; Change the                                                      ; Command 7
                       ; Command Operation                      btfsc   STATUS, Z
        movf    BS2Byte, w      ; Get the Command                nop     ; #### - Respond Command 7 Code
        xorlw   1               ; Compare to                     xorlw   0x88 ^ 0x87     ; Compare to Respond
                               ; Command 1                                               ; Command 8
        btfsc   STATUS, Z                                       btfsc   STATUS, Z
        nop     ; #### - Command 1 Code                         nop     ; #### - Respond Command 8 Code
        xorlw   2 ^ 1           ; Command 2?                     movlw   0xA5            ; Command Not
        btfsc   STATUS, Z
```

```
                   movwf     BS2Byte          ; Supported
                                              ; Bit Pattern to Show
                                              ; Connection
                   goto      BS2RRespond

; #### - Execute Respond Commands - Must
Execute within 1 ms

BS2RRespond:
   movlw     8                                ; Count Number of
                                              ; Bits to Send Out
   movwf     i

   movlw     HIGH (2500 + 256)
   movwf     j
   movlw     LOW (2500 + 256)
BS2RLoop4:                                    ; Wait for Data to
                                              ; become Active
   btfsc     ComClock
    goto     BS2RSkip6
   addlw     -1
   btfsc     STATUS, Z
    decfsz   j, f
     goto    BS2RLoop4
   goto      BS2RError

BS2RSkip6:                                    ; Output the Data Bit
   rrf       BS2Byte, f                       ; Put the LS Data Bit
                                              ; into Carry

   btfsc     STATUS, C
    bsf      ComData                          ; Data Bit High
   btfss     STATUS, C
    bcf      ComData                          ; Data Bit Low

   movlw     4
   movwf     j
BS2RLoop5:                                    ; Wait 15 msecs for
                                              ; Clock Pulse Active
   btfss     ComClock
    goto     BS2RSkip7
   decfsz    j, f
    goto     BS2RLoop5
   goto      BS2RError

BS2RSkip7:                                    ; Wait for Line Low
                                              ; to End
   movlw     HIGH (3000 + 256)
   movwf     j
   movlw     LOW (3000 + 256)
BS2RLoop6:                                    ; Wait 3 ms for Clock
                                              ; High
   btfsc     ComClock
    goto     BS2RSkip8
   addlw     -1
   btfsc     STATUS, Z
    decfsz   j, f
     goto    BS2RLoop6
   goto      BS2RError

BS2RSkip8:                                    ; More Bits to Send?
    decfsz   i, f
     goto    BS2RSkip6                        ; Yes, Loop Around
                                              ; Again
```

```
   goto      BS2RError          ; Finished, Return
                                ; and Keep Processing


end
```

To operate the application code, I modified asmBS2 If.asm to get asmBS2Test.asm by making the following changes:

1. "ComClock" and "DataClock" were changed to RA1 and RA0, respectively.

2. For "Command 1," replace the nop with "bsf PORTA, 4."

3. For "Command 2," replace the nop with "bcf PORTA, 4."

4. The xorwf/btfsc/nop instructions for the remaining six commands were deleted.

5. For "Respond Command 1," the "btfsc STATUS, Z/nop" was replaced with a "btfss STATUS, Z/goto InvalidReturn." And the following xorlw/btfsc/nop instructions were deleted.

6. After "goto InvalidReturn," the following instructions were added:

```
   movlw     0                  ; Return LED State
    btfsc    PORTA, 4
     movlw   1
   movwf     BS2Byte
   goto      BS2RRespond

InvalidReturn:
   movlw     0xA5               ; Command Not
                                ; Supported
   movwf     BS2Byte           ; Bit Pattern to Show
                                ; Connection
```

With these changes in place, the PIC12F675 interface connected to the breadboard with the circuit in Figure 10-8, and the PIC16F684 loaded with asmBS2 Test.asm, you can now remotely turn on and off an LED as well as query the LED's state. To turn on the LED, set the PIC12F675 interface's LCD to 1 and press the button—the LED connected to the PIC16F684 should turn on. Similarly, setting the PIC12F675 interface's LCD to 2 will turn off the LED. Setting the PIC12F675 interface's LCD to 0x81 and pressing the button will send a command response and return zero or one, indicating the state of the LED connected to the PIC16F684.

PC/ Simulator

PICkit™ 1

starter kit

Parts Bin

1	PIC16F684
2	TLC251
1	LED
1	Electret microphone
2	3.3M resistors
2	2.3M resistors
1	10k resistor
2	470Ω resistors
2	220Ω resistors
4	0.1 μF capacitors
4	0.01 μF capacitors

Tool Box

PIC12F675- or PIC16F684-
    based BS2 command simu-
    lator interface
DMM
Needle-nose pliers
Wire strippers
Breadboard
Wiring kit

Sound input recognition is an interesting and often useful sensor to add to your applications. With just a bit of signal-conditioning circuitry, you can add the ability to recognize sound to the PIC MCU. Before you start getting visions of adding voice control to different electronic devices around your house, I must warn you that the circuit used in this application is quite rudimentary and will not recognize sounds of different duration or frequencies. What you'll get after building this experiment is the front-end circuitry used in the "clapper" light switches that were popular a few years ago.

The application circuit in Figure 10-9 is probably a bit more complex than you would have expected. Along with the PIC16F684 with its BS2 sensor control interface, I have also included an LED, which flashes when sound is received, and a two-*operational-amplifier* (op-amp) circuit, which is used to filter out high-frequency sounds (defined as anything above 340 Hz) and amplify them to useful levels for the PIC MCU to recognize as logic level changes. The circuit fits easily on a small breadboard (see Figure 10-10). However, if you wanted to add any additional circuitry to the experiment, you would have to use a larger breadboard.

I don't want to go into a great deal of explanation about the operation of the TLC251 op-amps except to explain their basic operation. I should point out, however, that the component values are critical to correct operation of the circuit. If the resistor or capacitor values are changed, you will find that different parts of the circuit may go into oscillation. You should be particularly concerned about matching the impedance of the electret microphone to make sure that you maximize the voltage output. If you feel the need to change any of the component values, make sure that you understand the ramifications of what you are doing.

The electret microphone produces a relatively small-scale signal (in the tens of mV, as seen in Figure 10-11). This signal is passed through a dual-pole Butterworth low-pass filter, and then amplified an extreme amount. The amplification factor used in this application is on the order of ten thousand times for each op-amp. The reason for the extremely large amplification factor is that this application can amplify the signal only to the input voltage extremes. Anything greater is *clipped* and produces the signal seen in Figure 10-11. The signal from the microphone is the analog signal shown in the top waveform. By amplifying the signal

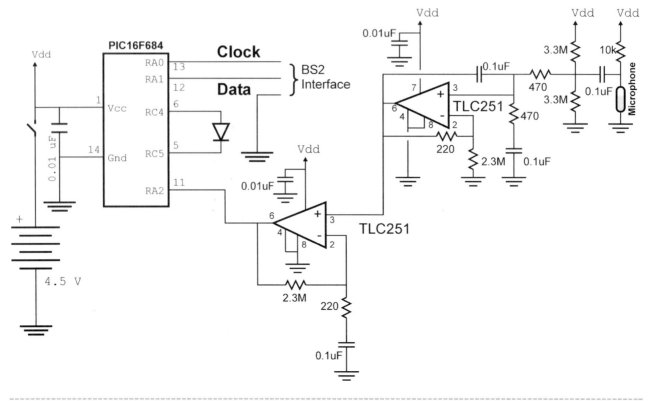

Figure 10-9   *Sound circuit*

by such an extreme amount, the more digital-looking signal is produced and can be passed to the PIC16F684's input directly.

There are many books available that will explain the operation of op-amps in greater detail than I will here. When you look at these texts, you will see that most of them use the LM741, not the TLC251 that I used in this application. I did not use the LM741 in this application due to its requirement of at least 12 volts to operate. The TLC251 works comfortably with the 5 volts used by this application.

There is one problem with this circuit: If the sound transition occurs when the BS2 interface is communicating with the PIC16F684, the PIC MCU cannot poll the incoming sound line. When I presented this circuit in *123 Robotics Experiments for the Evil Genius*, I passed the signal to a 74LS74 D-flip-flop clock with

Figure 10-10   *Sound input test circuit built on small breadboard with BS2 command interface providing signal wiring as well as power*

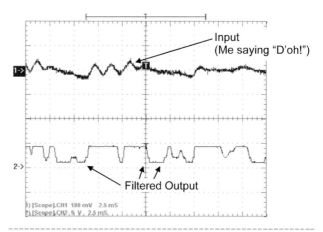

Figure 10-11   *Sound operation*

the expectation that the sound would cause a new value to be latched into the flip flop. To avoid this requirement now, I pass the conditioned sound output to the TMR0 input pin. Then, after setting TMR0 to 0xFF and resetting the T0IF bit to find out if sound has been detected, it is a simple matter of checking the T0IF bit as I do in the experiment's application code, asmSound.asm.

The simple ability to recognize that a sound has happened (even without the sound's characteristics) can be useful in a variety of different applications, ranging from the "clapper" to a robot's collision detector. The collision detector is actually a very useful sensor in robots and often more reliable than whiskers of the *infrared* (IR) or ultrasonic methods presented later in this section.

Experiment 88—Multiple Microswitch Debouncing

PC/ Simulator

PICkit™ 1

starter kit

Parts Bin

1 PIC16F684

4 10k resistors

1 0.01 μF capacitor

4 Breadboard-mountable push-button or SPST switches

Tool Box

PIC12F675 or PIC16F684-based BS2 command simulator interface

DMM

Needle-nose pliers

Wire strippers

Breadboard

Wiring kit

Previously, in this book, I have shown you how to debounce buttons, switches, and keypads (in both C and assembler). What I have not done is show how multiple, independent switches can be debounced in an environment that can be polled periodically. In this experiment, I will use the BS2 command interface periodically to poll a PIC16F684 wired with four buttons or switches that must be debounced before a value can be returned. This is very similar to what would be done in a robot that has several whiskers around its perimeter.

The circuit used for this experiment is quite simple and can be built on a small breadboard wired to a BS2 command simulator—either the PIC12F675 using a potentiometer command selection or the keypad interface (see Figure 10-12). When I was designing the application, I debated whether to use the pulled-up

pins on PORTA, but decided against it when I looked at how I would poll and debounce the buttons. Instead, I used the least significant four bits of PORTC, allowing me to come up with an efficient method of polling and debouncing the buttons. The application itself can be written out in C pseudo-code as follows:

```
j = 0;
while (1 == 1)
{
  for (i = 0; i < 216 us; i++);
// Loop takes 250 us
  if (0 == BS2Clock)
// Clock Low - BS2 Command
    BS2Read();
    if (0 == (PORTC & ( 1 << j)))
// Bit is Low
    if (0 == Value[j])
// Bit was Previously Low
      if (Count[j] < 20)
        Count = Count + 1;
      else;
// Count == 20, Do Nothing
    else
// Bit was Previously High
      {
        Value[j] = 0;
        Count[j] = 0;
      } // fi
    else
// Bit is high
  if (1 == Value[j])
// Bit was Previously High
    if (Count[j] < 20)
      Count = Count + 1;
```

```
  else;
// Count == 20, Do Nothing
  else
// Bit was Previously Low
  {
    Value[j] = 1;
    Count[j] = 0;
  } // fi
  j = (j + 1) % 4;
// Increment j to next value
  } // elihw
```

In this 250 μs loop, one of the four switches is polled, and if its value is different from the previous poll, the new value is saved and the counter is reset. If the value is the same, the counter is incremented to 20. With each loop taking 250 μs and executing four times between polls, a count of 20 will mean a 20 ms debounce period. The 216 μs delay loop, when added to the 34 μs polling code time, results in a 250 μs loop. The assembly code for this operation can be found in asmUSW.asm.

This code could be expanded quite easily for more buttons or reduced as required. (Six could be made available quite simply, by adding RC4 and RC5 to the

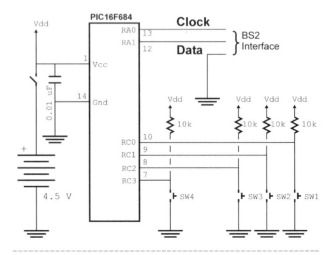

Figure 10-12   *BS2 uSwitch*

four already present.) The code executes in just a few instruction cycles, allowing you to add additional sensors or BS2 functions to the mix quite easily.

Experiment 89—Light Sensors

PC/ Simulator

PICkit™ 1

starter kit

Parts Bin

1 PIC16F684

3 10k resistors

2 10k light dependant resistors (R decreases with light)

1 10k potentiometer

1 0.01 μF capacitor

Tool Box

PIC12F675 or PIC16F684-based BS2 command simulator interface

DMM

Needle-nose pliers

Wire strippers

Breadboard

Wiring kit

In the previous experiment, I showed how you could time share between operations to check values while polling for BS2 commands. In this experiment, you will get the chance to experiment with different wiring of

light-dependant resistors (LDRs) or Cadmium Sulfide (CDS) cells as well as look at a method of reading each ADC individually while polling for the BS2 commands. This method of operation is not compatible with the previous, but either the microswitch polling method or the one in this experiment could be modified to work with the other.

To demonstrate the operation of the CDS cells along with how the ADC works for multiple devices, I used the circuit shown in Figure 10-13. The SPDT switch shows the high- and low-voltage level provided by a simple switch, and the LDR connections show how the voltage divider output voltage changes with the LDR resistance values. The in-circuit potentiometer should work exactly as you expect. That is, the

ADC value returned for the potentiometer is proportional to the position of the wiper.

The application code (asmLight.asm) for the experiment is very straightforward and reads each ADC value twice before saving the value. Reading the value twice is simply to ensure that the value to be read is correct. The code is called asmLight.asm. When you test this application, you should record the ADC values for the LDRs when they are exposed to ambient light and again when they are shielded from the light by a finger. (The ADC value for the LDR connected to Vdd should go down and visa versa for the LDR connected to the circuit's ground.)

The application should work quite well, with only one possible surprise: You may find that value of either the high- or low-voltage level returned for the SPDT switch is off by a few digits (a value of 0xFF or 0x00, respectively, is expected). This is due to resistances throughout the wires and breadboard, which cause small changes in the voltages at different points within the circuit (a single ADC digit represents 17 mV in this application). The reason for pointing this out is to rein-

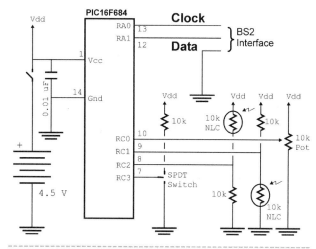

Figure 10-13 *BS2 light*

force the idea that nothing is absolute in electronics circuits, and you must always design your circuitry and software to accommodate these inconsistencies.

Experiment 90—Infrared (IR) Surface Sensor

PC/ Simulator

PICkit™ 1

starter kit

Parts Bin

1 PIC16F684

1 2N3904 NPN transistor

1 U-shaped IR opto-inter-rupter (see text)

1 LED

2 330Ω resistors

2 10k resistors

1 0.01 μF capacitor

1 Breadboard-mountable SPDT switch

1 Three-cell AA battery clip

3 AA alkaline batteries (see text)

Tool Box

DMM

Needle-nose pliers

Wire clippers

Wire strippers

Breadboard

Wiring kit

In *123 Robotics Experiments for the Evil Genius*, I demonstrated a basic line-following robot using two cut-apart U-shaped IR opto-interrupters. If you look back at those experiments (numbers 48 and 119 in that book), you'll see that by cutting apart the two halves of the opto-interrupter and placing them side by side, you

could make a fairly effective white/black surface sensor. The problem was the circuits required a fair amount of support circuitry due to the requirements of amplifying the current from the opto-interrupter's phototransistor and comparing it to an expected value. The built-in comparators of the PIC16F684 allow you to simplify the circuitry considerably as is shown in Figure 10-14. With just a few resistors and a transistor, you can produce a voltage that can be sensed by the PIC MCU and can be coupled with surprisingly simple software.

These parts go by a variety of names, including "slotted IR opto-interrupter" and "IR switch Opto-NPN." If you were looking in a catalog or online, be sure you see a picture of what you are getting, because the devices vary widely. Another issue that you should be aware of is that if you were to buy the parts from a distributor (e.g., Digi-Key, Mouser, Jameco), you'll probably be quite shocked at the price, especially considering I'm telling you to cut them apart (see Figure 10-15). There are also IR reflective sensors that perform the same task as the cut-apart opto-interrupter, but these too can be quite expensive. Rather than be subjected to the high prices of new opto-interrupters, I buy a few from electronics surplus bins for a just a few cents apiece.

Buying the parts from a surplus bin means that you will not have a part number or a datasheet for the part. Instead, you will have to build a circuit like the one in Figure 10-16 to determine which set of pins are for the IR LED and which set are for the IR *phototransistor* (PT) that make up the opto-interrupter. Assuming that the wires for each component come out at different ends, there are only four ways of wiring the circuit. You can do it by trial and error until the LED lights and turns off when something is put between the two sides of the opto-interrupter. Or you can spend a few minutes and come up with a strategy that simplifies the search. For example, to find the IR LED and its polarity, use the LED and a 330Ω resistor. Try each side with different polarities for the pins and the LED until the LED lights.

Once you have identified the components in each side of the opto-interrupter, record what each component is, record which wires belong to each, cut apart the

opto-interrupter, and then place the opto-interrupter in a breadboard (see Figure 10-15). This orientation will allow you to run a piece of paper with white and black marks over the IR LED and phototransistor. With this done, add the three-battery power supply, the resistors, and the transistor (see Figure 10-14). The two 10k resistors and the 2N3904 NPN transistor will produce an analog voltage that can be very easily sensed by the PIC MCU's comparators.

I decided to use comparators instead of the PIC16F684's ADC because I didn't expect there to be a need for processing analog values (white/black should be very binary), and the comparator outputs can be polled continuously, whereas the ADC must be initialized and then polled until the operation is complete. As can be seen in the application code, the task of polling the comparator and turning on the LED (if something is reflecting light from the IR LED to the IR phototransistor) is extremely simple.

Using your DMM, measure the voltage difference with something white in front of the IR LED/IR phototransistor pair and then with something black. I used a white sheet of paper that had large black areas printed on it. For testing the opto-interrupter, I used with the components specified, the voltage at the 2N3904 was 3.06 volts with the black printed paper in front of the phototransistor and 0.20 volts for plain white paper. For the comparator's VRef module, I choose 1/2 these extremes or 1.5 volts.

To calculate the value passed to the VRef module, I measured voltage at the PIC MCU's power pins (as the VRef module's output is proportional to Vdd) and found it to be 4.54 volts. For the VRef module, I decided to run it in low-range mode, which allows for

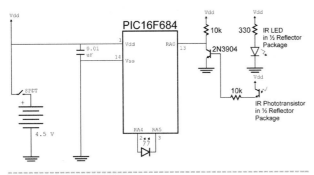

Figure 10-14 *IR white/black circuit*

Figure 10-15 *IR surface sensor circuit using a cut-apart U-shaped IR opto-isolator*

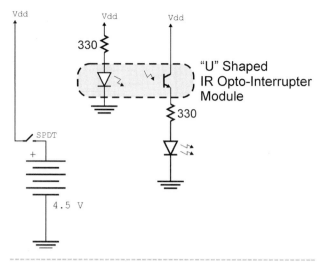

Figure 10-16 *IR phototransistor circuit*

voltages up to two-thirds of Vdd with a fraction denominator of 24. With this information, I calculated the VRef module's voltage specification value to be:

```
Value = (1.5 volts / 4.54 volts) × 24

= 7.93
```

As you can see in the software for this application (asmRwb.asm), I put in a value of 8 (b'1000') to the VRCON register:

```
title  "asmIRwb - Use Comparators with IR
White/Black Sensors"
;
;  This Program Polls the voltage coming from a
;  IR LED/Phototransistor and turns on an
;  LED when the voltage
;  in is greater than 1 Volt.  The
;  comparator is used for
;  this operation.
;
;  Hardware Notes:
;  PIC16F684 running at 4 MHz Using
;  the Internal Clock
;  4.45-4.5 Volt Power (3x "AA" Cells)
;  RA0 - Phototransistor Output
;  RA4 - LED Anode
;  RA5 - LED Cathode
;
```

```
;
;  Myke Predko
;  05.01.03
;
LIST R=DEC
 INCLUDE "p16f684.inc"


 __CONFIG _FCMEN_OFF & _IESO_OFF & _BOD_OFF &
_CPD_OFF & _CP_OFF & _MCLRE_OFF & _PWRTE_ON &
_WDT_OFF & _INTOSCIO


;  Variables
CBLOCK 0x020
ENDC

 PAGE
;  Mainline

 org      0

 nop                             ; For ICD Debug

 clrf     PORTA
 movlw    b'00001010'            ; Comparator with
                                 ; Vref Module
 movwf    CMCON0

 bsf      STATUS, RP0
 clrf     ANSEL ^ 0x80
 movlw    b'10101000'            ; Specify 1.5 Volt
                                 ; Transition
 movwf    VRCON ^ 0x80
 movlw    0x0F                   ; RC4/RC5 Outputs
 movwf    TRISA ^ 0x80
 bcf      STATUS, RP0

Loop:
 movlw    0                      ; Start with LED off
 btfsc    CMCON0, C1OUT          ; if C1Vin+ >
                                 ; Reference
  movlw   1 << 4                 ;   then turn on LED
 movwf    PORTA

 goto     Loop


 end
```

The asmRwb.asm code could be easily extended to up to four IR phototransistor modules by selecting different comparator inputs (two for each of the two comparators giving you a total of four). Although if you do this, I recommend you do your best to make sure the opto-interrupters have the same part number so you do not have to characterize the circuit repeatedly and provide different comparator (or component) values for each input.

PC/ Simulator

PICkit™ 1

starter kit

Parts Bin

1 PIC16F684

1 Sharp GP2D120 IR rang-
   ing module

1 10-LED bargraph display

1 0.01 μF capacitor

1 Breadboard-mountable
   SPDT switch

1 Three-cell AA battery
   clip

3 AA batteries

Tool Box

DMM

Needle-nose pliers

Wire strippers

Breadboard

Wiring kit

Sharp has a number of different IR object detection and ranging modules that you can interface very simply to any PIC MCU. In this experiment, I will introduce you to the GP2D120 module, which provides an analog signal roughly proportional to the distance between it and some light-reflecting object. The Sharp modules are designed to work using between 3 and 6 volts, with 4.5 to 5.4 volts being the range where they work most efficiently. Other than providing reasonably clean power to the modules, you have to connect only a single line to a PIC MCU to read distances.

The circuit that I used for this experiment is quite simple, as you will see in Figures 10-17 and 10-18, and it should not cause you too many problems in wiring it together. The basic GP2D120 module has a white plastic connector on the bottom, and to use it with a bread board, I simply soldered three short breadboard wiring kit wires to the three pins on the backside of the module (see Figure 10-18).

The code for the application is very straightforward and assumes that the maximum voltage put out by the GP2D120 is 3.0 volts. Instead of displaying the distance as a binary value, I use the bargraph as a 10-position scale with a single lit LED. The application code (asmGP2D12) tests the voltage output from the GP2D120 by repeatedly subtracting 15 from the

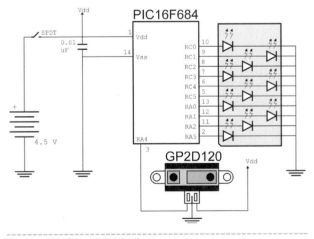

Figure 10-17 *GP2D12 circuit*

Figure 10-18 *Prototype circuit seen from the rear of the GP2D120 module*

returned ADC value. The reason for taking away 15 is that 3.0 volts measured in a maximum 255 range for a 5-volt circuit is 150. Therefore, 15 is equivalent to 300 mV.

I have been quite cavalier with the GP2D120 in this circuit because the circuit has only one LED active at any time and is powered by batteries, and the breadboard has a lot of built-in capacitance. If you were to use the GP2D120 (or any of the other Sharp IR object-detection modules) in an application, I would

recommend placing a fairly large (10 μF or greater) capacitor close to the module's power-supply pins. As I will show in the following experiments, IR receivers can be somewhat sensitive to electrical noise, and in an application that has electrical motors running (e.g., a robot), you will find that the output may be erratic. By adding the capacitor, you should be protecting the module from noise, which could prevent it from operating correctly.

Experiment 92—Do-It-Yourself IR Object Sensor

PC/ Simulator

PICkit™ 1

starter kit

Parts Bin

1	PIC16F684
1	38 KHz IR TV remote control receiver
1	IR LED
1	LED
1	10k resistor
1	1k resistor
1	470Ω resistor
1	100Ω resistor
1	1k Breadboard-mountable potentiometer
1	0.01 μF capacitor
1	47 μF electrolytic capacitor
1	5mm black heat-shrink tubing 1 to 1.25 inches (2.5 to 3 cm) in length
1	Breadboard-mountable SPST switch
1	Three-cell AA battery clip
3	AA batteries

Tool Box

DMM

Needle-nose pliers

Breadboard

Breadboard wiring kit

When Ben Wirz and I designed the "TAB Electronics Build Your Own Robot Kit," we spent about a month struggling with coming up with a way that objects could be detected using simple wires that, when moved, would close or open circuits. An enormous amount of work was done to come up with a design that was sensitive (would detect an object with enough time for the robot to stop), robust (required no bending of wires back into shape after a collision), reliable (made no "false" hits), and inexpensive. The solution that we came up with was a noncontact method of detecting objects by using a modulated IR light beam and a TV remote-control receiver. You would probably be surprised to discover that this method outperformed mechanical switches in each area previously

listed in this book and it was cheaper to implement. In this and the experiments that follow, I will present you with some noncontact object-detection methodologies with which you can experiment.

When you use a remote control to turn a TV on or off or change the station or the volume, the remote control in your hand is usually sending out a modulated stream of IR light pulses. You may have

---

discovered that the remote control can "bounce" its output signal off different objects in the room (e.g., the wall behind you) and the TV will respond as if you were pointing the remote control directly at it. This is the theory behind this experiment; as you can see in Figure 10-19, light is bounced from an IR LED to a TV remote control via some other object. Earlier in this section, an IR white/black sensor performed a similar function, but the distance at which it works is limited to a half-inch (1 cm) or so. The circuit presented here can detect objects as far as several feet away and will even give you some idea of its distance, as you will see in the next experiment.

Considering it costs just a few cents, the TV remote-control receivers (such as the Sharp GP1UD series) are amazingly complex devices. They continually monitor the incoming IR light and respond to a signal that is modulated (turned on and off) at 38 kHz. Inside the TV remote-control receiver, the input signal is processed, and if a 38 kHz signal is encountered, the open collector output is pulled low, which allows multiple receivers on the same circuit. Amazingly it will filter out any constant signals and continually adapt to its environment. What I discovered with the TAB Electronics robot is important: It will learn to filter out any continuous 38 kHz signal as noise if it is left active for more than a few milliseconds. This is why in Figure 10-19, I have drawn a PWM signal that allows the 38 kHz signal to be passed to the IR LED only periodically. Figure 10-20 shows the signal sent to an IR LED and the response from a TV remote-control receiver; the delay of several hundred microseconds in the output is simply the TV remote-control receiver recognizing the incoming signal and then recognizing the signal has stopped.

To demonstrate using an IR LED and TV remote-control receiver for object detection, I came up with the circuit shown in Figure 10-21. It's wired with the IR LED bent over and pointing in the same direction as the TV remote-control receiver (see Figure 10-22). The

IR LED has a 1-inch (2.5 cm) piece of black heat-shrink tubing placed over it to direct the IR waveform away from the TV remote control. The circuitry shown in Figure 10-22 is actually the circuitry for the next experiment. The four LEDs are used to determine the range from the circuit to another object. Only one LED is required for this application.

The 1k resistor and potentiometer used in the application is to limit the amount of current being passed to the IR LED and therefore to limit the light output from the LED, which will help set the point where objects will be detected. This scheme is often used in robots to limit the detection to 1 foot (30 cm) or so, even though it is not really a recommended method to limit the detection distance. A better way will be shown in the next experiment.

The first application for testing this circuit is asmIR.asm and simply delays for 50 ms the sending of 10 pulses to the IR LED. As the signal is being sent, the TV remote-control output is being polled, and if the signal is active for two cycles, then the application acknowledges an object is in front of the circuit.

```
title   "asmIR - Roll Your Own IR Object
Detector"
;
;   This Program Outputs a 38 KHz Signal
;   (26 µs Period) signal
;   for 10 Cycles and Monitors a
;   IR TV Remote Control Receiver
;   for reflections.
;
;   Hardware Notes:
;   PIC16F684 running at 4 MHz
;   Using the Internal Clock
;   RC4 - IR Receiver Input
;   RC0 - Indicator LED
;   RC5 - IR LED Output
;
;
;   Myke Predko
```

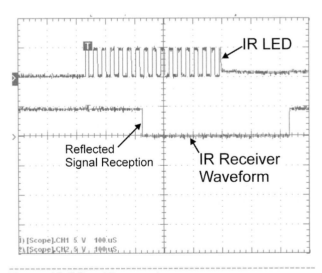

Figure 10-20   *IR operation*

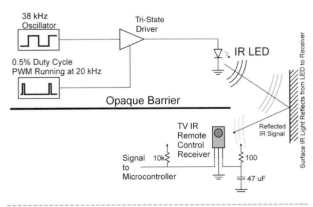

Figure 10-19   *IR detector theory*

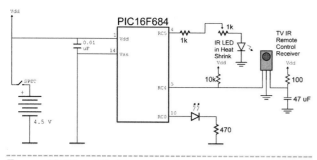

Figure 10-21   *IR detect circuit*

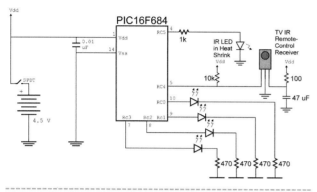

Figure 10-22   *Breadboard circuit used to detect objects and measure the distance to them*

```
;   05.01.03
;
  LIST R=DEC
  INCLUDE "p16f684.inc"


  __CONFIG _FCMEN_OFF & _IESO_OFF & _BOD_OFF &
_CPD_OFF & _CP_OFF & _MCLRE_OFF & _PWRTE_ON &
_WDT_OFF & _INTOSCIO

; Variables
  CBLOCK 0x020
i, j
Dlay
PORTShadow
  ENDC

  PAGE
; Mainline

  org     0

  nop                         ; For ICD Debug

  clrf    PORTA
  clrf    PORTC
  movlw   7                   ; Turn off
                              ; Comparators
  movwf   CMCON0

  bsf     STATUS, RP0
  clrf    ANSEL ^ 0x80
  bcf     TRISC ^ 0x80, 0     ; Visible Light LED
                              ; Output
  bcf     TRISC ^ 0x80, 5     ; IR LED Output
  bcf     STATUS, RP0

  clrf    PORTShadow          ; Clear the PORTC
                              ; Output Value
```

```
Loop:
  movlw   HIGH ((50000 / 5) + 256)
  movwf   Dlay
  movlw   LOW ((50000 / 5) + 256)
  addlw   -1                  ; Wait to Repeat the
                              ; Test
  btfsc   STATUS, Z
  decfsz Dlay, f
  goto    $ - 3               ; 5 Cycle Delay Loop
                              ; for 20 ms

  clrf    j                   ; "j" counts the
                              ; Active IR
  movlw   20                  ; Send 10 Cycles, 1/2
                              ; Cycle at
  movwf   i                   ;  a time
IRLoop:                       ; Output 8 Cycles and
                              ; Count Low Rx
  goto    $ + 1               ; Delay Full 13
                              ; Cycles in Loop
  goto    $ + 1
  btfss   PORTC, 4            ; Poll IR Receive
  incf    j, f                ; If Reset, Increment
                              ; Counter
  movlw   1 << 5              ; Toggle RC5
  xorwf   PORTShadow, w
  movwf   PORTShadow
  movwf   PORTC
  decfsz  i, f                ; Loop Again?
  goto    IRLoop

  movlw   4                   ; Four Polls Low?
  subwf   j, w                ; Carry Set if Four
                              ; are Low
  movf    PORTShadow, w
  andlw   0xFF ^ 1            ; Turn on LED?
  btfsc   STATUS, C
  iorlw   1                   ; Carry Set, j >= 4
  movwf   PORTShadow
  movwf   PORTC

  goto    Loop


  end
```

Although asmIR.asm works perfectly, I wanted to come up with a way of performing the same operation using the CCP's PWM circuitry. Instead of having to come up with the 38 kHz signal for the IR LED in software, the PWM can do it and tell me when it is complete by using the TMR2IF flag and the 16-times postscaler. This has some advantages that will become apparent presently. The changed application is called asmIR 2.asm.

```
title  "asmIR 2 - Roll Your Own IR Object
Detector"
;
;  This Program Outputs a 38 KHz Signal
;   (26 µs Period) signal
;  for 16 Cycles using the ECCP PWM
;  and Monitors a IR TV
;  Remote Control Receiver for reflections.
;
;  Hardware Notes:
;   PIC16F684 running at 4 MHz Using
;   the Internal Clock
;   RC4 - IR Receiver Input
;   RC0 - Indicator LED
```

```
;   RC5 - IR LED Output
;
;
;   Myke Predko
;   05.01.03
;
   LIST R=DEC
   INCLUDE "p16f684.inc"

   __CONFIG _FCMEN_OFF & _IESO_OFF & _BOD_OFF &
_CPD_OFF & _CP_OFF & _MCLRE_OFF & _PWRTE_ON &
_WDT_OFF & _INTOSCIO

;   Variables
   CBLOCK 0x020
i
Dlay
PORTShadow
   ENDC

   PAGE
;   Mainline

   org     0

   nop                         ; For ICD Debug

   clrf    PORTC
   movlw   7                   ; Turn off
                               ; Comparators
   movwf   CMCON0

   movlw   b'00001100'         ; Enable PWM Mode of
                               ; ECCP
   movwf   CCP1CON
   movlw   13                  ; Set the PWM Output
                               ; Reset Control
   movwf   CCPR1L

   bsf     STATUS, RP0
   clrf    ANSEL ^ 0x80
   movlw   26                  ; Set TMR2 Period
   movwf   PR2 ^ 0x80
   bcf     TRISC ^ 0x80, 0     ; Visible Light LED
                               ; Output
   bcf     STATUS, RP0

   clrf    PORTShadow          ; Clear the PORTC
                               ; Output Value

Loop:
   movlw   HIGH ((50000 / 5) + 256)
   movwf   Dlay
   movlw   LOW ((50000 / 5) + 256)
   addlw   -1                  ; Wait to Repeat the
                               ; Test
   btfsc   STATUS, Z
   decfsz  Dlay, f
    goto   $ - 3               ; 5 Cycle Delay Loop
                               ; for 20 ms
   clrf    TMR2                ; Clear TMR2
   bcf     PIR1, TMR2IF        ; Reset Interrupt
                               ; Request Flag
```

```
   movlw   b'01111100'         ; Run TMR2 for 16
                               ; Cycles
   movwf   T2CON

   bsf     STATUS, RP0
   bcf     TRISC ^ 0x80, 5     ; IR LED Output ON
   bcf     STATUS, RP0

   clrf    i                   ; "j" counts the
                               ; Active IR
IRLoop:                        ; Output 8 Cycles and
                               ; Count Low Rx
   btfss   PORTC, 4            ; If RC4 High, Don't
                               ; Increment Count
    incf   i, f
   btfss   PIR1, TMR2IF        ; Wait for TMR2 to
                               ; Time Out
    goto   IRLoop

   bsf     STATUS, RP0
   bsf     TRISC ^ 0x80, 5     ; IR LED Output Off
   bcf     STATUS, RP0

   movlw   b'01111000'         ; Turn TMR2 Off
   movwf   T2CON

   movlw   30                  ; Thirty Polls Low?
   subwf   i, w                ; Carry Set if Four
                               ; are Low
   movf    PORTShadow, w
   andlw   0xFF ^ 1            ; Turn on LED?
   btfsc   STATUS, C
    iorlw  1                   ; Carry Set, j >= 4
   movwf   PORTShadow
   movwf   PORTC

   goto    Loop


   end
```

By using the PIC MCU's built-in PWM, I have come up with a method of performing this task that does not need counted assembly language statements to create the $26\mu s$-period wave sent to the IR LED. Because these assembly language statements are not required, the application could be written in C (as demonstrated in cIR.c).

In cIR.c, you may have noticed that I check the poll count for only a value of 15 instead of 30, as in the asmIR 2.asm routine. The reason for this was my use of the 16-bit variable i as the counter instead of an eight-bit variable. I assumed that the code would take approximately twice as long to execute with a 16-bit counter as it would with an eight-bit counter. The assumption must be reasonable, as the code works without any issues.

Experiment 93—IR Object-Ranging Sensor

PC/ Simulator

PICkit™ 1

starter kit

Parts Bin

1 PIC16F684

1 38 KHz IR TV remote-control receiver

1 IR LED

4 LEDs

1 10k resistor

1 1k resistor

4 470Ω resistors

1 100Ω resistor

1 0.01 µF capacitor

1 47 µF electrolytic capacitor

1 5mm black heat-shrink tubing 1 to 1.25 inches (2.5 to 3 cm) in length

1 Breadboard-mountable SPST switch

1 Three-cell AA battery clip

3 AA batteries

Tool Box

DMM

Needle-nose pliers

Breadboard

Breadboard wiring kit

In the previous experiment, I used a 1k potentiometer to limit the amount of current that passed the IR LED. By limiting the current, I dimmed the LED's output and shortened the detection distance. The problem with this method is that it is not very reliable and would be difficult to reproduce in a manufacturing setting. If you look at a TV remote-control receiver's datasheet, you would discover that as the IR LED's modulating frequency changes, so does the sensitivity of the TV remote-control receiver. If you assume that a brighter signal is required for the TV remote-control receiver to recognize it (when the signal was different than 38 kHz), then you would also assume that a closer object would produce a brighter reflected signal and would only be detected at a closer distance. The purpose of this experiment is to test this hypothesis.

To indicate the basic distance to an object, I modified the circuitry from the previous experiment (see Figure 10-23). The actual differences are minor; the 1k potentiometer has been removed and three additional visible light LEDs and three 470Ω current-limiting resistors have been added. These additional LEDs are used to indicate whether an object is detected at different modulating frequencies.

The base for this experiment's code (asmIRDist.asm) was taken from asmIR 2.asm. One of the reasons for using the PWM to generate the modulating IR signal is that it allows changes to the values passed to it conditionally. By taking the delay and IR object detect code of the previous experiment and placing it in a macro, I was able to come up with a code

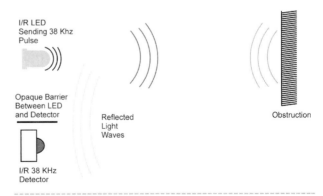

Figure 10-23   *IR ranging circuit*

base that could test different modulating frequencies easily as well as output object detections at these frequencies on different LEDs.

```
title    "asmIRDist - Measure Distance with the
IR Object Detector"
;
;  This Program Outputs a multi-period
;   (starting at 38 kHz Signal,
;   26 µs Period and going to a 31 kHz
;   Signal, 32 µs Period) signal
;   for 16 Cycles using the ECCP PWM
;   and Monitors a IR TV
;   Remote Control Receiver for reflections.
;    Depending on which
;   Signal Produced a reflection, a Specific
;   LED will be lit.
;
;  The actual frequency, Percent off nominal
;  and Receptive Percentage
;   is listed in the Macro invocations
;
;  Note: Macros are used extensively in
;  this application
;
;  Hardware Notes:
;   PIC16F684 running at 4 MHz Using
;   the Internal Clock
;   RC4       - IR Receiver Input
;   RC3:RC0 - Indicator LEDs
;   RC5       - IR LED Output
;
;
;  Myke Predko
;  05.01.07
;
  LIST R=DEC
  INCLUDE "p16f684.inc"

  __CONFIG _FCMEN_OFF & _IESO_OFF & _BOD_OFF &
_CPD_OFF & _CP_OFF & _MCLRE_OFF & _PWRTE_ON &
_WDT_OFF & _INTOSCIO

; Variables
CBLOCK 0x020
i
Dlay
PORTShadow
 ENDC

; Macros
Dlay50ms Macro                 ; 50 ms Delay
  movlw   HIGH ((50000 / 5) + 256)
  movwf   Dlay
  movlw   LOW ((50000 / 5) + 256)
  addlw   -1                   ; Wait to Repeat the
                               ; Test
  btfsc   STATUS, Z
   decfsz Dlay, f
    goto  $ - 3                ; 5 Cycle Delay Loop
                               ; for 20 ms
  endm

IRTest Macro period, led
  movlw   b'00001100'          ; Enable PWM Mode of
                               ; ECCP
  movwf   CCP1CON
  movlw   period / 2           ; Square Wave Output
  movwf   CCPR1L

  bsf     STATUS, RP0
```

```
  movlw   period               ; Set TMR2 Period
  movwf   PR2 ^ 0x80
  bcf     STATUS, RP0

  clrf    TMR2                 ; Clear TMR2
  bcf     PIR1, TMR2IF         ; Reset Interrupt
                               ; Request Flag
  movlw   b'01111100'          ; Run TMR2 for 16
                               ; Cycles
  movwf   T2CON

  bsf     STATUS, RP0
  bcf     TRISC ^ 0x80, 5      ; IR LED Output ON
  bcf     STATUS, RP0

  clrf    i                    ; "i" counts the
                               ; Active IR
  btfss   PORTC, 4             ; If RC4 High, Don't
                               ; Increment Count
   incf   i, f
  btfss   PIR1, TMR2IF         ; Wait for TMR2 to
                               ; Time Out
   goto   $ - 3

  bsf     STATUS, RP0
  bsf     TRISC ^ 0x80, 5      ; IR LED Output Off
  bcf     STATUS, RP0

  movlw   b'01111000'          ; Turn TMR2 Off
  movwf   T2CON

  movlw   30                   ; Thirty Polls Low?
  subwf   i, w                 ; Carry Set if Four
                               ; are Low
  movf    PORTShadow, w
  andlw   0xFF ^ (1 << led)    ; Clear LED Bit
  btfsc   STATUS, C
   iorlw  (1 << led)           ; Carry Set, j >= 30
                               ; so Turn on LED
  movwf   PORTShadow
  movwf   PORTC
endm


  PAGE
; Mainline

  org     0

  nop                          ; For ICD Debug

  clrf    PORTC
  movlw   7                    ; Turn off
                               ; Comparators
  movwf   CMCON0

  bsf     STATUS, RP0
  clrf    ANSEL ^ 0x80
  movlw   0x30                 ; Bottom 4 PORTC
                               ; Bits are
  movwf   TRISC ^ 0x80         ;  Visible Light LED
                               ; Outputs
  bcf     STATUS, RP0

  clrf    PORTShadow           ; Clear the PORTC
                               ; Output Value

Loop:
  Dlay50ms                     ; 38.46 kHz -
                               ; 1.2%/100%
  IRTest 26, 0                 ; Period = 26 us,
                               ; LED0
```

```
Dlay50ms                    ; 35.71 kHz -
                              6.0%/90%
IRTest 28, 1                ; Period = 28 us,
                              LED1

Dlay50ms                    ; 33.33 kHz -
                              12.2%/60%
IRTest 30, 2                ; Period = 30 us,
                              LED2

Dlay50ms                    ; 31.25 kHz -
                              20.0%/35%
IRTest 32, 3                ; Period = 32 us,
                              LED3

    goto    Loop

    end
```

After building the circuit, I tried it with both Sharp and Liteon IR receivers. I was surprised at the nonlinear distances the different frequencies produced for both manufacturers' products. With a faster PWM base frequency (i.e., the PIC MCU's clock frequency), modeling of the application could take place, and a better understanding of the distance detection versus PWM frequency could be developed.

Experiment 94—Ultrasonic Distance-Range Sensor

PC/ Simulator

PICkit™ 1

starter kit

Parts Bin

1 PIC16F684

1 Polaroid 6500 ultra-sonic ranger

1 10-LED bargraph display

1 10k resistor

1 0.01 μF capacitor

1 1,000 μF electrolytic capacitor (see text)

1 6-volt lantern battery (see text)

Tool Box

DMM

Needle-nose pliers

Breadboard

Breadboard wiring kit

Although the IR Light provided a basic level of object ranging to a sensor, the best noncontact method for measuring the distance to an object is to measure the *flight time* of an ultrasonic signal. A number of demonstration circuits are available that you could look at, but the Polaroid 6500 is very commonly used, as it is fairly available and very easy to interface to. In this experiment, I will show you how to add a Polaroid 6500 ultrasonic ranger to a PIC MCU and use it to measure distances.

The Polaroid 6500 is a high-energy device. When it is active, it requires 1 ampere of current. This is why I specified a 6-volt lantern battery for this experiment instead of the traditional AA batteries. This battery will provide several amps of current for a surprisingly long period of time. I chose this battery because it could also be used to power the controlling PIC MCU without a voltage regulator. When the Polaroid 6500 is active, it creates on-board voltages in the neighborhood of 400 volts (DC), which will give you a surprising shock (and I am speaking from experience) if you are not careful. When you are working with the Polaroid 6500, make sure you never touch a metal object that could provide a current path through your body, and definitely do not touch the metal transducer (the circular perforated metal piece seen in Figure 10-24) when it is in operation.

The circuit used for this experiment (see Figure 10-25) is quite simple, although a couple of things should be noted when you are laying out your own circuit. First, you'll notice there is no on/off switch in this application. When I used the traditional EG1903 breadboard-mountable switch (rated for 200 mA, not the 1 amp of this application), I found there was a

Figure 10-24 *Polaroid 6500 ultrasonic ranging module with circular transducer disk (from which distances are measured) along with the breadboard experiment circuit and 6-volt lantern battery*

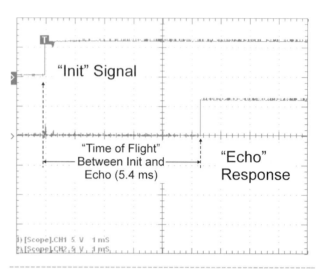

Figure 10-26 *Ultra waveform*

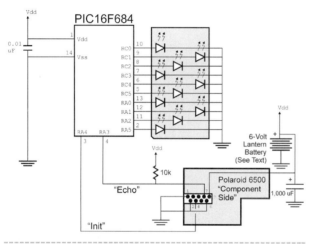

Figure 10-25 *Ultra circuit*

significant voltage drop through it and the breadboard traces that the power passed through. I put the Vdd wire directly into the Vcc rail and kept the power pin of the Polaroid 6500 very close to it. In addition to the lack of a switch and the need to keep power resistances to a minimum, there is a 1,000 μF capacitor across the application's Vdd and Vss. The purpose of this large capacitor is to prevent any large draws from

the Polaroid 6500 from affecting any of the other circuitry (i.e., the PIC MCU) in the application.

Although wiring the Polaroid 6500 into a circuit presents some unique challenges, interfacing it to a PIC MCU is very easy. As shown in Figure 10-26, the Init pin (pin 4) is driven high, and then the time required for the ultrasonic signal to be sent from and return to the transducer is the time it takes for the Echo pin (pin 7) to become active.

Assuming that sound travels at 1,127 feet per second, it takes 73.94 μs to travel 1 inch. When determining the *time of flight*, you have to assume that the signal actually takes 147.88 μs for each inch, to account for the time out and the time back. In the application code (asmUltra.asm), I rounded up the time of flight to 148 μs per inch to measure the time it takes from when the Init pin becomes active to when the Echo pin acknowledges the signal.

In asmUltra.asm, when you look at the LED bargraph, remember that the first LED indicates the distance from the transducer is between 0 and 1 foot. The second LED indicates 1 foot and 2 feet, and so on. If the distance is greater than 10 feet (or an echo is not returned), the first and last LEDs are turned on to indicate the error condition.

Experiment 95—Robot IR Tag

PC/ Simulator

PICkit™ 1

starter kit

Parts Bin

1 PIC16F684

2 38 KHz IR TV remote-control receivers

1 IR LED

3 High-intensity white LEDs

1 Yellow/Red LED

1 4.7k resistor

2 100Ω resistors

2 47 ΩF electrolytic capacitors

1 Length of 5mm black heat-shrink tubing, 1 to 1.25 inches (2.5 to 3 cm) long

Tool Box

BS2-based robot

DMM

Breadboard wiring kit

I don't go out of my way to watch them, but anytime "Robot Wars," "Battle Bots," or other robot combat shows are on, I love to tape them. It's a lot of fun watching two robots duke it out with parts flying everywhere, although I shudder at the amount of work that is being destroyed in just a few minutes. I would love to take part in the competitions, but I would like somebody else to rebuild (and pay for) the robots. In this experiment, you can have many of the thrills of robot fighting, without the mess, cost, and effort of rebuilding. This experiment will let two, or more, robots shoot at each other with beams of light and record the number of "hits."

The circuit and software presented here were originally designed for the "Tab Electronics Sumo-Bot Kit," which I codeveloped, but they could be transferred directly to any other microcontroller-based robot fairly easily. In the photograph of the circuit mounted on a Sumo-Bot (see Figure 10-27), you can see the IR LED cannon, enclosed in a piece of 5mm heat-shrink tubing, along with the IR receivers and four indicator LEDs. The operation of the circuit is quite simple, by pressing the one-bar button on the robot's remote control, a specially coded signal is sent from the IR LED cannon to anything in front of the robot (e.g., another robot). If the robot receives a hit from another robot (the operation of the application prevents a robot from hitting itself), it registers the hit and increments a counter. This counter can be set for one, five, or an unlimited number hits until the indicator LEDs flash, letting the world know that the robot has been "killed." Normally, three high-intensity white LEDs flash in sequence to allow light-seeking, autonomous robots to find and shoot at a human-controlled robot. When a robot has been hit five times, the counter can be reset from the BS2.

Figure 10-27 *Front view of TAB Electronics Sumo-Bot showing off its IR LED "cannon," receivers, LEDs, and controlling PIC16F684*

With the robot controlling the motors and providing basic guidance, the circuit in Figure 10-28 takes care of the business of shooting and registering hits. If you have a Sumo-Bot or a Parallax Boe-Bot, you might think the circuit will be difficult to wire on the small breadboards that come with the robots, but this is really not the case (see Figure 10-29).

There are a few things you should be aware of when you are assembling the circuit:

- The bright white LEDs should be arranged 120 degrees apart. This is important to allow autonomous robots a chance to "see" their competition and know when to fire on them.

- The IR LED must have a piece of 5mm heat-shrink tubing placed over it. If the heat-shrink tubing is not in place, the robot's cannon will have a much wider than 5-to-10-degree field of fire and be able to hit its competitors very easily.

- The IR receivers' leads should not be trimmed so they stick up high above the robot. As I have outlined the circuit layout, they should be placed roughly back to back to ensure they receive signals from all around them.

- Before starting a tournament, I highly recommend that the two robots are placed facing each other, and that one shoots at the other until the other starts flashing. Then reverse the process. This is less to ensure that the receivers are in place than to make sure the cannons are properly aimed. Because the cannons are going into breadboards, they can be easily dislodged and put back into the breadboards incorrectly. Before the competition starts, everybody should be confident that his or her robots can fire. This process will also ensure that a competitor has not commanded the PIC16F684 to turn off receiving hits from the two IR receivers.

The PIC MCU software has resided on a PIC16C505, PIC16F630, and PIC16F684. It is quite versatile and easily ported to different devices. I have taken out the conditional assembly code for the low-end (PIC16C505) execution to try and make it as readable as possible. (You may think this was a losing battle.) However, I want to point out that the conditional assembly code was used only for disabling the comparators and ADC (which the PIC16C505 does not have), changing PORTA in the PIC16F630 and PIC16F684 to PORTB in the PIC16C505, and taking advantage of the TRIS instruction of the PIC16C505. This last change was the most significant and potentially the most confusing when you read the code; to provide a one-instruction access to the TRISA register, I placed its address in the FSR register rather than setting and resetting the RP0 bit of the STATUS register.

Table 10-2 lists the different BS2 commands that are built into the PIC16F684 for controlling the IR tag circuit. These commands were chosen to give the user as much flexibility as possible. Note that the only BS2 command that returns values is 16, not 0x81 that I use in the BS2 interface applications presented in this book. I created this application (asmIRTag.asm) before starting this book, and I followed the BS2 interface built into the Sumo-Bot rather than thinking of a more generic standard. Second, notice that the robot will fire only once a second, and if the hit-counter value has been reached, it can no longer fire until it is reset.

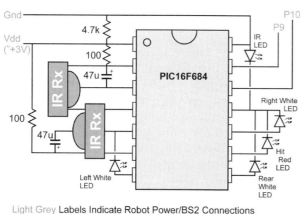

Light Grey Labels Indicate Robot Power/BS2 Connections

Figure 10-28   *IR tag circuit*

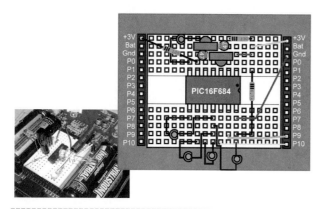

Figure 10-29   *IR tag layout on a Tab Electronics Sumo-Bot Breadboard*

I also have listed a sample Sumo-Bot application (called Max 5 Tournament.bs2) for a five-hit tournament competition. This code is loaded into the robot's BS2 and allows the robot to be controlled by the robot's remote control as well as fire at its opponent.

This circuit (and software) has been tested on a variety of different Sumo-Bots and has been built by a fair number of people. Through quite a bit of testing, we have found that the five-hit tournament seems to be optimal, giving a minute and a half of combat and an opportunity for somebody to come from behind and win. The tournaments are very exciting and as fun as watching other people's robots destroy each other on TV.

Table 10-2

BS2 Commands Supported by the PIC16F684 IR Tag Microcontroller

Command	Value	Comments
TagFire	1	Fire once per second. No firing when robot "killed"
TagIROff	2	Turn off the IR sensors
TagIROn	3	Turn on the IR sensors (default condition)
TagWhiteOff	4	All three white LEDs are turned off
TagWhiteAll	5	All three white LEDs are turned on
TagLeftWhite	6	Toggle the state of the left white LED
TagRightWhite	7	Toggle the state of the right white LED
TagRearWhite	8	Toggle the state of the rear white LED
TagWhiteCycle	9	White LEDs cycle (default condition)
TagIndOn	10	Indicator LED on
TagIndOff	11	Indicator LED off (default condition)
TagHitClear	12	High counter clear
TagMaxHit1	13	Maximum 1 hit before LEDs flash/firing stops
TagMaxHit5	14	Maximum 5 hits before LEDs flash/firing stops
TagNoMaxHit	15	No maximum hits (default condition)
TagGetHitNum	16	RETURN the number of hits in the counter

Motor Control

PC/ Simulator

PICkit™ 1

starter kit

Tool Box

PIC12F675 or PIC16F684- based BS2 command simulator interface

DMM

Soldering iron

Solder

Needle-nose pliers

Wire clippers

Breadboard

Wiring kit

Scissors

Krazy Glue

Required Parts

1 PIC16F684

1 L293D motor driver chip

4 2N3904 NPN bipolar transistors

2 2N3906 PNP bipolar transistors

4 1N914 (1N4148) silicon diodes

1 10-LED bargraph display

1 Servo connector (see text)

1 470Ω 10-pin resistor SIP

4 100Ω resistors

1 10k breadboard-mountable potentiometer

1 0.01 μF capacitor

1 Breadboard-mountable SPST switch

2 DC motors

1 Bipolar stepper motor

1 Unipolar stepper motor (see text)

2 Radio control servos

1 Four-cell AA battery clip

1 Three-cell AA battery clip

4 AA alkaline batteries

1 Six-pin 0.100-inch header (see text)

1 Four-pin 0.100-inch header (see text)

1 Cardboard

I expect that you feel you understand small DC motors very well; you've been exposed to them since you were a small child, in toys and different devices. You've probably seen how you can reverse the direction of motor rotation by switching the polarity of the battery that drives. And, you surely know that you can change the speed of the motor by adding more batteries. It doesn't seem like there is much more to understand.

As it turns out, there's quite a bit more, especially when you consider that the basis of all types of motors is the *inductor* (or *coil*). To devise different methods of controlling current and power to the inductor, you need to be aware of the moderate to high current that passes through it and of the "kickback" produced when the amount of current passing through the

inductor is changed. An inductor stores energy as a magnetic field. When the current passing through the coil is changed, the coil resists this change, as it will cause a change in the amount of energy stored in the magnetic field. This *resistance to change* takes the form of large voltage spikes and current transients.

Figure 11-1 shows a simple NPN transistor control of an inductor. The diode beside the inductor is a kickback diode, designed to absorb the voltage and current transients produced when the transistor turns on and off. When using a bipolar transistor for motor control, this diode must always be present. This circuit can be used as the basis of a DC motor control by changing the inductor with a relay and then wiring the application, as shown in Figure 11-2.

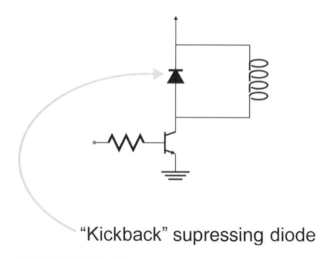

"Kickback" supressing diode

Figure 11-1   *Coil control*

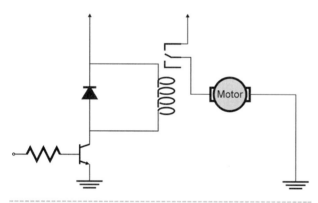

Figure 11-2   *Relay control*

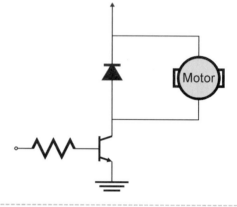

Figure 11-3   *Motor control*

The relay motor-control circuit of Figure 11-2 is the first approach many people use to controlling a motor. This is disappointing because it is redundant, especially compared with the single-transistor motor control of Figure 11-3. There are reasons for using a relay in some applications; for example, the available driver transistors cannot switch enough current to effectively power the motor or the voltage drop through the transistor is too high for the motor selected for the application to work at full power. But for the most part, you should be using the simple single-transistor motor control. Not only does the circuit control the basic operation of turning on and off the motor, it also allows very fast switching of the motor, which in turn allows you to use high-speed *pulse width modulation* (PWM) for motor-speed control, rather than attempting to vary the voltage level, which can be very problematic.

There is a basic problem with the motor control shown in Figure 11-3: The direction of the motor cannot be reversed. Figure 11-4 shows a *full H-bridge circuit* in which the direction of current passing through the motors can be changed by selecting different

transistors. In this section you will be building a full H-bridge circuit using discrete bipolar transistors, and you will see a common full H-bridge chip that can be used for controlling two DC motors simultaneously.

A modification to the full bridge is the *half H-bridge circuit* (see Figure 11-5). Although requiring literally twice the power supply of the full H-bridge, the half H-bridge requires only half the number of driver transistors. This circuit has advantages over the full H-bridge, as it does not require complementary PNP or P-Channel MOSFETs to the NPN or N-Channel MOSFET drivers.

In this section, I will experiment with a number of different DC motors, their drivers, and PIC® MCU software control methodologies. I am going to assume that you understand basic transistor theory and the operation of PWMs. By doing so we can focus on how the PIC MCU can be used to control DC motors and what kind of built-in features are available to take advantage of.

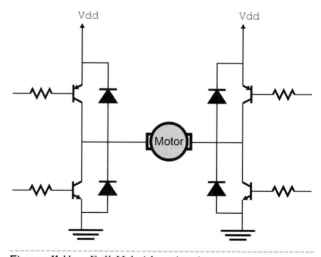

Figure 11-4   *Full H-bridge circuit*

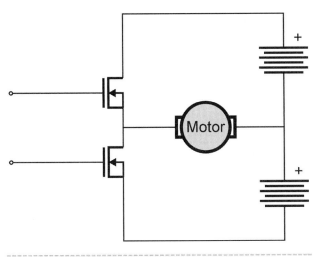

I specified the particular parts for motor drivers in the following experiments because these parts should (a) be easy to find, and (b) work with virtually all motors. The parts are not optimized for specific motors, and you will find cases where the motors will not develop full torque, will drain your batteries quickly, and/or will get warm. The drivers, motors, and PIC MCUs used in the experiments shouldn't become uncomfortably hot to the touch. The purpose of this section is to demonstrate how the PIC MCU is used to control motors, not how to design motor drivers.

Figure 11-5   *Half H-bridge circuit*

Experiment 96—DC Motor Driven Using the CCP PWM and Using a Potentiometer Control

PC/ Simulator

PICkit™ 1

starter kit

Parts Bin

Tool Box

DMM

Soldering iron

Solder

Needle-nose pliers

Wire clippers

Breadboard

Wiring kit

1 PIC16F684

2 2N3904 NPN bipolar transistors

2 2N3906 PNP bipolar transistors

4 1N914 (1N4148) silicon diodes

4 100Ω resistors

1 10k breadboard-mountable potentiometer

1 0.01 μF capacitor

1 Breadboard-mountable SPST switch

1 DC motor (see text)

1 Three-cell AA battery clip

3 AA alkaline batteries

Programming a transistor to turn a motor on and off is a very simple application for the PIC MCU. I would expect that at this point, you would be able to create an application that sent a high output to an I/O pin that was connected to a current-limiting resistor and a PNP transistor. In this application, I want to jump right to creating a fairly complex control application that allows forwards and reverses, as well as speed control of a DC motor using the full H-bridge circuit (see Figure 11-6). The direction and speed of the motor is specified by the potentiometer. When the potentiometer is

centered, the motor is stopped. When the potentiometer is turned toward one extreme, the motor will start turning and speed up as the potentiometer reaches the end of its travel. If the potentiometer is turned in the opposite direction from center, the motor will turn in the opposite direction, starting off slowly and moving much faster as the wiper approaches the other stop. This is a fairly intuitive interface and one that will allow you to see the operation of the H-bridge as well as the PWM signal that controls the motor's speed.

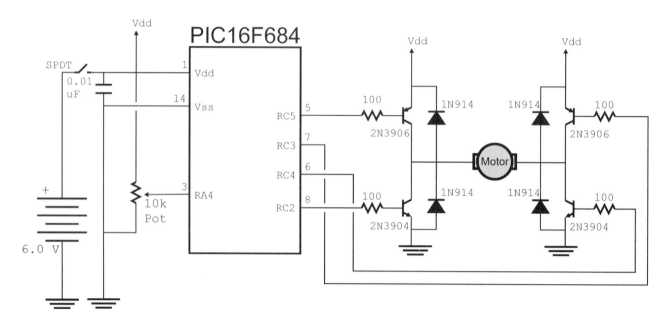

Figure 11-6   *DC motor circuit*

The circuit (see Figure 11-6) was designed for using the PWM capabilities of the PIC16F684's ECCP, which requires RC5:RC2 to interface to four motor drivers and run a single motor in both directions with PWM speed control. My prototype circuit was built on a small breadboard as shown in Figure 11-7, and I used a small hobby motor rated at 4 volts, which I bought at an electronics store. The stranded wire used to connect power to the motor was soldered to clipped leads so that they could be plugged in directly to a breadboard.

As I indicated previously, one of the purposes of this experiment is to control the motors using the built-in ECCP PWM and a potentiometer, which is polled once every 100 ms for controlling the speed and direction of the motor. A PWM frequency of 15 kHz was chosen, as it allows a 64-step PWM control output and still runs above the range of most people's hearing. Generally, a PWM control output of 20 kHz is desired, as it is above the range of hearing and will not affect other devices. To produce a PWM of 20 kHz, a TMR2 reset value of 48 would have to be produced, which requires passing one-third of the value of the ADC, rather than one-half as I have done in this application.

The ECCP PWM is quite easy to set up but does not always work properly when PNP bipolar transistors are used, as in this experiment. The high output of the PIC16F684 is less than the Vdd applied to the PNP transistor, and when the PIC MCU output was high (which turned off the PNP transistor), some current flow remained. To eliminate this current flow, as well as any potential current flow to the NPN transistors when

Figure 11-7   *Single DC motor control circuit using discrete transistors*

they are supposed to be off, I made the pin connected to the base of the unused transistor an input. The application code is called asmMotor.asm:

```
title  "asmMotor - Controlling a DC Motor from
a PIC16F684"
;
;  This Program Monitors a Pot at RA3 (RA3)
;   and moves a
;   DC Motor Accordingly.  Values less than
;   0x80 move the
;   motor in reverse while Values greater
;   than 0x80 move
;   the motor forwards.  When the Pot is
;   at an extreme, the
;   ECCP PWM moves at full speed.
;
;  Hardware Notes:
;   PIC16F684 running at 4 MHz Using
;   the Internal Clock
```

```
;     RA4 - Pot Command
;     RC5/P1A - Motor Forwards High
;     RC4/P1B - Motor Reverse PWM (on Low)
;     RC3/P1C - Motor Reverse High
;     RC2/P1D - Motor Forwards PWM (on Low)
;
;
;  Myke Predko
;  05.01.09
;
  LIST R=DEC
  INCLUDE "p16f684.inc"


  __CONFIG _FCMEN_OFF & _IESO_OFF & _BOD_OFF &
_CPD_OFF & _CP_OFF & _MCLRE_OFF & _PWRTE_ON &
_WDT_OFF & _INTOSCIO


; Variables
  CBLOCK 0x020
Dlay
ADCValue
  ENDC


  PAGE
; Mainline

  org     0

  nop                         ; For ICD Debug

  clrf    PORTC
  movlw   7                   ; Turn off
                              ; Comparators
  movwf   CMCON0
  movlw   b'00001101'         ; Enable ADC on RA4
  movwf   ADCON0
  clrf    TMR2                ; Using TMR2 as a PWM
                              ; Generator
  movlw   b'00000100'         ; Enable TMR2
  movwf   T2CON
  clrf    CCPR1L              ; Nothing Moving
                              ; (Yet)

  bsf     STATUS, RP0
  movlw   1 << 3              ; RA4 (AN3) ADC Input
  movwf   ANSEL ^ 0x80
  movlw   b'00010000'         ; Select ADC Clock as
                              ; Fosc/8
  movwf   ADCON1 ^ 0x80
  movlw   64                  ; Setup the Limit for
                              ; the PWM
  movwf   PR2 ^ 0x80          ; 15 kHz Frequency
  movlw   b'000011'           ; RC5:RC2 Outputs
  movwf   TRISC ^ 0x80
  bcf     STATUS, RP0

  movlw   b'01001110'         ; Enable PWM Forwards
                              ; with
  movwf   CCP1CON

Loop:
  movlw   HIGH ((100000 / 5) + 256)
  movwf   Dlay                ; Delay 100 ms
                              ; Between ADC Polls
  movlw   LOW ((100000 / 5) + 256)
  addlw   -1
  btfsc   STATUS, Z
   decfsz Dlay, f
    goto  $ - 3

  bsf     ADCON0, GO          ; Start ADC
  btfsc   ADCON0, GO          ; Wait for it to
                              ; finish
   goto   $ - 1

  movf    ADRESH, w           ; Read the ADC Value
  movwf   ADCValue
  movlw   0x80                ; Forwards or
                              ; Reverse?
  subwf   ADCValue, w
  btfss   STATUS, C           ; Less than 0x80 go
                              ; in Reverse
   goto   MotorReverse
MotorForwards:
  movwf   ADCValue            ; Divide ADValue / 2
                              ; for PWM Value
  movlw   b'01001110'         ; Enable PWM Forwards
                              ; with
  movwf   CCP1CON             ; P1A/P1C/P1B/P1D
                              ; Active High
  bsf     STATUS, RP0
  bsf     TRISC ^ 0x80, 3     ; Turn Off Low Side
                              ; PNP Driver
  bsf     TRISC ^ 0x80, 4     ; Turn Off High Side
                              ; NPN Driver
  bcf     TRISC ^ 0x80, 5     ; Turn On High Side
                              ; PNP Driver
  bcf     TRISC ^ 0x80, 2     ; Turn ON Low Side
                              ; NPN Driver
  bcf     STATUS, RP0
  bcf     STATUS, C
  rrf     ADCValue, w
  movwf   CCPR1L              ; Have the New ADC
                              ; Limit
  goto    Loop                ; Wait 100 ms to
                              ; Resample


MotorReverse:
  movlw   0x7F                ; Cheesy 7 Bit
                              ; Negation of the
  xorwf   ADCValue, f         ; Reverse CCPR1L
                              ; Value
  movlw   b'11001110'         ; Enable PWM Reverse
                              ; with
  movwf   CCP1CON             ; P1A/P1C/P1B/P1D
                              ; Active High
  bsf     STATUS, RP0
  bsf     TRISC ^ 0x80, 5     ; Turn Off Low Side
                              ; PNP Driver
  bsf     TRISC ^ 0x80, 2     ; Turn Off High Side
                              ; NPN Driver
  bcf     TRISC ^ 0x80, 3     ; Turn On High Side
                              ; PNP Driver
  bcf     TRISC ^ 0x80, 4     ; Turn ON Low Side
                              ; NPN Driver
  bcf     STATUS, RP0
  bcf     STATUS, C
  rrf     ADCValue, w
  movwf   CCPR1L              ; Have the New ADC
                              ; Limit
  goto    Loop                ; Wait 100 ms to
                              ; Resample

  end
```

Along with writing the application in assembler, I translated it into C as cMotor.c as follows:

```c
#include <pic.h>
/*  cMotor - Control a DC Motor using a
Potentiometer
```

```
This Program Monitors a Pot at RA3 (RA3) and
moves a DC Motor Accordingly. Values less than
0x80 move the motor in reverse while Values
greater than 0x80 move the motor forwards.
When the Pot is at an extreme, the ECCP PWM
moves at full speed.


   Hardware Notes:
     PIC16F684 running at 4 MHz Using the Internal
Clock
     RA4 - Pot Command
     RC5/P1A - Motor Forwards High
     RC4/P1B - Motor Reverse PWM (on Low)
     RC3/P1C - Motor Reverse High
     RC2/P1D - Motor Forwards PWM (on Low)


myke predko
05.01.10

*/

__CONFIG(INTIO & WDTDIS & PWRTEN & MCLRDIS &
UNPROTECT \
   & UNPROTECT & BORDIS & IESODIS & FCMDIS);


int  Dlay;
             // LED Time on Delay Variable
char ADCValue;

main()
{

  PORTC = 0;
  CMCON0 = 7;
             // Turn off Comparators
  ANSEL = 1 << 3;
             // RA4 (AN3) is the ADC Input

  ADCON0 = 0b00001101;
             // Turn on the ADC
             // Bit 7 - Left Justified Sample
             // Bit 6 - Use VDD
             // Bit 4:2 - RA4
             // Bit 1 - Do not Start
             // Bit 0 - Turn on ADC
  ADCON1 = 0b00010000;
             // Select the Clock as Fosc/8

  TMR2 = 0;
             // TMR2 Provides PWM Period
  PR2 = 64;
             // 15 kHz PWM Frequency
  T2CON = 0b00000100;
             // Enable TMR2
  CCPR1L = 0;
             // 0 Duty Cycle to Start Off

  while(1 == 1)
  {
    NOP();
    for (Dlay = 0; Dlay < 6666; Dlay++);
             // 100 ms between
    NOP();
             // Samples

    GODONE = 1;
             // Read Pot Value
    while (GODONE);

    ADCValue = ADRESH;
             // Read in ADC Value
    if (ADCValue > 0x80)
             // go Forwards
    {
      CCPR1L = (ADCValue - 80) >> 1;
      CCP1CON = 0b01001110;
      TRISC = 0b011011;
             // RC5/RC2 Output, RC3/RC4 Input
    }
    else
             // Go in Reverse
    {
      CCPR1L = (ADCValue ^ 0x7F) >> 1;
      CCP1CON = 0b11001110;
      TRISC = 0b100111;
// RC5/RC2 Output, RC3/RC4 Input
    } // fi
  } // elihw
} // End cMotor
```

You should be aware of a few things before beginning this experiment. First, the circuit is meant to work with a wide variety of different small hobby motors. You may find that the circuit does not have a very long battery life, the transistors get warm due to excessive current passing through them, or the motors do not produce a lot of torque. As indicated previously, the motor drivers are general-purpose circuits. Therefore by understanding your motor's parameters and specified power supply, and by using driver transistors with appropriate current, voltage, and resistance parameters, your circuits will be more likely to perform at maximum efficiency. The four kickback suppression diodes are not optional, even if you have a very small motor. And, if your circuit acts strangely, you may wish to add a 10 to 47 μF electrolytic capacitor to the PIC MCU's power pins and a 0.1 μF capacitor across the motor to help reduce the electrical noise produced by the motor.

Do not disassemble this circuit when you are finished; it will be required for the next experiment.

Experiment 97—DC Motor Control with Simple TMR0 PWM

PC/Simulator

PICkit™ 1

starter kit

Tool Box

DMM
Needle-nose pliers
Wire clippers
Breadboard
Wiring kit

Parts Bin

1 PIC16F684

2 2N3904 NPN bipolar transistors

2 2N3906 PNP bipolar transistors

4 1N914 (1N4148) silicon diodes

4 100Ω resistors

10k breadboard-mountable potentiometer

1 0.01 µF capacitor

1 Breadboard-mountable SPST switch

1 DC motor (see text)

1 Three-cell AA battery clip

3 AA alkaline batteries

This experiment is a repeat of the previous one, except instead of using the built-in ECCP PWM circuitry, I used TMR0 to overflow every 1,024 cycles, providing a timebase for the PWM. By delaying for 32 cycles, a 30 Hz PWM is produced. Although in the previous experiment, I used the built-in PWM hardware to produce a signal above the range of human hearing, by using TMR0 I have created a PWM signal *below* human hearing. Several reasons exist for doing this. First, small hobby and toy motors often work more efficiently at very low PWM frequencies. Another reason is to allow a basic timebase in the software so other events or elements (including multiple motors) can be synchronized. This is a very important consideration when you are designing a robot. Finally, TMR0-based timebase allows the use of low-cost PIC MCUs that don't have built-in PWM generators.

The application circuit for this experiment is the same as in the previous experiment (see the schematic in Figure 11-6 and prototype wiring in Figure 11-7).

The application code for this experiment is called asmMotor 2.asm. Along with the reduced PWM interval, I also added a very simple state machine that alternatively sets the GO/DONE bit of ADCON0 and reads the value in ADRESH. In the previous experiment, I could poll the state of GO/DONE continuously, as the PWM operates in the background without any software intervention (other than to disable or change the PWM duty cycle). Another difference is that the pin values of RC5:RC2 do not change, but to change how the H-bridge operates, I changed the pin bit input/output mode values in the TRISC register.

```
  title   "asmMotor 2 - DC Motor Control using a
TMR0 Timebase"
;
;   This Program Monitors a Pot at RA3 (RA3)
;   and moves a DC Motor Accordingly.  Values
;   less than 0x80 move the motor in reverse
;   while Values greater than 0x80 move
;   the motor forwards.  When the Pot is at an
;   extreme, the Software PWM moves at full
;   speed.
;
;   Hardware Notes:
;   PIC16F684 running at 4 MHz Using
;   the Internal Clock
;   RA4 - Pot Command
;   RC5/P1A - Motor Forwards High
;   RC4/P1B - Motor Reverse PWM (on Low)
;   RC3/P1C - Motor Reverse High
;   RC2/P1D - Motor Forwards PWM (on Low)
;
;
;   Myke Predko
;   05.01.10
;
  LIST R=DEC
  INCLUDE "p16f684.inc"

  __CONFIG _FCMEN_OFF & _IESO_OFF & _BOD_OFF &
_CPD_OFF & _CP_OFF & _MCLRE_OFF & _PWRTE_ON &
_WDT_OFF & _INTOSCIO
```

```
; Variables
 CBLOCK 0x020
Direction
ADCState, ADCValue
PWMDuty, PWMCycle
 ENDC

 PAGE
; Mainline

 org    0

 nop                            ; For ICD Debug

 movlw  b'010111'               ; Set PORTC According
                                ; to
 movwf  PORTC                   ; Operating Parameters
 movlw  7                       ; Turn off Comparators
 movwf  CMCON0
 movlw  b'00001101'             ; Enable ADC on RA4
 movwf  ADCON0
 clrf   TMR0                    ; Using TMR0 as a PWM
                                ; Base

 bsf    STATUS, RP0
 movlw  b'11010001'             ; 1:4 Prescaler to TMR0
 movwf  OPTION_REG ^ 0x80;
 movlw  1 << 3                  ; RA4 (AN3) ADC Input
 movwf  ANSEL ^ 0x80
 movlw  b'00010000'             ; Select ADC Clock as
                                ; Fosc/8
 movwf  ADCON1 ^ 0x80
 bcf    STATUS, RP0

 bcf    INTCON, T0IF            ; Wait for TMR0 to
                                ; Overflow
 clrf   ADCState
 clrf   PWMDuty                 ; Not Moving at First
 clrf   PWMCycle                ; Start at the
                                ; Beginning
 clrf   Direction               ; Moving Forwards

Loop:
 btfss  INTCON, T0IF            ; Wait for Timer
                                ; Overflow
 goto   $ - 1
 bcf    INTCON, T0IF            ; Reset and Wait for
                                ; Next

 btfsc  ADCState, 0             ; Start or Read ADC?
 goto   ADCRead
 bsf    ADCON0, GO              ; Start ADC Read
 bsf    ADCState, 0             ; Next State
 goto   ADCDone
ADCRead:                        ; Read the ADC Value
 movf   ADRESH, w
 movwf  ADCValue
 clrf   ADCState                ; Original State
ADCDone:                        ; Process ADC Value

 movlw  0x80                    ; Forwards or Reverse?
 subwf  ADCValue, w
 btfss  STATUS, C               ; Less than 0x80 go in
                                ; Reverse
 goto   MotorReverse

MotorForwards:
 movwf  PWMDuty
 bcf    STATUS, C               ; Convert the Value
                                ; from 7 Bits
 rrf    PWMDuty, f              ; to 5
 bcf    STATUS, C
 rrf    PWMDuty, f
```

```
 clrf   Direction               ; Move Forwards
 goto   MotorUpdate

MotorReverse:
 movlw  0x7F                    ; Cheesy 7 Bit Negation
                                ; of the
 xorwf  ADCValue, w            ; Reverse CCPR1L Value
 movwf  PWMDuty
 bcf    STATUS, C               ; Convert the Value
                                ; from 7 Bits
 rrf    PWMDuty, f              ; to 5
 bcf    STATUS, C
 rrf    PWMDuty, f
 movlw  1
 movwf  Direction               ; Move Backwards

MotorUpdate:                    ; Check to Update the
                                ; Motor
 movf   PWMDuty, w              ; If Duty > Cycle, then
                                ; Off
 subwf  PWMCycle, w
 movlw  b'011011'               ; TRISC Forwards
 btfsc  Direction, 0           ; Forwards or Reverse?
 movlw  b'100111'               ; TRISC Reverse
 btfsc  STATUS, C
 movlw  b'111111'               ; Turn off Motors?
 bsf    STATUS, RP0
 movwf  TRISC ^ 0x80
 bcf    STATUS, RP0

 incf   PWMCycle, f             ; Increment the PWM
                                ; Cycle Count
 bcf    PWMCycle, 5             ; Maximum of 32 States

 goto   Loop                    ; Finished, Loop Around
                                ; Again


 end
```

I created a C version of this application called cMotor 2.c.

```
#include <pic.h>
/*  cMotor 2 - Control a DC Motor using a
Potentiometer

  This Program Monitors a Pot at RA3 (RA3)
    and moves a
  DC Motor Accordingly.  Values less
    than 0x80 move the
  motor in reverse while Values greater
    than 0x80 move
  the motor forwards.  A 1 ms loop (produced
    by TMR0) is
  used to provide a 30 Hz PWM for toy motors


  Hardware Notes:
  PIC16F684 running at 4 MHz Using the Internal
Clock
  RA4 - Pot Command
  RC5/P1A - Motor Forwards High
  RC4/P1B - Motor Reverse PWM (on Low)
  RC3/P1C - Motor Reverse High
  RC2/P1D - Motor Forwards PWM (on Low)


myke predko
05.01.10

*/
```

```
___CONFIG(INTIO & WDTDIS & PWRTEN & MCLRDIS &
UNPROTECT \
    & UNPROTECT & BORDIS & IESODIS & FCMDIS);


char ADCState = 0;
char ADCValue;
char Direction = 0;
char PWMDuty = 0;
char PWMCycle = 0;


main()
{

 PORTC = 0b010111;
            // PORTC to Control Values
 CMCON0 = 7;
            // Turn off Comparators
 ANSEL = 1 << 3;
            // RA4 (AN3) is the ADC Input

 ADCON0 = 0b00001101;
            // Turn on the ADC
            // Bit 7 - Left Justified Sample
            // Bit 6 - Use VDD
            // Bit 4:2 - RA4
            // Bit 1 - Do not Start
            // Bit 0 - Turn on ADC
 ADCON1 = 0b00010000;
            // Select the Clock as Fosc/8

 TMR0 = 0;
            // Use TMR0 for a 1 ms Delay
 OPTION = 0b11010001;
            // 1:4 Prescaler to TMR0

 T0IF = 0;
            // Use Interrupt Flag for 1 ms

 while(1 == 1)
  {
  while(!T0IF);
            // Wait for TMR0 to Overflow
   T0IF = 0;
            // Reset for Next 1 ms Delay

   if (0 == ADCState)
            // Start or Read ADC?
    {
    GODONE = 1;
            // Start ADC
    ADCState = 1;
    }
   else
    {
    ADCValue = ADRESH;   // Read ADC
    ADCState = 0;
            // Reset State Machine
    } // fi

   if (ADCValue >= 0x80)
            // Forward Command?
    {
    PWMDuty = (ADCValue - 0x80) >> 2;
    Direction = 0;
    }
   else
            // Reverse
    {
    PWMDuty = (ADCValue ^ 0x7F) >> 2;
    Direction = 1;
    } // fi

   if (0 == Direction)
    TRISC = 0b011011;
            // Enable Forward Bits
   else
    TRISC = 0b100111;
            // Enable Reverse Bits

   if (PWMCycle >= PWMDuty)
    TRISC = 0b111111;
            // Stop Motors

   PWMCycle = (PWMCycle + 1) % 32;

  } // elihw
} // End cMotor 2
```

This code is much easier to follow than the assembly code version. But if you look closely, you will discover that the C code specifying the on/off of the motor works differently, and even if the PWM has a duty cycle of 0 percent (fully off), the motor driver will be on for a very short time (just a few μs) before the software determines it should be turned off completely. The reason for the difference is primarily an oversight on my part. I discovered it only when I compared the operation of the two programs line by line after seeing the very short pulses on my oscilloscope when I looked at the operation of the two applications.

Experiment 98—Controlling Multiple Motors with PWM and BS2 Interface

PC/Simulator

PICkit™ 1

starter kit

Parts Bin

1 PIC16F684

1 L293D motor driver chip

1 Breadboard-mountable SPST switch

2 DC motors (see text)

1 Four-cell AA battery clip

4 AA alkaline batteries

Tool Box

PIC12F675 or PIC16F684-based BS2 command simulator interface

DMM

Needle-nose pliers

Wire clippers

Breadboard

Wiring kit

One of the points I made in the previous experiment was that multiple motors and complex control software could be added to the 30 Hz motor control PWM software. This is due to the relatively large number of cycles the code spends waiting for TMR0 to overflow before updating the motor PWM code. In this experiment, a second motor is added to the control code that executes while waiting for TMR0 to overflow and that polls for the BS2 instrument interface that was created in the previous section.

Instead of repeating the build of the four-transistor H-bridge used in the previous two experiments, I decided to use the L293D chip, which consists of four two-output-level motor drivers as shown in Figure 11-8. The chip includes kickback diodes on the outputs, so you do not have to add them to your circuit, and is designed for driving two motors with an extra PWM output control (pins 1 and 9). But for most applications, I simply tie these inputs high and control the operation of the motor by changing the level of the control outputs. The chip runs on a 4.5- to 6.0-volt power input (Vcc on pin 16) and can switch a higher or lower voltage input on Mot Powr pin (pin 8). This makes the L293D useful in a lot of applications.

There are three things you should be aware of with the chip (1) The 0.7-volt difference between Vcc and Gnd and the selected output level is due to the bipolar components used in the manufacture of the chip; (2) The chip will dissipate a fair amount of heat when a

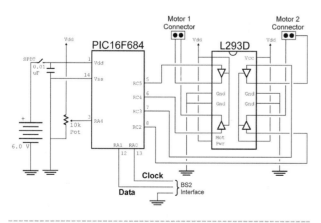

Figure 11-8 *Dual BS2 motor*

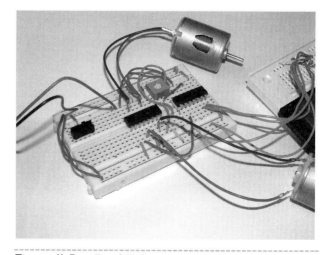

Figure 11-9 *Dual DC motor H-bridge*

motor is running (caused by the 1.4-volt drop through the high and low pins multiplied by the current passing through the motors). The total amount of power the chip can dissipate safely is 3 watts. It is a good idea to provide copper PCB heat sinking to the chip via large *fill* or *flood areas* on your PCB connected to the ground pins to wick away as much heat as possible. (3) The L293D has had spotty availability over the past five years, with some manufacturers dropping shipments of the chip while others have been ramping up. The internal kickback diodes make the chip a lot more useful than does the 754410 (also known as the L293), which is pinout compatible but does not have the diodes on the outputs.

Once you have wired the circuit shown in Figure 11-08 on a breadboard (see Figure 11-9) and connected it to a BS2 or simulator application, you can burn asmBS2Motor.asm into a PIC16F684 and test out the operation of the two motors.

The BS2 command and response operation is quite simple with values 0x00 to 0x3F being used for Motor 1 and values 0x40 to 0x7F used for Motor 2. Like the previous applications, the middle values (0x20 and 0x60 for Motor 1 and Motor 2, respectively) stop the motors, values greater than the middle will cause them to turn forward, and values less than the middle will cause the motors to turn in the reverse direction. Like the previous experiment's code, the PWMs built into this application will have higher duty cycles when the motor values are further away from the middle point.

The motor software shown here could be used as is for a BS2-controlled differentially driven (two motor, one on each side) robot or be the basis of a differentially driven robot that is controlled using the single PIC16F684. The motor control code and PWM cycle-count increment take only 24 cycles, leaving a thousand cycles for sensor and control operations of the robot.

Experiment 99—Bipolar Stepper Motor Control

PC/ Simulator

PICkit™ 1

starter kit

Parts Bin

1 PIC16F684
1 L293D motor driver chip
1 10k breadboard-mountable potentiometer
1 0.01 µF capacitor
1 Breadboard-mountable SPST switch
1 Bipolar stepper motor (see text)
1 Four-cell AA battery clip
4 AA alkaline batteries
1 Four-pin 0.100-inch header (see text)
1 Cardboard

Tool Box

DMM
Soldering iron
Solder
Needle-nose pliers
Scissors
Krazy Glue
Wire clippers
Breadboard
Wiring kit

Stepper motors are very popular devices for a variety of applications because they can be turned a specified amount, do not require external gearing, and are generally very simple for which to design driver circuitry.

In this experiment and the next, I will introduce you to the two most common types of stepper motors, unipolar and bipolar, and then to the driver circuitry and PIC MCU software that will first simply turn them and then control them using a potentiometer, as I did with the DC motors at the start of the section. To help you understand how the software works, I will first present you with C code for driving the stepper motors followed by assembler code.

Stepper motors have a few notable characteristics: They consist of four (or more) coils arranged perpendicularly to each other (see Figure 11-10). These four coils surround a magnetized shaft, which will be either attracted or repelled when the coils are energized. To turn the stepper motor, the coils are energized in a pattern that will cause it to turn in one direction or another. Because of the time required to energize the coils and because of the inertia (as well as any load resistance) of the shaft and reduction gearing placed on the shaft output, the speed of the stepper motor is much more limited than that of the DC motor. The reduction gearing reduces the movement of the motor output from 45 or 90 degrees for each change in position of the shaft to just a couple of degrees or so to maximize the torque output of the motor. Along with the slower speed of the stepper motor, the need to keep at least one coil energized at any one time will draw more current than does the DC motor. On the plus side, stepper motors can be moved a precise amount, and they produce much more torque than does a DC motor.

There are two types of stepper motors commonly in use. The bipolar stepper motor will be presented here and consists of four coils around the magnetized shaft (see Figure 11-10). The bipolar stepper motor can most easily be identified by four wires coming out of the body of the motor with pairs of wires connected together through a small resistance (which is caused by the coils). The high-current push-pull drivers of the H-bridge are used to alternatively turn the two sets of coils on and off as well as turn them to different polarities. In Table 11-1, I have listed the polarities for the different coils to move the shaft by 45 degrees at a time. This is known as *half stepping* and requires that

one or two coils be energized at any time. *Full stepping* moves the shaft by 90 degrees at a time, and only one set of coils is energized at any time.

Table 11-1

Half-Step Coil Energization Pattern for a Bipolar Stepper Motor

Step	Up-Down Coil	East-West Coils
1	South	**Off**
2	South	**South**
3	**Off**	South
4	**North**	South
5	North	**Off**
6	North	**North**
7	**Off**	North
8	**South**	North

In Table 11-1, the North and South specifications are arbitrary and are used to indicate that the polarity of the coils' magnetic fields changes over the course of the sequence. Also note that in Table 11-1, I have emphasized the text of changing coil (one coil changes in each step).

To test out the information in Table 11-1, I created a circuit (see Figure 11-11) to drive a bipolar stepper motor and wired it on a breadboard (see Figure 11-12). With the stepper motor that I used, the connector attached it to a double in-line connector that could be plugged into the breadboard (similar to the ones used in the servo experiments elsewhere in the book). Chances are you will not be so lucky and you will have to solder the individual pins to a single pin in-line header that can be plugged into the breadboard.

Before burning the PIC16F684 with the following software, I suggest you cut a sliver of cardboard as a pointer and Krazy Glue it to the end of the stepper

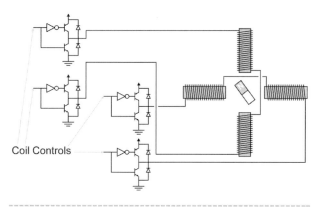

Figure 11-10 *Bipolar stepper motor control*

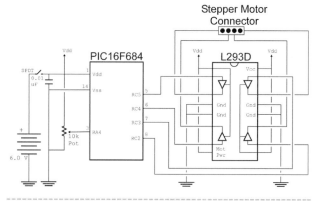

Figure 11-11 *Bipolar stepper circuit*

Figure 11-12   *Bipolar stepper motor test circuit*

motor's output shaft (see Figure 11-12) so that you can clearly observe the movement of the stepper motor.

When the circuit is built, you can burn a PIC16F684 with cStepper.c, which takes the information from Table 11-1 and uses it to create a simple table for half-step driving the bipolar stepper motor. In between steps, there is a quarter second delay, and if your application is wired correctly, you will see the pointer you glued to the stepper motor shaft turning through 360 degrees (a degree or so at a time). If you do not see this pattern, you will have to rearrange the wires on the breadboard until the motor starts working correctly. (Do not desolder and resolder the leads on the 0.100-inch header.)

```
#include <pic.h>
/*  cStepper.c - Turn a Stepper Motor

This Program is based on "asmStepper.asm".

Hardware Notes:
PIC16F684 Running at 4 MHz with Internal
Oscillator
RC5:RC2 - Stepper Motor Outputs

myke predko
05.01.15

*/

__CONFIG(INTIO & WDTDIS & PWRTEN & MCLRDIS &
UNPROTECT \
  & UNPROTECT & BORDIS & IESODIS & FCMDIS);


unsigned int i = 0, j;
const char StepperTable[] = {0b011100, 0b010100,
                             0b000100, 0b100100,
                             0b100000, 0b101000,
                             0b111000, 0b011000};

main()
{
```

```
PORTC = 0;
CMCON0 = 7;
        // Turn off Comparators
ANSEL = 0;
        // Turn off ADC
TRISC = 0b000011;
        // RC5:RC2 Outputs

while(1 == 1)
        // Loop Forever
  {
  NOP();
  for (j = 0; j < 21000; j++);
  NOP();

  PORTC = StepperTable[i];

  i = (i + 1) % 8;

  }  // elihw
}  // End cStepper
```

The assembly language version of cStepper.c is asmStepper.asm as follows. Knowing that the code would never be longer than 255 instructions, I decided to use a basic, not the general-case, table for the stepper positions to simplify the coding.

```
 title  "asmStepper - PIC16F684 Bipolar Stepper
Motor Control"
;
;  This Program Outputs a new Bipolar Stepper
;  Motor Sequence once every 250 ms.
;
;  Hardware Notes:
;   PIC16F684 running at 4 MHz Using
;   the Internal Clock
;   Internal Reset is Used
;   RC5:RC2 - L293D Stepper Motor Control
;
;
;  Myke Predko
;  05.01.14
;
  LIST R=DEC
  INCLUDE "p16f684.inc"

  __CONFIG _FCMEN_OFF & _IESO_OFF & _BOD_OFF &
_CPD_OFF & _CP_OFF & _MCLRE_OFF & _PWRTE_ON &
_WDT_OFF & _INTOSCIO

; Variables
CBLOCK 0x20
Dlay, i
  ENDC

  PAGE
; Mainline

  org     0

  nop                        ; For ICD Debug

  movlw   1 << 2             ; Start with Bit 2
                             ; Active
  movwf   PORTC
  movlw   7                  ; Turn off Comparators
  movwf   CMCON0
  bsf     STATUS, RP0        ; Execute out of Bank 1
  clrf    ANSEL ^ 0x080      ; All Bits are Digital
  movlw   b'000011'          ; RC5:RC2 are Outputs
```

```
        movwf    TRISC ^ 0x080
        bcf      STATUS, RP0        ; Return Execution to
                                    ; Bank 0

        clrf     i

Loop:                              ; Return Here for Next
                                   ; Value
        movlw    HIGH ((250000 / 5) + 256)
        movwf    Dlay
        movlw    LOW ((250000 / 5) + 256)
        addlw    -1                ; 250 ms Delay
        btfsc    STATUS, Z
         decfsz  Dlay, f
          goto   $ - 3

        movf     i, w
        call     SwitchRead
        movwf    PORTC

        incf     i, f             ; i = (i + 1) % 8;
        bcf      i, 3

        goto     Loop

SwitchRead:
        addwf    PCL, f           ; Staying in First 256
                                  ; Instructions
        dt       b'011100', b'010100', b'000100',
                 b'100100'
        dt       b'100000', b'101000', b'111000',
                 b'011000'


        end
```

Once I got the stepper moving consistently in one direction, I then expanded the application so that a potentiometer would control the operation of the bipolar stepper motor in the same way as it would control the DC motors. cStepper 2.c was created to read the potentiometer wired to RA4 and then to change the direction and the delay between steps in the range of 257 to 2 ms.

```
#include <pic.h>
/* cStepper 2.c - Control a Stepper Motor Using
a Pot

This Program is based on "asmStepper 2.asm".

Hardware Notes:
PIC16F684 Running at 4 MHz with Internal
Oscillator
RC5:RC2 - Stepper Motor Outputs
RA4 - Potentiometer Control

myke predko
05.01.15

*/

__CONFIG(INTIO & WDTDIS & PWRTEN & MCLRDIS &
UNPROTECT \
    & UNPROTECT & BORDIS & IESODIS & FCMDIS);


unsigned int i = 0, j;
unsigned char Period;
```

```
const char StepperTable[] = {0b011100, 0b010100,
                             0b000100, 0b100100,
                             0b100000, 0b101000,
                             0b111000, 0b011000};

const int Onems = 83;

main()
{

  PORTC = 0;
  CMCON0 = 7;
// Turn off Comparators
  ANSEL = 1 << 3;
// RA4 (AN3) is the ADC Input

  ADCON0 = 0b00001101;
// Turn on the ADC
        // Bit 7 - Left Justified Sample
        // Bit 6 - Use VDD
        // Bit 4:2 - RA4
        // Bit 1 - Do not Start
        // Bit 0 - Turn on ADC
  ADCON1 = 0b00010000;
// Select the Clock as Fosc/8
  TRISC = 0b000011;
// RC5:RC2 Outputs

  while(1 == 1)                    // Loop Forever
   {
     NOP();
     for (j = 0; j < Onems; j++);
     NOP();
     GODONE = 1;                   // Start ADC
     for (j = 0; j < Onems; j++);
     Period = ADRESH;             // Read Value
     if (0x80 != Period)
// Only Move if Something There
      {
      if (0x80 < Period)          // Forwards
       {
        Period = (Period - 0x80) ^ 0x7F;
        i = (i + 1) % 8;
       }
      else
// Reverse - "Period" OK
        i = (i - 1) % 8;

        PORTC = StepperTable[i];
// Move Stepper

        while (0 != Period)
// Delay at New Position
         {
         for (j = 0; j < Onems; j++);
         Period = Period - 1;
         }  // elihw
      }  // fi
    }  // elihw
}  // End cStepper 2
```

asmStepper 2.asm is the assembly language version of cStepper 2.c and performs exactly the same function.

Due to the nature of stepper motors (i.e., having to delay a set amount between each step), I have not included a BS2 interface example. It is quite simple, due to the ability of the code to poll the BS2 clock line while it is delaying between steps.

PC/ Simulator

PICkit™ 1

starter kit

Parts Bin

1 PIC16F684

4 2N3904 NPN bipolar transistors

4 1N914 (1N4148) silicon diodes

4 100Ω resistors

1 10k breadboard-mountable potentiometer

1 0.01 μF capacitor

1 Breadboard-mountable SPST switch

1 Unipolar stepper motor (see text)

1 Four-cell AA battery clip

4 AA alkaline batteries

1 Six-pin 0.100-inch header (see text)

1 Cardboard

Tool Box

DMM

Soldering iron

Solder

Needle-nose pliers

Wire clippers

Scissors

Krazy Glue

Breadboard

Wiring kit

The unipolar stepper motor is subtly different from the bipolar motor in how its four coils are wired. Instead of current passing through two coils at a time, the unipolar stepper motor is designed with a common connection between each set of parallel coils that allows each coil to be turned on or off individually. The normal wiring configuration of the unipolar stepper motor is shown in Figure 11-13, with the common wires connected to positive power and the individual coil wires being passed to open collector drivers, turning the coils on in sequence and drawing the shaft toward one coil at a time.

The unipolar stepper motor is usually differentiated by the number of wires coming from it. Although in Figure 11-13 I imply that five wires come from the unipolar stepper motor, there are usually six. Each pair of coils has its own common wire. Some stepper motors have five wires coming out of them; these seem to be hybrid unipolar/bipolar stepper motors, where each pair of coils is wired differently. In these cases, you should treat the stepper motors simply as bipolar. If you compare Figures 11-13 and 11-10, you should realize that the unipolar stepper motor can be used in exactly the same applications as the bipolar motor when the common wires are disconnected and the four wires leading to an individual coil are used.

The differences you should be aware of between the two types of stepper motors are few. Because the unipolar motor has only one coil active instead of two, as in the bipolar motor, it doesn't have the same torque as the bipolar motor. On the plus side, the bipolar motor control circuitry is a lot simpler to program, simply pulsing each coil in sequence.

Figure 11-14 shows the circuit I came up with for testing the unipolar servo motor, and Figure 11-15 shows it wired to a breadboard. In Figure 11-14, you

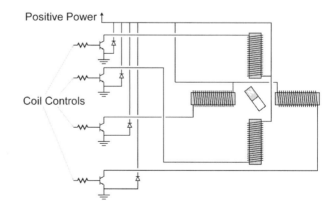

Figure 11-13   *Unipolar stepper motor control*

can see the six-pin header to which the six unipolar stepper motor wires are soldered, with the common wires (found with a DMM resistance check) being placed in the middle of the connector. Along with soldering the stepper motor wires to a header, you should also Krazy Glue a cardboard pointer to the shaft of the stepper motor to observe its motion when you test it.

Testing the unipolar stepper motor is accomplished in exactly the same way as testing the bipolar stepper motor: cStepper 3.c will sequence through the coils, hopefully moving the cardboard pointer continuously. Again, if it doesn't, move the wires to the control transistors until it does. A simpler way of testing and decoding the wiring is to touch the base connection (through the 100Ω resistor) of each transistor to the Vdd rail of the breadboard. This will tell you which circuit is wired to which pin. And then you can begin attaching them in sequence, starting at:

```
#include <pic.h>
/*  cStepper 3.c - Turn a Unipolar Stepper Motor

This Program is based on "asmStepper.asm".

Hardware Notes:
PIC16F684 Running at 4 MHz with Internal
Oscillator
RC5:RC2 - Stepper Motor Outputs

myke predko
05.01.15

*/

__CONFIG(INTIO & WDTDIS & PWRTEN & MCLRDIS &
UNPROTECT \
    & UNPROTECT & BORDIS & IESODIS & FCMDIS);


unsigned int j;
unsigned char OutputVal = 1 << 2;
```

Figure 11-15 *Unipolar stepper motor circuit using NPN transistors as motor control drivers*

```
main()
{

  PORTC = 0;
  CMCON0 = 7;              // Turn off Comparators
  ANSEL = 0;               // Turn off ADC
  TRISC = 0b000011;        // RC5:RC2 Outputs

  while(1 == 1)            // Loop Forever
  {
  NOP();
  for (j = 0; j < 21000; j++);
  NOP();

  OutputVal = (OutputVal & 0x3C) << 1;
  if ((1 << 6) == OutputVal)
    OutputVal = 1 << 2;
  PORTC = OutputVal;

  }  // elihw
}  // End cStepper 3

"asmStepper 3.asm" is the translation of
"cStepper 3.c".
```

As with the previous experiment, I wrote code to control the movement of the unipolar stepper motor using a potentiometer. The control software is almost identical to that used in the previous experiment with a 2 to 257 ms delay in the stepper motor movements. The C language version is called cStepper 4.c:

```
#include <pic.h>
/*  cStepper 4.c - Control a Unipolar Stepper
Motor Using a Pot

This Program is based on "asmStepper 2.asm".

Hardware Notes:
PIC16F684 Running at 4 MHz with Internal
Oscillator
RC5:RC2 - Stepper Motor Outputs
RA4 - Potentiometer Control
```

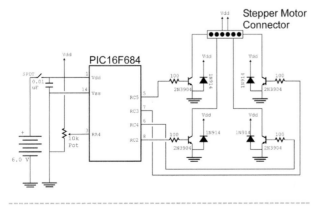

Figure 11-14 *Unipolar stepper circuit*

```
myke predko
05.01.15

*/

__CONFIG(INTIO & WDTDIS & PWRTEN & MCLRDIS &
UNPROTECT \
  & UNPROTECT & BORDIS & IESODIS & FCMDIS);


char OutputVal = 1 << 2;
unsigned int j;
unsigned char Period;
const int Onems = 83;

main()
{

    PORTC = 0;
    CMCON0 = 7;        // Turn off Comparators
    ANSEL = 1 << 3;  // RA4 (AN3) is the ADC
                       Input

    ADCON0 = 0b00001101;
            // Turn on the ADC
            // Bit 7 - Left Justified Sample
            // Bit 6 - Use VDD
            // Bit 4:2 - RA4
            // Bit 1 - Do not Start
            // Bit 0 - Turn on ADC
    ADCON1 = 0b00010000;
            // Select the Clock as Fosc/8
    TRISC = 0b000011;
            // RC5:RC2 Outputs

    while(1 ==                 // Loop Forever
    {
        NOP();
        for (j = 0; j < Onems; j++);
        NOP();
        GODONE = 1;            // Start ADC
        for (j = 0; j < Onems; j++);
        Period = ADRESH;       // Read Value
        if (0x80 != Period)
                // Only Move if Something There
        {
            if (0x80 < Period)  // Forwards
            {
                Period = (Period - 0x80) ^ 0x7F;
                OutputVal = (OutputVal & 0x3C)
<< 1;

                if ((1 << 6) == OutputVal)
                    OutputVal = 1 << 2;
            }
            else
                // Reverse - "Period" OK
            {
                OutputVal = (OutputVal & 0x3C)
>> 1;
```

```
                if ((1 << 1) == OutputVal)
                    OutputVal = 1 << 5;
            }    // fi

        PORTC = OutputVal;   // Move Stepper

        while (0 != Period)
                // Delay at New Position
        {
            for (j = 0; j < Onems; j++);
            Period = Period - 1;
        }    // elihw
    }    // fi
}    // elihw
}    // End cStepper 4
```

asmStepper 4.asm is the assembly language version of asmStepper 4.asm.

You might be wondering why an intermediate value is used for storing and shifting the PORTC unipolar stepper motor position value. Instead of:

```
rlf     OutputVal, w ; Shift the Saved Value
andlw   b'111100'    ; Clear Out Shifted up
                     ; Bits
btfsc   PORTC, 5     ; Roll Over?
 iorlw  1 << 2
movwf   PORTC        ; Store New Active Coil
                     ; Bit
movwf   OutputVal    ; Save New Set Bit
```

you might think that:

```
rlf     PORTC, w     ; Shift Up the Active
                     ; Coil Bit
andlw   b'111100'
btfsc   PORTC, 5     ; If Bit 5 Was Set,
                     ; Rolling over
 iorlw  1 << 2       ; to Bit 2
movwf   PORTC        ; Save New Stepper
                     ; Value
```

is more efficient, as it saves an instruction and a variable. The problem with the second method is that the voltage across the pin is at the one-half Vdd threshold (i.e., the threshold that determines if the pin is high or low and when it is read), and therefore the value is indeterminate. By putting a larger resistor on the base of the driver transistor, there should be a noticeably high voltage at the output pin. But I wanted to keep the circuit as general as possible and provide the maximum current switching possible so I left the 100Ω resistor and added the variable.

Experiment 101—Radio-Control Model Servo Control

PC/Simulator

Tool Box

PICkit™ 1

starter kit

Parts Bin

DMM
Oscilloscope
Needle-nose pliers
Wiring kit

1 PIC16F684
1 0.01 µF capacitor (any type)
1 10k breadboard-mountable potentiometer
1 470Ω 10-pin resistor SIP
1 10-LED bargraph display
1 Servo connector (see text)
1 Radio control servo
1 Breadboard-mountable SPST switch
1 Three-cell AA battery clip

When you are first getting started with robots, I highly recommend using radio control servos for drive motors. Servos are quite inexpensive ($10 or less, and cheaper than many DC motor driver kits or gearhead motors), surprisingly powerful, and very easy to wire into an application as you can see from the schematic circuit for this experiment (see Figure 11-16), which I built on a breadboard (Figure 11-17). Electrically, all you need is a 4.5- to 6-volt power supply, a line from a PIC MCU to drive it, along with a servo connector built out of a couple of three-position, 0.100-inch, in-line connectors (see Figure 11-18).

If you are looking to use servos for driving a mobile robot, they will probably have to be modified for continuous rotation. In *123 Robotics Experiments for the Evil Genius*, I go through the basic steps required to modify a servo for continuous rotation. A quick Google search should find the information for your servos very quickly. Each make and model of servo requires slightly different modifications, and with several hundred different servos available, it would be impossible to try and list the modifications for each one here.

The control signal passed to the R/C servo is known as a *digitally proportional signal*, which is quite a mouthful for something that is really a modified PWM signal. The period of the signal is 20 ms with a pulse and a duty cycle that ranges from 1 ms to 2 ms, with the time between being the set position of the servo.

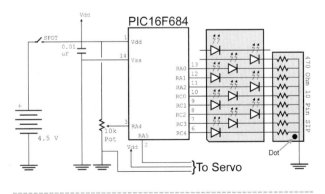

Figure 11-16 *Servo circuit*

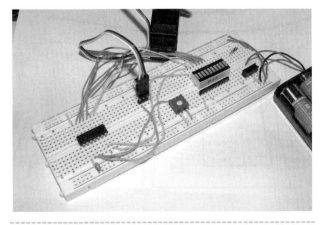

Figure 11-17 *Servo control circuit built on a breadboard*

Figure 11-18 *Servo-to-breadboard connector made from two three-pin in-line connectors soldered together*

Some servos require a 1 to 1.5 ms pulse, but you should be able to identify these quite easily using this circuit. And, if you have such a servo, modifying the signal generated by this experiment is quite easy.

In this experiment, I use a potentiometer and the PIC16F684's ADC to specify the position of the servo. The value returned from the potentiometer is displayed on eight of the 10 LEDs built into an LED display bargraph. This function was originally put in to debug the ADC operation, but I left it in because I liked seeing the LEDs move with the servo. The code for this application is asmServo.asm.

```
  title  "asmServo - Controlling a Servo from a
PIC16F684"
;
;   This Program Monitors a Pot at RA3
;   (RA3) and moves a
;   Servo at RA5 Accordingly.  LEDs Indicate
;   the Position of the Pot.
;
;   Hardware Notes:
;   PIC16F684 running at 4 MHz Using the
;   Internal Clock
;   RA4 - Pot Command
;   RA5 - Servo Connection
#define ServoPin PORTA, 5
;   RC4:RC0 - Bits 7:3 of LED Output
;   RA2:RA0 - Bits 2:0 of LED Output
;
;
;   Myke Predko
;   04.12.26
;
  LIST R=DEC
  INCLUDE "p16f684.inc"

  __CONFIG _FCMEN_OFF & _IESO_OFF & _BOD_OFF &
_CPD_OFF & _CP_OFF & _MCLRE_OFF & _PWRTE_ON &
_WDT_OFF & _INTOSCIO

; Variables
CBLOCK 0x020
Temp, Dlay
ServoCount
ServoState
  ENDC
```

```
  PAGE
;  Mainline

  org      0

  nop                                   ; For ICD Debug

  clrf    PORTA
  clrf    PORTC
  movlw   7                             ; Turn off
                                        ; Comparators
  movwf   CMCON0
  movlw   b'00001101'                   ; Enable ADC on RA4
  movwf   ADCON0

  bsf     STATUS, RP0
  movlw   0xD1                          ; Enable TMR0 with
                                        ; 4x Prescaler
  movwf   OPTION_REG ^ 0x80
  movlw   1 << 3                        ; RA4 (AN3) ADC
                                        ; Input
  movwf   ANSEL ^ 0x80
  movlw   b'00010000'                   ; Select ADC Clock
                                        ; as Fosc/8
  movwf   ADCON1 ^ 0x80
  movlw   b'011000'                     ; RA4/RA3 As Inputs
  movwf   TRISA ^ 0x80
  clrf    TRISC ^ 0x80                  ; All PORTC Outputs
  bcf     STATUS, RP0

  clrf    ServoState                    ; Use Simple Servo
                                        ; State M/C

  movlw   0x80;                         ; Start with Servo
                                        ; in Middle
  movwf   ServoCount

  movlw   HIGH ((20000 / 5) + 256)
  movwf   Dlay
  movlw   LOW ((20000 / 5) + 256)
  addlw   -1                            ; Wait for ADC Input
                                        ; to be Valid
  btfsc   STATUS, Z
  decfsz  Dlay, f
  goto    $ - 3                         ; 5 Cycle Delay Loop
                                        ; for 20 ms

Loop:
  bsf     ServoPin                      ; Output a Servo
                                        ; Signal
  clrf    TMR0
  bcf     INTCON, T0IF                  ; Wait for Overflow
  bsf     INTCON, T0IE
  btfsc   ServoState, 0                 ; Calculate Value in
                                        ; 1 ms Pulse
  goto    ReadADC
StartADC:
  bsf     ADCON0, GO                    ; Start ADC
  bsf     ServoState, 0
  goto    ADCDone
ReadADC:                               ; Read ADC Value
  movf    ADRESH, w
  movwf   ServoCount
  andlw   b'00000111'                   ; Display the ADC
                                        ; Value
  iorlw   1 << 5                        ; Make Sure ServoPin
                                        ; Stays High
  movwf   PORTA
  rlf     ServoCount, w                 ; Need Top 5 Bits
  movwf   Temp
  swapf   Temp, w
  andlw   0x0F
```

```
 btfsc   STATUS, C
  iorlw  0x10             ; Add Top Bit
 movwf   PORTC
 bcf     ServoState, 0    ; Repeat
ADCDone:
 movf    ServoCount, w    ; Get Read with
                          ; Servo Value

 btfsc   STATUS, Z
  movlw  1                ; If Zero, Make 1
 sublw   0                ; Take it away from
                          ; 256

 btfss   INTCON, T0IF
  goto   $ - 1
 movwf   TMR0
 bcf     INTCON, T0IF     ; Wait for Overflow
 movf    ServoCount, w    ; Repeat to get 2 ms
                          ; Delay

 btfss   INTCON, T0IF
  goto   $ - 1
 bcf     ServoPin         ; Finished with the
                          ; Servo

 movwf   TMR0
 bcf     INTCON, T0IF     ; Wait for Overflow
 btfss   INTCON, T0IF
  goto   $ - 1

 movlw   HIGH ((18000 / 5) + 256)
 movwf   Dlay
```

```
 movlw   LOW ((18000 / 5) + 256)
 addlw   -1               ; Want 20 ms Loop
 btfsc   STATUS, Z
  decfsz Dlay, f
   goto  $ - 3            ; 5 Cycle Delay Loop
                          ; for 20 ms
 goto    Loop             ; Repeat


 end
```

There isn't a lot to this application that should surprise you. I use TMR0 to create the delays or 1,024 cycles for the 1 ms delay, and then I use it twice to create the servo position delay. Reading the ADC takes place in the first 1 ms delay, as the operation takes about 15 μs. So there is no chance of it taking longer than the delay, and this makes the servo control operation completely *self-contained*. (Self-contained operations will be discussed in the next experiment.) The three TMR0 delays end up lasting a total of 2 ms, which means I simply have to delay for an additional 18 ms to get a total PWM period of 20 ms.

Experiment 102—Multiple Servo Control
Software Structure

PC/ Simulator

PICkit™ 1

starter kit

Parts Bin

1 PIC16F684

1 0.01 μF capacitor (any type)

1 10k breadboard-mountable potentiometer

1 470Ω 10-pin resistor SIP

1 10-LED bargraph display

2 Servo connectors

2 Radio control servos

1 Breadboard-mountable SPST switch

1 Three-cell AA battery clip

3 AA batteries

Tool Box

DMM

Oscilloscope

Needle-nose pliers

Wiring kit

As I said in the previous experiment, servos are very easy to wire into a circuit. To add a second servo to the previous experiment's circuit, all you have to do is add an additional servo connector as shown in Figure 11.18.

In addition, the additional code is surprisingly simple, especially considering that the second servo operates completely independently of the first.

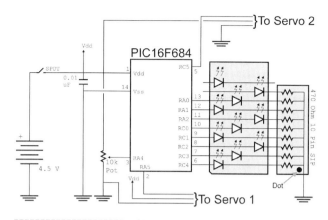

Figure 11-19 *Two-servo circuit*

The code for this experiment (asmServo 2.asm) is a modification of the single-servo application, with code added to hold the servo at one extreme for 600 ms, move it to the other extreme, hold it there for 600 ms, and then move it back to the original extreme. This operation is carried out using a software state machine that is built into the 1 ms initial pulse delay of the servo pulse, just like the potentiometer reading for the first servo. As both servo operations take 2 ms each, the final delay is reduced to 16 ms, so the loop (and each servo's PWM) period remains at 20 ms.

```
title  "asmServo 2 - Controlling two a Servos
from a PIC16F684"
;
;   This Program Monitors a Pot at RA3
;   (RA3) and moves a Servo at RA5 Accordingly.
;   LEDs Indicate the Position of the Pot. A
;   second servo, connected to RC5 will move back
;   and forth under a simple pre-programmed
;   routine.
;
;   Hardware Notes:
;   PIC16F684 running at 4 MHz Using the
;   Internal Clock
;   RA4 - Pot Command
;   RA5 - Servo 1 Connection
#define Servo1Pin PORTA, 5
;   RC5 - Servo 2 Connection
#define Servo2Pin PORTC, 5
;   RC4:RC0 - Bits 7:3 of LED Output
;   RA2:RA0 - Bits 2:0 of LED Output
;
;
;   Myke Predko
;   04.12.26
;
  LIST R=DEC
  INCLUDE "p16f684.inc"


  __CONFIG _FCMEN_OFF & _IESO_OFF & _BOD_OFF &
_CPD_OFF & _CP_OFF & _MCLRE_OFF & _PWRTE_ON &
_WDT_OFF & _INTOSCIO

; Variables
CBLOCK 0x020
Temp, Dlay
Servo1Count, Servo1State
```

```
Servo2Count, Servo2State, Servo2Dlay
  ENDC

    PAGE
;  Mainline

  org     0

  nop                              ; For ICD Debug

  clrf    PORTA
  clrf    PORTC
  movlw   7                        ; Turn off Comparators
  movwf   CMCON0
  movlw   b'00001101'              ; Enable ADC on RA4
  movwf   ADCON0

  bsf     STATUS, RP0
  movlw   0xD1                     ; Enable TMR0 with 4x
                                   ; Prescaler
  movwf   OPTION_REG ^ 0x80
  movlw   1 << 3                   ; RA4 (AN3) ADC Input
  movwf   ANSEL ^ 0x80
  movlw   b'00010000'              ; Select ADC Clock as
                                   ; Fosc/8
  movwf   ADCON1 ^ 0x80
  movlw   b'011000'                ; RA4/RA3 As Inputs
  movwf   TRISA ^ 0x80
  clrf    TRISC ^ 0x80             ; All PORTC Outputs
  bcf     STATUS, RP0

  clrf    Servo1State              ; Use Simple Servo
                                   ; State M/C
  movlw   0x80;                    ; Start with Servo in
                                   ; Middle
  movwf   Servo1Count

  clrf    Servo2State
  movlw   0x20                     ; Stay in Position for
                                   ; 600 ms
  movwf   Servo2Dlay

  movlw   HIGH ((20000 / 5) + 256)
  movwf   Dlay
  movlw   LOW ((20000 / 5) + 256)
  addlw   -1                       ; Wait for ADC Input
                                   ; to be Valid
  btfsc   STATUS, Z
  decfsz  Dlay, f
   goto   $ - 3                    ; 5 Cycle Delay Loop
                                   ; for 20 ms

Loop:
  bsf     Servo1Pin                ; Output a Servo
                                   ; Signal
  clrf    TMR0
  bcf     INTCON, T0IF             ; Wait for Overflow
  bsf     INTCON, T0IE
  btfsc   Servo1State, 0           ; Start Poll?
   goto   ReadADC
StartADC:
  bsf     ADCON0, GO               ; Start ADC
  bsf     Servo1State, 0
  goto    ADCDone
ReadADC:                           ; Read ADC Value
  movf    ADRESH, w
  movwf   Servo1Count
  andlw   b'00000111'              ; Display the ADC
                                   ; Value
  iorlw   1 << 5                   ; Keep Servo Control
                                   ; High
  movwf   PORTA
  rlf     Servo1Count, w           ; Need Top 5 Bits
  movwf   Temp
  swapf   Temp, w
```

```
        andlw   0x0F
        btfsc   STATUS, C
        iorlw   0x10             ; Add Top Bit
        movwf   PORTC
        bcf     Servo1State, 0   ; Repeat
ADCDone:
        movf    Servo1Count, w   ; Get Read with Servo
                                 ; Value
        btfsc   STATUS, Z
        movlw   1                ; If Zero, Make 1
        sublw   0                ; Take it away from
                                 ; 256
        btfss   INTCON, T0IF
        goto    $ - 1
        movwf   TMR0
        bcf     INTCON, T0IF     ; Wait for Overflow
        movf    Servo1Count, w   ; Repeat to get 2 ms
                                 ; Delay
        btfss   INTCON, T0IF
        goto    $ - 1
        bcf     Servo1Pin        ; Finished with the
                                 ; Servo
        movwf   TMR0
        bcf     INTCON, T0IF     ; Wait for Overflow
        btfss   INTCON, T0IF
        goto    $ - 1

        bsf     Servo2Pin        ; Servo2 Code Based on
                                 ; Servo 1
        clrf    TMR0
        bcf     INTCON, T0IF     ; Wait for Overflow
        bsf     INTCON, T0IE
        movf    Servo2State, 0   ; Four Servo 2 States
        xorlw   0
        btfsc   STATUS, Z
        goto    Servo2State_0
        xorlw   1 ^ 0
        btfsc   STATUS, Z
        goto    Servo2State_1
        xorlw   2 ^ 1
        btfsc   STATUS, Z
        goto    Servo2State_2
Servo2State_3:                   ; Going to Original
                                 ; Extreme
        bcf     STATUS, C
        rlf     Servo2Dlay, w    ; Multiply by 16
        movwf   Temp
        rlf     Temp, f
        rlf     Temp, f
        rlf     Temp, f
        movf    Temp, w
        btfsc   STATUS, C        ; At Extreme
        movlw   0xFF
        movwf   Servo2Count      ; Otherwise Store
        decfsz  Servo2Dlay, f
        goto    Servo2Done
        clrf    Servo2State      ; Wait At Extreme
        movlw   0x20
        movwf   Servo2Dlay
        goto    Servo2Done
Servo2State_1:                   ; Going to Other
                                 ; Extreme
        bcf     STATUS, C
        rlf     Servo2Dlay, w    ; Multiply by 16
        movwf   Temp
        rlf     Temp, f
        rlf     Temp, f
        rlf     Temp, f
        movf    Temp, w
        btfsc   STATUS, C        ; At Extreme
        movlw   0xFF
        sublw   0                ; Negate it
        movwf   Servo2Count      ; Otherwise Store
        decfsz  Servo2Dlay, f
```

```
        goto    Servo2Done
        incf    Servo2State, f   ; Wait At Extreme
        movlw   0x020
        movwf   Servo2Dlay
        goto    Servo2Done
Servo2State_0:                   ; At Original Point
        clrf    Servo2Count
        decfsz  Servo2Dlay, f
        goto    Servo2Done
        movlw   0x10             ; Move to Other
                                 ; Extreme
        movwf   Servo2Dlay
        incf    Servo2State, f
        goto    Servo2Done
Servo2State_2:                   ; At Extreme
        movlw   0xFF
        movwf   Servo2Count
        decfsz  Servo2Dlay, f
        goto    Servo2Done
        movlw   0x10             ; Move to Other
                                 ; Extreme
        movwf   Servo2Dlay
        incf    Servo2State, f
Servo2Done:
        movf    Servo2Count, w   ; Get Read with Servo
                                 ; Value
        btfsc   STATUS, Z
        movlw   1                ; If Zero, Make 1 to
                                 ; Avoid Negation Error
        sublw   0                ; Take it away from
                                 ; 256
        btfss   INTCON, T0IF
        goto    $ - 1
        movwf   TMR0
        bcf     INTCON, T0IF     ; Wait for Overflow
        movf    Servo2Count, w   ; Repeat to get 2 ms
                                 ; Delay
        btfss   INTCON, T0IF
        goto    $ - 1
        bcf     Servo2Pin        ; Finished with the
                                 ; Servo
        movwf   TMR0
        bcf     INTCON, T0IF     ; Wait for Overflow
        btfss   INTCON, T0IF
        goto    $ - 1

        movlw   HIGH ((16000 / 5) + 256)
        movwf   Dlay
        movlw   LOW ((16000 / 5) + 256)
        addlw   -1               ; Want 20 ms Loop
        btfsc   STATUS, Z
        decfsz  Dlay, f
        goto    $ - 3            ; 5 Cycle Delay Loop
                                 ; for 20 ms
        goto    Loop             ; Repeat


        end
```

This application demonstrates that you can create surprisingly sophisticated operations in the PIC MCU even when you have a limited amount of time to create them. The second servo state machine requires around 33 instruction cycles to control the servo; about 30 times fewer cycles than are available. With many robot applications, it isn't unusual to see a variety of different functions "tucked inside" the servo delay loops and using the PIC MCU's built-in timers.

Experiment 103—Two-Servo Robot Base with BS2 Interface

PC/Simulator

PICkit™ 1

starter kit

Parts Bin

1 PIC16F684

1 0.01 μF capacitor (any type)

2 Servo connectors

2 Radio control servos

1 Breadboard-mountable SPST switch

1 Three-cell AA battery clip

3 AA batteries

1 Breadboard

Tool Box

PIC12F675 or PIC16F684-based BS2 command simulator interface

DMM

Oscilloscope

Needle-nose pliers

Wiring kit

In the previous two experiments, the full eight-bit value of the ADC was used to specify the position of a servo, and this worked quite well. As will be shown in this experiment, the BS2 interface, as designed, cannot send an eight-bit position value along with a command value, which limits how the controlling servos can move. Along with the restriction on absolute number of bits that can be sent, there is also a restriction on when data can be sent to the receiver. This further reduces how data can be sent to the servos. In this experiment, I will demonstrate one way of solving these issues and discuss how to decide on the best method of implementing this function in your own applications.

Using the program base of the previous experiment, I changed the application so that if a BS2 command came in when the servos were being written to, the command would be ignored. If the servo command was processed while the servos were active, the pulse would be extended, which would result in an unexpected move command. In Figure 11-20, this is indicated by the shaded Windows in which received commands are accepted. The problem with this method occurs when the BS2 command is active, the PIC MCU software has finished sending the control pulses to the servos, and a low BS2 clock is encountered. The protocol used to send the data to the servo-controlling PIC device must

be able to sense that the command is not complete and should be ignored.

The basic BS2 command-interface software does handle a good portion of this function by terminating reception if a timeout happens while waiting for additional clock pulses. In addition to this basic capability, a method is required to determine whether or not the data is valid and to feed back to the BS2 an indicator of whether or not the command has been accepted.

The way I solved these issues was to send only a response command to the PIC16F684, with certain restrictions. The first restriction is that as a response command, bit 7 of the eight bits sent, is always set. If it is low, the command will have to be ignored. Because there are two servos, a bit must be allocated to specify which servo the command is being sent to (I used bit 8 of the incoming data packet). Finally, to ensure the data is valid, the last bit must be zero (the reasons why will be explained in a moment). This reduces the number of position bits to five and provides the servos with only 32 different position states.

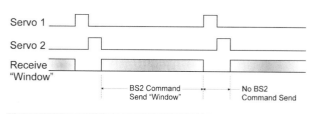

Figure 11-20 *Servo available*

The reason for making the least significant bit zero is to handle the case where data is partially sent when the data-read window becomes active and some of the bits are read. After sending eight bits, the BS2 will put its data pin into input mode, but the PIC MCU will still be reading the primary command data, and its data pin will be in input mode also. Because the PIC MCU data pin is pulled up, a 1 would be received as the least significant bit, which is invalid. Therefore the PIC16F684 controlling the application would reject the command.

I decided to use the response command instead of the straight eight-bit command because if the received command was invalid, the PIC16F684 would not respond (because the command had bit 7 reset) or because bit 0 was set. In the case where eight valid bits were received, the PIC16F684 would attempt to respond with the eight times the servo-position value sent or with 0xA5 to indicate that the command was not received. These responses (i.e., anything other than eight times the servo position) indicate to the sending BS2 or BS2 simulator that that the command was not received and the servo position information has not changed.

This method, while seemingly complex, worked very well on the prototype circuit (see Figure 11-21) used with the asmBS2Servo.asm software and connected to

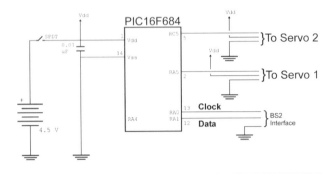

Figure 11-21   *BS2 two-servo circuit*

a BS2 simulator. If the position information was invalidly received, the LCD display on the BS2 simulator would either have an invalid value or a value of 0xA5.

When deciding how to implement a communications protocol like this one, a number of issues should be considered, and, as I will discuss in the next section, a number of different options should be reviewed. In Table 11-2, I have listed four options for implementing a BS2 command and their comments. While the method presented here isn't what I would consider the best, it was the easiest to implement with an unmodified BS2 simulator on a workbench.

Table 11-2

Comparing Different Methods of Sending Servo Position Data from a BS2 or Simulator

Method	Operation Sequence	Comments
Expand the data packet	16-bit data word from BS2	Will require custom software instead of shiftin and shiftout commands. The resulting waveform will not be compatible with existing software.
Add a "handshaking" line from the servo controller to the BS2	After servo commands complete, enable "BS2Receive" pin until 5 ms before next set of servo pulses	Disable "BS2Receive" 5 ms before next servo pulses to avoid changing software timing. Eliminates reset bit 0 and gives a maximum of seven servo position bits. Adds requirements for an additional pin to be used.
Send five-bit servo position along with servo and reset bit 0	Send servo position, expect position * 8 returned	Method used here: Only five position bits but could be used for mobile robot driving servos.
Send two packets with eight bits of position data for specified servo	1. Send high four bits of servo position 2. Send low four bits of servo position 3. Send response command to verify servo position information received	Longest packet length but properly received four bit values could be saved to avoid resending data. This is the most difficult method to test on the workbench with the BS2 simulators.

Solving Programming Problems in PIC® Microcontroller Assembly Language

PC/ Simulator

A good test of your ability to develop assembly-language applications is to attempt to develop programs in assembly that perform functions that are normally programmed in high-level languages. The class of programs that might be most effective for this experiment includes traditional mathematical programming problems and conversions that don't require a lot of learning to understand how they work, but may require a bit of thinking to solve them effectively.

The problems and solutions listed in this section were originally developed for grade 12 (senior year) high school students to learn about assembly-language programming. The assignments themselves are quite simple. None require more than 50 assembly source code lines. Each student was given one of the assignments listed below and the task of researching the requirement and then coming up with the "best" solution for it. Despite the advanced level of the students and the simplicity of the final applications, the students had a surprising amount of difficulty coming up with the solutions to the assignments.

I believe the reasons for the difficulties are as follows:

- Uncertainty with PIC MCU assembly language and the MPLAB® IDE
- Difficulty translating mathematical concepts into programming algorithms
- Inexperience in evaluating programming algorithms

There is a certain irony with the first point; the point of the exercise was to help the student better understand how to program the PIC MCU in assembly language. In the previous sections, the PIC MCU instructions and basic programming building blocks were presented and should have been fairly easy to work through successfully. The programming assignments listed here, although using the skills developed in the previous sections, force the developer to choose the best instructions and code snippets to implement the applications. This uncertainty can be daunting.

The best way to become more familiar with PIC MCU assembly-language programming is to actually do it. This means you should simply start writing code. It may not work properly, but as you think through the process of putting down instructions that carry out specific tasks and see them execute, you'll learn from your mistakes. You'll learn how to recognize that it isn't working, be able to identify the instructions that are producing the wrong result, and then change them so the application code performs the task that you wanted. These skills do not come from reading a book and following an example; you have to force yourself to come up with some code on your own and get it to work.

The MPLAB IDE combines the editor, compiler/ assembler, and simulator in one package and is a very powerful development tool that you will have to become familiar with. So far in this book, I have introduced you to its basic capabilities and given you a few hints on how to use it for application development, but you'll have to learn how to use and customize it so it works most efficiently for you. None of the applications presented in this section (and, as I have argued, none of the applications in the book can be accomplished without taking advantage of the MPLAB IDE simulator).

Translating mathematical concepts into programming algorithms requires an explicit understanding of both the requirements and the potential solution. For these applications, you will have to do some research and be able to understand what is being presented. I recommend keeping the mathematics and programming textbooks handy and using a search engine like Google to research different explanations of the mathematical concepts. Do not fall into the trap of using only one source of information. Although one source may be good for some information, there will definitely be better sources for other information. And as with any research, don't treat everything you find (whether it's from books or the Internet) as being 100 percent correct. Check your sources and get more than one source if possible.

When you understand both the application's requirements and the theory behind the solution, look for multiple ways of implementing the solution. Use this as your basis for determining what is best. When I am given a task, I try to come up with three different ways of accomplishing it. Chances are one solution will become immediately obvious, but if I push myself to look for multiple solutions, I'll either come up with a better solution, or an improvement to your original solution.

Before going on to the assignments and their solutions, I wanted to say a few words about what makes a solution the "best" solution. Try to avoid simplistic measurements like "the shortest program" or "the fastest program," because they don't reflect the dynamic nature of real-world application programming. Instead, try to articulate your requirements as clearly as possible in a defined area as a *requirement continuum*, like the one shown in Figure 12-1. By defining this area, you'll be able to more clearly and easily define execution speed, amount of program memory, and variables that are available to the prob-

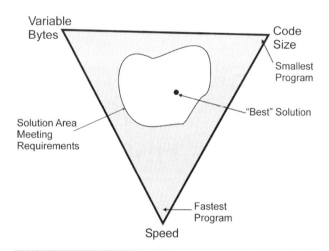

Figure 12-1 *Best triangle*

lem. The solutions that lie outside this area can be immediately rejected, and the best solution can be more easily identified.

Before working through each experiment of this section, I suggest you try to come up with your own solution to the problem (even going as far as writing your own solution and comparing it to mine). In most of the experiment write-ups, I have presented multiple ways the problem can be solved as well as suggested some additional experiments you can write. By doing this, you will get a better understanding of how the solution works and you'll become more familiar with PIC MCU assembly-language programming and the MPLAB IDE. Chances are you'll also come up with better solutions than the ones that I came up with.

Experiment 104—Eight-Bit Multiplication with a 16-Bit Product

PC/ Simulator

The first assignment, multiplying two eight-bit numbers together to form a 16-bit product, can be accomplished a number of different ways. The most obvious way of doing this is to go back to the definition of

multiplication as *repeated addition* and add the multiplicand multiplier number of times as I show in the example below:

```
title   "asmMultiply 1 - Multiply Using Repeated
Addition"
;
; This program uses repeated addition to
; implement Multiplication.
;
; "C" Equivalent Code:
;
; Product = 0;              // 16-Bit Product Value
; Multiplier = Value1;
; Multiplicand = Value2;
;
; while (Multiplier != 0)  // Add Each Value
```

```
; {
;  Product = Product + Multiplicand;
;                 // Add Multiplicand to Product
;  Multiplier = Multiplier - 1;
;                 // Decrement Multiplier Count
; }  // rof
;
; while (1 == 1); // Finished, Loop Forever
;
; Hardware Notes:
;  PIC16F684 running at 4 MHz in Simulator
;
;
; Myke Predko
; 04.08.18
;
  LIST R=DEC
 INCLUDE "p16f684.inc"

; Variables
 CBLOCK 0x20
Product:2                  ; Finished Result
Multiplicand, Multiplier   ; Values for
                           ; Processing
 ENDC

 PAGE
; Mainline of Multiply 1

 org     0

  clrf    Product          ; Initialize
                           ; Variables
  clrf    Product + 1
  movlw   47               ; Value1 <- 47
  movwf   Multiplier
  movlw   35               ; Value2 <- 35
  movwf   Multiplicand
  movf    Multiplier, f    ; Multiplier == 0?
  btfsc   STATUS, Z
   goto   EndLoop          ; Yes, Product = 0
Loop:
  addwf   Product, f       ; Add Multiplicand to
                           ; 16-Bit Product
  btfsc   STATUS, C
   incf   Product + 1, f
  decfsz  Multiplier, f    ; Decrement Multiplier
                           ; and Repeat
   goto   Loop             ; while !=0

EndLoop:
  goto    $                ; Finished, Loop
                           ; Forever

 end
```

This solution could be considered best by some measurements because it requires only 15 instructions. But a fundamental problem exists with this method: The number of cycles needed to multiply two values together ranges from 10 to 1,539 depending on the value of multiplier. For some applications, this is not a problem. But for many applications, especially those involved with real-time operations (such as robots), this variability and potentially long execution time could be a significant problem.

When you first learned to multiply, you may have done it like the example application has done it, but after a while, you used a more sophisticated method in which the multiplicand was shifted to the left accord-

ing to the multiplier digit by which it is being multiplied. This method is actually very easy to implement in PIC MCU assembly if you assume that each digit is a bit. The product of a multiplier digit (or bit) and the multiplicand can be either the multiplicand or zero, and shifting the multiplicand by the multiplier digit is accomplished by simply shifting the multiplicand. For example, to multiply 12 (B′1100′) by 5 (B′0101′) in binary, it would look like:

```
    B'1100' <dot> 1 = B'1100'      <- Bit 0 of
Multiplier (Set)
    B'11000' <dot> 0 = B'00000'    <- Bit 1 of
Multiplier (Reset)
    B'110000' <dot> 1 = B'110000'  <- Bit 2 of
Multiplier (Set)
+ B'1100000' <dot> 0 = B'0000000'  <- Bit 3 of
Multiplier (Reset)
-----------------------------------------------
                  B'0111100' <- Product
(Decimal 60)
```

"asmMultiply 3.asm" shows how this is implemented.

If you were to test the second application with different multipliers, you would discover that it generally executes in 90 to 110 instruction cycles. This is a huge improvement in speed and predictability, with only six instructions (and no file register variables) added to the first example.

I wanted to see if I could improve upon the execution speed of the second multiplication example by multiplying larger digits. Four bits, or hexadecimal digits, seemed like a natural progression, and a table was built with the products for all the possible four-bit multipliers and multiplicands. With this table in place, I then multiplied the two eight-bit numbers by breaking up the multiplier and multiplicand so that they look like the binomial expansion:

```
Product = Multiplier x Multiplicand
        = ((Multiplier & 0x00F) + (Multiplier &
          0x0F0)) x ((Multiplicand & 0x00F) +
          (Multiplicand & 0x0F0))
```

and then multiplying them together using *first, outside, inside, last* (FOIL). The program that I came up with is "asmMultiply 2.asm."

This last program requires about 12 times the number of instructions as the previous examples and one more file register byte for the Temp variable. What it has in its favor is that it *always* executes in 104 instruction cycles. The conditional assembly statements (i.e., the if, else, and endif directives) select between setting the PCLATH register for the table jump or delaying a cycle so no difference exists in how the code executes based on where it is located.

Despite requiring many times more instructions and not being substantially faster than the shifting multiplication algorithm, you should remember this method because the theory behind it can be used for processors that have a built-in eight-bit multiplication instruction (such as the Microchip PIC18 series of microcontrollers). By using the methodology shown in this example, you can create a 16-bit multiplication routine that executes in just a few instructions.

After creating these three applications, we can now ask the question: which one is "best"? For virtually all applications, the second example (asmMultiply 2) would be considered best in terms of execution speed and instruction optimization. The last example would be considered best only for applications where the constant speed is required or where it was known beforehand that size of the multiplier and multiplicand never exceeds 15 (requiring no more than four bits). In the latter case, the eight-bit product would be found in a constant 13 cycles.

Experiment 105—Division of a 16-Bit Value by an Eight-Bit Value

PC/ Simulator

Programming for division routines has many of the same characteristics as programming for multiplication. There are several ways of doing it, and each method has different characteristics that make them best for certain applications. In this experiment, I will look at a number of different ways of implementing division operations and comment on their effectiveness. The routines that are presented here are for *positive, nonzero integers*. I will discuss making these routines work in the general case at the end of the experiment.

If multiplication can be described as repeated addition, then, as its inverse operation, *division* can be described as repeated subtraction, subtraction being the inverse of addition. A simple division routine can be built from a subtraction routine that repeatedly subtracts the divisor into the dividend until the dividend is less than the divisor. The *quotient* is incremented during each loop, and at the end of the routine is equal to the number of times the divisor can be taken away from the dividend. When the dividend is less than the divisor, it is passed back to the calling routine as the *remainder*. A sample program that implements division in this way follows:

```
title   "asmDivide 1 - Divide a 16 Bit Value by
an 8 Bit Value"
;
; This program finds the quotient and
; remainder for dividing a 16-bit
; value from an 8-bit value by
; repeated subtraction.
;
; "C" Equivalent Code:
;
; Remainder = Dividend = 12345;
; Quotient = 0;   // Quotient is 16 Bits
; Divisor = 47;   // Divisor is 16 Bits
;
; while(Remainder >= Divisor)
; {
;   Remainder = Remainder - Divisor;
;                 // Take away Divisor Value
;   Quotient = Quotient + 1;
;                 // Increment Quotient
; }  // elihw
;
; // Quotient and Dividend are correct
;
; while (1 == 1);
;                 // Finished, Loop Forever
;
; Hardware Notes:
;   PIC16F684 running at 4 MHz in Simulator
;
;
; Myke Predko
; 04.04.13
;
  LIST R=DEC
  INCLUDE "p16f684.inc"

; Variables
  CBLOCK 0x20
Remainder:2, Quotient:2
Dividend:2
Divisor
Temp
  ENDC
```

```
        PAGE
; Mainline of Divide

  org     0

  movlw   high 12345      ; Initialize
                          ; Variables
  movwf   Dividend + 1
  movwf   Remainder + 1
  movlw   low 12345
  movwf   Dividend
  movwf   Remainder
  movlw   47
  movwf   Divisor

  clrf    Quotient + 1
  clrf    Quotient

DivideLoop:
  movf    Remainder + 1, w  ; Subtract Divisor
                            ; from Remainder and
  movwf   Temp              ; see if >= 0
  movf    Divisor, w
  subwf   Remainder, w
  btfss   STATUS, C         ; If Carry Set, Low
                            ; Byte of Remainder
   decf   Temp, f           ; >= Divisor
  btfsc   Temp, 7           ; If MSB of (Remainder
                            ; - Divisor) != 1
                            ; then Subtract
   goto   DivideEnd         ; Else, if set then
                            ; Finished

  movwf   Remainder         ; Save Results of
                            ; Subtraction
  movf    Temp, w
  movwf   Remainder + 1

  incf    Quotient, f       ; Increment the
                            ; Quotient
  btfsc   STATUS, Z
   incf   Quotient + 1, f

  goto    DivideLoop

DivideEnd:
  goto    $                 ; Finished, Loop
                            ; Forever

  end
```

When you look through the division by a repeated subtraction subroutine, you'll see that I subtract the divisor from the dividend to see if the result is negative (and the dividend can then be referred to as the remainder). If the result is positive, this result is saved as the new dividend instead of calculating a new dividend.

Repeatedly subtracting the divisor from a number is potentially the slowest method of implementing division, just as repeated addition is a fairly poor way of implementing multiplication. Division can be implemented in a similar manner to multiplication by first shifting the divisor up so it is greater than the dividend and then shifting down and testing to see if the shifted divisor is less than the dividend. If the shifted divisor is

less than or equal to the dividend, then the divisor is taken away from the dividend, and the shifted amount is added to the quotient. I find it is easier to understand how the algorithm works by coding it out as I have in "asmDivide 2.asm."

This method and code for division is probably the most efficient and suitable for virtually all applications. Even though it requires twice the number of instructions as the repeated subtraction method requires, it executes reasonably quickly and accurately.

Going back to your grade school mathematics, you probably remember there is another way of dividing one number into another: That is, write the number as a fraction and multiply the inverse of the denominator with the numerator:

$$A \text{ divided by } B = A / B = A \times (1 / B)$$

And although you may remember it, you may not see its applicability immediately. The PIC MCU's processor does not have the floating-point number capability required to calculate the reciprocal of B (or 1/B). This is true, but a positive number can be produced if, instead of dividing the divisor into 1, it is divided into a much larger number—such as 256. This would turn the previous equation into:

$$A \text{ divided by } B = A / B = [A \times (1 / B)]$$
$$\times (256/256) = [A \times (100 / B)] \times 256$$

Although this is an improvement, you may also be wondering why I would go through the extra work of dividing the divisor into some number and then multiplying the reciprocal of the divisor by the dividend. The answer: This method makes the most sense when you are working with a constant divisor. Calculating the reciprocal of B can be implemented by the assembler's built-in constant calculator, and once the reciprocal has been calculated as a constant, it can then be multiplied with the dividend, as I show in "asmDivide 3.asm."

This method uses about 60 percent of the instructions of the full division routine and executes in less than a quarter of the number of cycles. Note that division by 256 is accomplished by simply taking the upper byte of a two-byte number. This routine is simply not as accurate as the other methods; no remainder is produced and the quotient is typically off by 5 percent. This method is suited for applications where the divisor is a constant and known, and the dividend goes through only the division processing step. Situations where this division routine is best suited include converting user and sensor input data for storage/ processing.

Remember, the case where the divisor is zero is generally considered invalid. In many processors with built-in division circuitry, if a divisor equal to zero is encountered, an error is flagged. As noted at the start of this experiment, none of the three methods of division shown here will work for this case (and will either never end or cause a build-time error).

To check a single-byte divisor, you can simply run the value through the PIC MCU's Algorithm/Logic Unit (ALU) status check and go to a flagging routine if it is zero:

```
movf    Divisor, f    ; Run Divisor through
                       ; status check
                       ;  without changing
                          contents of WREG
btfsc   STATUS, Z     ; If Zero flag set,
                       ; jump to Zero
   goto DivisionByZero ;    Divisor handler
```

To check a 16-bit value, you can OR the two bytes together and see if the result is zero:

```
movf    Divisor, w    ; Two Bytes of Divisor
                       ; == 0?
iorwf   Divisor + 1, w
btfsc   STATUS, Z     ; If Zero flag set,
                       ; jump to Zero
   goto DivisionByZero ;    Divisor handler
```

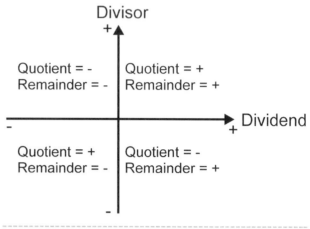

Figure 12-2   *Division signs*

Finally, you may have an application that has to handle positive *and* negative dividends and divisors. For these cases, I use the Division Sign Chart shown in Figure 12-2. Before executing your division routine, mark flags indicating if the dividend and divisor are positive or negative, convert the negatives to positive values, execute the division routine, and, upon exit, change the quotient and remainder to negatives if appropriate.

Experiment 106—Squaring a Number Using Finite Difference Theory

When you are data processing, you will discover cases where you need mathematical operations other than multiplication and division. You might be thinking of sine functions and logarithmic operations and shuddering at the thought of trying to calculate values for them in the simple PIC MCU's processor. But remember, these calculations were performed literally for centuries before the invention of the digital "computer" or handheld calculator.

I have put the word *computer* in quotations in the previous paragraph because it was the name given to individuals that were responsible for calculating values *computers*. These people spent literally months working through a series of calculations to come up with the answer to a problem. As you would imagine, the major problems encountered by doing complex mathematical operations using humans included the time required as well as the accuracy of the final result. The seemingly simple operation of producing sine and cosine values for a book of tables for navigation was fraught with many problems, due simply to human error, that often resulted in the loss of ships and lives.

In order to minimize the amount of time required to perform calculations (and the opportunities for errors), a number of different mathematical tools were developed over the years. One of the most powerful tools was given the name *finite differential theory*, in which simple operations were analyzed to find easily recognizable differences between calculation values. The most-often-used example of how finite differential theory works is the calculation of squares. Looking at squares may seem like a strange place to start, but it will make more sense at the end of this experiment.

To show how finite differential theory works, consider Table 12-1, in which I have listed different integers, their squares, and the differences of the squares to the values of the adjacent squares. You can see that when I have calculated the differences in the values between squares, they can be expressed simply by adding two to the previous difference.

It should be obvious that by simply knowing that the square of zero is zero and the square of one is one, that you can calculate the square of any integer quite quickly. The code that I came up for calculating the square of 47 is named asmDifference and is listed here:

Table 12-1
Integers, Their Squares, and the Differences Between Them

Integer	Square	Difference to Previous	Difference to Difference
0	0	N/A	N/A
1	1	1	N/A
2	4	3	2
3	9	5	2
4	16	7	2
5	25	9	2
6	36	11	2
7	49	13	2
8	64	15	2
9	81	17	2
10	100	19	2

```
title  "asmDifference - Squares using Finite
Difference Theory"
;
; Using Finite Difference Theory, this
;  Program finds the Square for
;  the given 8 Bit Integer (with a
;  16 Bit Result).
;
; "C" Equivalent Code:
;
; Value = 47      // Number to Start With
; Square = 0;     // Current Square Value
; Difference = 1; // Start with 1 and Add
;                 // Difference
; while (Value != 0)
;                 // Calculate Square for "Value"
; {
;  Square = Square + Difference;
;                 // Calculate Updated Square
;  Difference = Difference * 2;
;                 // Add to the Difference
;  Value--;
;                 // Decrement the Value Counter
; }  // elihw
;
; while (1 == 1);
;                 // Finished, Loop Forever
;
; Hardware Notes:
;  PIC16F627A running at 4 MHz in Simulator
;
;
; Myke Predko
; 04.04.14
;
   LIST R=DEC
   INCLUDE "p16f627a.inc"

; Registers

; Variables
   CBLOCK 0x20
```

```
Value, Square:2
Difference:2
 ENDC

   PAGE
; Mainline of Difference

 org     0

   movlw   47              ; Initialize Variables
   movwf   Value
   clrf    Square + 1
   clrf    Square
   clrf    Difference + 1
   movlw   1
   movwf   Difference

Loop:                      ; Return Here until
                           ; [Value] == 0
   movf    Difference, w   ; Add "Difference" to
                           ; "Square" to get
   addwf   Square, f       ; its new value
   movf    Difference + 1, w
   btfsc   STATUS, C
    incf   Difference + 1, w
   addwf   Square + 1, f
   movlw   2               ; Increase Difference
                           ; by 2
   addwf   Difference, f
   btfsc   STATUS, C
    incf   Difference + 1, f
   decfsz  Value, f        ; Decrement the Value
                           ; Counter
    goto   Loop

   goto    $               ; Finished, Loop
                           ; Forever

   end
```

In addition to squaring numbers (finding the value of a number to the power two), finite difference theory can be used to calculate other powers. You may want to spend a bit of time reading about Babbage's *Difference Engine* to understand how other powers are calculated and maybe to try to come up with an application to do it on your own. Hint: As you go to larger powers, you will discover that more than one difference value changes for each new integer.

Once you are comfortable with using finite difference theory for calculating powers of numbers, you can use this knowledge to calculate different values. The series formulas listed here will help:

$$\sin(x) = x - x^3/3! + x^5/5! - x^7/7! + \dots$$

$$\cos(x) = 1 - x^2/2! + x^4/4! - x^6/6! + \dots$$

$$e^x = 1 + x + x^2/2! + x^3/3! + x^3/3! + \dots$$

$$\log_e x = (x - 1)^1/1 - (x - 1)^2/2 + (x - 1)^3/3 - \dots$$

Experiment 107—Find the Square Root of a 16-Bit Number

PC/ Simulator

I can honestly say that the only thing I did not understand in high school mathematics was how to manually calculate square roots. Looking back, I suspect that it was due primarily to my poor understanding of how repeating algorithms worked (I had not yet learned any programming) and secondarily to my reluctance to learn the material because of the $\sqrt{}$ on my calculator. Unfortunately in most simple microprocessors and microcontrollers, no built-in square root instruction exists—which means that when I have to implement these functions, I wish I had paid more attention in high school math.

When faced with the challenge of calculating the square root of a value, I usually fall on what I know—finite difference theory. It can be used to find the square of a value, so by reversing the process (subtracting the difference from the value) and recording the number of iterations, the square root of a number can be found.

```
title   "asmSqRoot 2 - Find the Square Root of
the 16 Bit Value"
;
; This program takes advantage of the
;   property of square roots that
```

```
;   are the number of odd numbers, who's
;   sum is less than or equal to
;   the number the square root is being
;   calculated for.
;
; This is the reverse of the Finite
;   Difference Squaring Program.
;
; "C" Equivalent Code:
;
; Value = 12345;      // Value to Square Root
;
; DigitCount = 1;     // Square Root Value
; Odd = 1;            // Difference Value
;
; while (Odd <= Value)
;                     // Take Away Difference
; {                   // Until >Remainder
;     Value = Value - Odd;
;     Odd = Odd + 2;
;                     // Increase the Difference
;     DigitCount = DigitCount + 1;
;                     // Increment Square Root Value
; }  // elihw
;
; while (1 == 1);     // Finished, Loop Forever
;
; Hardware Notes:
;   PIC16F684 running at 4 MHz in Simulator
;
;
; Myke Predko
; 04.04.13
;
    LIST R=DEC
    INCLUDE "p16f684.inc"

; Variables
    CBLOCK 0x20
Value:2                      ; Value the Square Root
                             ; is to be found
Odd:2                        ; Processing Variables
DigitCount                   ; Square Root of the
                             ; Value
    ENDC
```

```
        PAGE
; Mainline of SqRoot

  org     0

    movlw   high 12345      ; Initialize Variable
                            ; to have the Square
    movwf   Value + 1       ; Root Found for it
    movlw   low 12345
    movwf   Value

    clrf    Odd + 1         ; Initialize the
                            ; Different Values
    movlw   1
    movwf   Odd
    clrf    DigitCount

Loop:                       ; Repeat Here until
    movf    Odd + 1, w      ; Calculate Value - Odd
    subwf   Value + 1, f    ; Note, Result Stored
                            ; Back in "Value"
    movf    Odd, w
    subwf   Value, f
    btfss   STATUS, C
     decf   Value + 1, f

    btfsc   Value + 1, 7    ; If MSB of "Value" Set,
     goto   SqRootDone      ; Value is Negative

    movlw   2               ; Go to the Next Odd
                            ; Number
    addwf   Odd, f
    btfsc   STATUS, C
     incf   Odd + 1, f

    incf    DigitCount, f   ; Increment Value of
                            ; the Square Root

    goto    Loop

SqRootDone:                 ; Completed
    goto    $               ; Finished, Loop
                            ; Forever

    end
```

If you compare the equivalent C code to this assembly code, you'll notice that there is quite a big difference; the assembly code calculates the value minus the difference (Odd) and stores the result in "Value" before checking to see if "Value" is negative (the most significant bit is set). The correct C equivalent code is:

```
while ((Value = Value - Odd) > 0)
                // Take Away Difference Until
{               // Remainder is Less than Zero
    Odd = Odd + 2;
                // Increase the Difference
    DigitCount = DigitCount +1;
                // Increment Square Root Value
}   // elihw
```

I did not note this difference in the original code, because, as I have said, if you are new to C programming, the ability to embed assignment statements in conditional statements and check their values is not intuitively obvious. By calculating the value for "Value" instead of doing the compare, a full subtrac-

tion operation is eliminated, making the application both smaller and faster.

The method shown here works reasonably well, but an arguably better method would be using Newton's method of zero finding. This method is based on the function:

$$f(x) = G^2 - A$$

Where A is the original value and G is its square root. When $f(x)$ is equal to zero, it should be obvious that G is equal to the square root of A. To calculate G, the approximation formula:

$$G' = [(A/G) + G] / 2$$

is used. G' (G-prime) is the next value of G and is calculated by dividing A by its square root and then averaging the difference with G. The formula is executed repeatedly until G and G' are essentially the same. Table 12-2 shows how this formula is used to calculate the integer square root of 12,345 (the same value as the example code). For the initial value of G, I usually use $1/3$ of the value to find the square root of. The closer you make the initial value of G to the actual square root of the number, the fewer iterations are required to find the actual square root, but any value other than zero can be used.

In Table 12-2, you can see that eight iterations were needed to calculate the square root of 12,345, but two issues should be noted. First, although this method is quite simple, it does require a division capability in the processor, which will add to the size of the square root routine. This is not a major problem, at least not compared to the other issue. And second, when I calculated the values for G' in Table 12-2, I used a calculator and rounded up fractional values. If you were to create this

Table 12-2
Newton's Method for Calculating Integer Square Roots

Iteration	G	G' = [(A/G) + G]/2
1	4115	2059
2	2059	1032
3	1032	522
4	522	273
5	273	159
6	159	118
7	118	111
8	111	111

application in assembly language (or even in C), you would have to somehow perform this rounding operation.

Unless, of course, you used the floating-point capabilities of the PICC Lite™ compiler, as I have done in cSqRoot.c and cSqRootN.c, which I created quickly to test the PIC MCU's ability to perform a square root with a floating-point result. In cSqRoot.c, I used the built-in C sqrt function:

```
#include <pic.h>
#include <math.h>
/*  cSqRoot.c - Test PICC Lite Square Root
Function

This program will simply execute a single square
root function to see how big it is and how long
it takes to execute.

myke predko
05.01.19

*/

double Number, SquareRoot;


main()
{

    Number = 12345.0;  // Establish the Number
    SquareRoot = sqrt(Number);
                       // Get its Square Root

    while (1 == 1);

}  // End cSqRoot
```

and in cSqRootN.c, I came up with a simple implementation of Newton's method:

```
#include <pic.h>
#include <math.h>
/*  cSqRootN.c - Use Newton's Method to Find
Square Root

This program will simply Execute Newton's Square
Root Function until the new value is the same as
the old.

myke predko
05.01.19

*/

double Number, SquareRoot, NSquareRoot;


main()
{

  Number = 12345.0;    // Establish the Number
  NSquareRoot = Number / 3;
                       // Get Seed Value
  do
  {
    SquareRoot = NSquareRoot;
    NSquareRoot = (SquareRoot +
                  (Number / SquareRoot)) / 2;
  } while (SquareRoot != NSquareRoot);

    while (1 == 1);

}  // End cSqRootN
```

cSqRoot.c required 585 instructions and 37 bytes of variable memory, although cSqRootN.c required only 292 instructions and 21 bytes of variable memory. This is an improvement in size of about half. The question you should be asking is why isn't Newton's method used in the C sqrt function? The reason seems to be that Newton's method is quite a bit slower. When I plugged in different values to find their square roots, I discovered that Newton's method could take as much as 300 percent of the time that the sqrt function took.

This is another example of how different code, performing the same function will have different operating characteristics. The standard C sqrt function takes up considerably more space than Newton's method, but is quite a bit faster. Although I should note that even though it's faster, the average calculation time is on the order of 7 to 9 milliseconds. In either case, the square root function will take up a substantial amount of program memory, so its use must be chosen judiciously.

Experiment 108—Converting a Byte into Three Decimal, Two Hex, or Eight Binary ASCII Bytes

PC/ Simulator

When I first introduced you to programming the PIC MCU, I used the high-level capabilities of the C programming language to facilitate converting numeric data to ASCII. These operations are quite easy to do in a high-level language, but you might find it daunting to write them in assembly language. It may be surprising to you to discover that creating these functions is assembly language is not very difficult and is usually quite a bit more efficient than writing them in C.

The trick is recognizing that ASCII values are numbers and can be manipulated in just the same way as regular values are manipulated in assembly language. For example, if you had a numeric value of seven and wanted to convert it into the ASCII character code for 7, you could add seven to the ASCII character code for 0. If you had the value of seven in the WREG, you could convert it to the ASCII character code for 7 with the single instruction:

```
addlw    '0'
```

Like in C, a character's ASCII code is substituted when the character, enclosed in single quotes, is encountered.

The first program, asmByte2Bin, converts a byte value to eight ASCII bit values (1 or 0) by looping through each bit and testing them to see whether or not they are set or reset. In this program (like in the others), I start the conversion with the high-order bits so they can be seen easily in the MPLAB IDE simulator's File Register display.

```
 title   "asmByte2Bin - Number to ASCII"
;
; This application converts a single
;  8-Bit (Byte) Value into eight
;  ASCII Binary characters.
;
; "C" Equivalent Code:
;
; Temp = Number;        // Save the Number
;
```

```
; for (i = 0; i < 8; i++)
;                       // Loop through all 8 Bits
; {
;     if ((Temp & 0x080) != 0)
;         Digit[i] = '1';
;                       // Bit Set
;     else
;         Digit[i] = '0';
;                       // Bit Reset
;     Temp = Temp << 1;
;                       // Shift in New Bit
; }  // rof
;
; while(1 == 1);        // Done, Loop Forever
;
;
; Hardware Notes:
;   PIC16F684 running at 4 MHz in Simulator
;
;
; Myke Predko
; 04.08.29
;
  LIST R=DEC
  INCLUDE "p16f684.inc"

; Variables
  CBLOCK 0x20
Digit: 8                ; Number Variables
Temp, i
  ENDC

  org     0

  movlw   123           ; Use "123" to Test the
                        ; Program
  movwf   Temp

  movlw   Digit         ; FSR to Point to "Digit"
                        ; Variable
  movwf   FSR

  movlw   8             ; Want to Do 8 Bits
  movwf   i
Loop:
  movlw   '0'           ; Want to Save Either '0'
                        ; or '1'
  btfsc   Temp, 7       ; Is Bit 7 of "Temp" Set?
   addlw  1             ; Yes, Convert WREG to '1'
                        ; from '0'
  movwf   INDF          ; Save the ASCII Digit
  rlf     Temp, f       ; Shift up tested Temp bit
  incf    FSR, f        ; Point to the Next Digit
   decfsz i, f          ; Repeat 8x
    goto  Loop

  goto    $             ; Finished, Loop Forever

  end
```

Converting a byte's value to hexadecimal is also quite simple due to the swapf instruction built into the PIC MCU's processor. This instruction swaps the two nybbles (high four bits and low four bits) to facilitate converting these four bits to a hexadecimal value using a subroutine. The most obvious method to convert a byte to two hexadecimal ASCII characters is to use a table as I do in "asmByte2Hex 1:"

```
title    "asmByte2Hex 1 - Number to Hex ASCII"
;
; This application converts a single
;  9-Bit (Byte) Value into Two ASCII
;  Hex characters using a table.
;
; "C" Equivalent Code:
;
; HexTable[] = {'0', '1', '2', '3', '4',
;  '5', '6', '7',
;  '8', '9', 'A', 'B', 'C', 'D', 'E', 'F'};
;
; Digit[0] = HexTable[(Number >> 4) & 0x00F];
; Digit[1] = HexTable[Number & 0x00F];
;
; while(1 == 1);     // Done, Loop Forever
;
;
; Hardware Notes:
;  PIC16F684 running at 4 MHz in Simulator
;
;
; Myke Predko
; 04.08.29
;
  LIST R=DEC
  INCLUDE "p16f684.inc"

; Variables
CBLOCK 0x20
Digit:2               ; Number Variables
Temp
  ENDC

  org     0

  movlw   123         ; Use "123" to Test the
                      ; Program
  movwf   Temp

  swapf   Temp, w     ; Get the High Nybble
  call    HexTable
  movwf   Digit + 0

  movf    Temp, w     ; Get the Low Nybble
  call    HexTable
  movwf   Digit + 1

  goto    $           ; Finished, Loop Forever

HexTable:             ; Return Hex ASCII Char
                      ; for LSNybble
  clrf    PCLATH      ; in WREG
if ((_HexTable & 0x100) != 0)
  bsf     PCLATH, 0
endif
if ((_HexTable & 0x200) != 0)
  bsf     PCLATH, 1
endif
if ((_HexTable & 0x400) != 0)
  bsf     PCLATH, 2
endif
  andlw   0x0F        ; Just want lower 4 bits
  addlw   low _HexTable ; Calculate the correct
                      ; offset
  btfsc   STATUS, C
```

```
  incf    PCLATH, f
  movwf   PCL
_HexTable:
  dt      "0123456789ABCDEF"

  end
```

A potentially more efficient method of performing this conversion is to calculate the ASCII hex character algorithmically, replacing HexTable in the program above with the following subroutine:

```
GetHex:                  ; Return Hex ASCII Char
                         ; for LSNybble
                         ; in WREG
  andlw   0x00F          ; Just want lower 4 Bits
  addlw   6              ; is the Hex Digit > 9?
  btfsc   STATUS, DC
  addlw   7              ; Yes, Move Value from
                         ; Digits to Chars
  addlw   '0' - 6
  return
```

This code tests the value in WREG to see if it is greater than or equal to 0x00A, and if it is, adds the difference between the ASCII character codes for 9 and A. Once this is done, the ASCII character code for 0 is added to convert the numeric value to an ASCII character. To test out the operation of this subroutine, create a new program called asmByte2Hex 2 and replace HexTable with GetHex, and then test it out. In terms of efficiency, I would generally use GetHex over HexTable because it can be located anywhere in the PIC MCU's memory without concern and is short enough that you could consider placing it in line rather than making it a subroutine. For example, a general-case byte to hex conversion could be written as and takes about the same number of instructions as calling the HexTable to convert a Nybble once.

```
Byte2Hex:                ; Convert Value in WREG
                         ; to two Hex
  movwf   Temp           ; characters at location
  swapf   Temp, w        ; pointed to by FSR
  andlw   0x00F          ; Do High Nybble First
  addlw   6              ; Nybble > 9?
  btfsc   STATUS, DC
  addlw   7              ; Yes - Put in "A" to "F"
                         ; range
  addlw   '0'            ; Convert to ASCII
                         ; Character
  movwf   INDF           ; Save ASCII Character
  incf    INDF, f        ; Point to the Next
                         ; Destination Byte
  movf    Temp, w        ; Convert the LSNybble
  andlw   0x00F          ; Do High Nybble First
  addlw   6              ; Nybble > 9?
  btfsc   STATUS, DC
  addlw   7              ; Yes - Put in "A" to "F"
                         ; range
  addlw   '0'            ; Convert to ASCII
                         ; Character
  movwf   INDF           ; Save ASCII Character
  incf    INDF, f        ; Point to the Next
                         ; Destination Byte
  return                 ;Return to Caller
```

The GetHex subroutine is a good example of working with a single nybble (four bits) and using the digit carry (DC) flag instead of the standard carry flag.

The last program that I am going to present for this experiment is the conversion of a byte to a decimal value (asmBin2Dec1.asm). To find the hundreds and tens digits, I repeatedly take away values (division by repeated subtraction) and add the quotient to the ASCII character code for 0 to produce the correct characters. After taking away hundreds and tens, the remainder is the number of decimal ones in the number, which is simply added to the ASCII character code for 0.

This isn't a terrible method of converting a byte to decimal, but it could be improved by noting that the least significant digit is the sum of the least significant nibble and 16 times the most significant nibble. The asmBin2Dec 3.asm routine takes advantage of this property and goes one better: instead of adding 16 times the high nibble, it adds six times the high nibble to the low nibble, and it adds the 10-times value to the high nibble. The code itself is almost twice the length of asmBin2Dec 1, but it executes in a constant 38 instructions.

Experiment 109—Produce the Even Parity Values for a Byte

One of the dying technologies of computer interfacing is RS-232. This serial interface was one of the first methods of connecting computers and peripherals and required a number of different skills to implement successfully. In *123 Robotics Experiments for the Evil Genius*, I discussed some of the issues regarding creating the voltages needed for the data transfer and how systems are connected together. In this experiment, I want to look at how the error correction code was created and see how efficiently it can be done.

The traditional format for an RS-232 data packet is shown in Figure 12-3. It contains two bits after the data: the stop bit is an indicator that the data packet has been sent, and gives the sender and receiver time to process the current byte and prepare for the transmission of the next one.

The second bit, the parity bit, is used to detect whether any one of the transmitted bits was received in error. This bit, when summed with the total of all the previous bits will result in either an even or odd value. There are five different types of parity: mark, space, even, odd, and zero, in which a 1, 0, even sum, odd sum, and no bits (respectively) are sent. In most modern communications, the no parity bit is most often used due to the reliability of modern communications. If a parity bit is required for error detection, then it is most often even because it is calculated as the sum of all the bits that have been received in a packet and should be an even value (i.e., bit 0 of the sum should be zero). The following program demonstrates the most obvious way of calculating the even parity bit—the value of each bit of a byte is summed together.

```
title   "asmParity 1 - Calculate Even Parity
Value for a Byte"
;
; This application calculates the
;   even parity for a byte by summing
;   each bit.
;
; "C" Equivalent Code:
;
; Temp = 123;              // Set the Byte
;
; Parity = 0;
; for (i = 0; i < 8; i++)
;                          // Read through each bit
; {
;     if ((Temp & 1) == 1)
;         Parity = Parity + 1;
;     Temp = Temp >> 1; // Shift 1 Bit Down
; } // rof
;
```

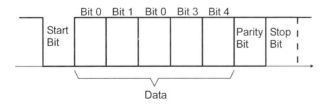

Figure 12-3   *RS-232 data packet*

```
; Parity = Parity & 1;
;                        // Parity is a Single Bit
;
; while(1 == 1);         // Done, Loop Forever
;
;
; Hardware Notes:
;  PIC16F684 running at 4 MHz in Simulator
;
;
; Myke Predko
; 04.04.07
;
  LIST R=DEC
  INCLUDE "p16f684.inc"

; Variables
 CBLOCK 0x20
Temp, i
 ENDC

  PAGE
  org     0

  movlw   123           ; Initialize Variables
  movwf   Temp
  movlw   8
  movwf   i
  clrw                  ; Use WREG as Parity
                        ; Counter

Loop:                   ; Loop Here for Each Bit
  btfsc   Temp, 0       ; LSB of Value Set?
   addlw  1
  rrf     Temp, f
  decfsz i, f           ; Repeat for each bit
   goto  Loop

  andlw  1              ; Only LSB is Parity Bit

  goto   $              ; Finished, Loop Forever

  end
```

This method works acceptably well, but is another case where a bit of knowledge or binary numbers and

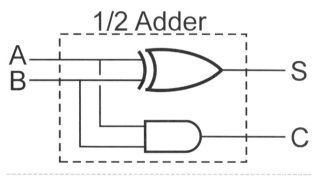

Figure 12-4   *Half adder*

a bit of experimentation comes in handy. Remember that a binary adder consists of an XOR gate and an AND gate (see Figure 12-4) in which the sum (S) bit is the XOR result of the inputs. So if we want to add two values together and get the single-bit sum, we simply require an XOR gate.

In asmParity 2.asm, this function of the half adder is exploited (along with shifting bits) to produce a parity bit in the same number of instructions, executing in one-seventh the number of instructions and using one less file register variable than the original.

Despite improving how the parity calculation software works, as a tool, it's pretty limited. The parity bit will detect only one bit in error in the packet. With two bits, the parity bit will indicate the data is correct. Additionally, when an error is detected, the incorrect bit is not indicated, nor is there any correction information included. These functions are critical features of modern *error detection/correct codes* (ECC) used in high-speed communications and peripheral interfacing.

Experiment 110—Sort a List of 10 Eight-Bit Values Using the Bubble-Sort Algorithm

After the time and effort spent in Section 4 describing different ways to implement a bubble sort in C, you may be hoping that I won't have a lot to say about creating a sort in assembly language. You'll be happy to hear this is true: The assembly-language bubble sort (shown as follows) is a quite simple and faithful translation of the C language algorithm.

```
        title    "asmbSort 1 - Sort 10 8-Bit Values"
;
; This program uses the Bubble Sort Algorithm to
Sort 10 Values
;
; "C" Equivalent Code:
;
; for (i = 0; i < 10; i++)
;     SortList[i] = ReadValue;
;               // Read in the Ten Byte Values
;
; for (i = 0; i < 9; i++)
;               // Outside Sort Loop
;     for (j = 0; j < (9 - i); j++)
;               // Inside Sort Loop
;         if (SortList[i] > SortList[i + 1])
;         {     // Have to Swap Values
;             Temp = SortList[i + 1];
;             SortList[i + 1] = SortList[i]
;             SortList[i] = Temp;
;         } //fi
;
; Hardware Notes:
;   PIC16F684 running at 4 MHz in Simulator
;
;
; Myke Predko
; 04.04.05
;
  LIST R=DEC
  INCLUDE "p16f684.inc"

; Variables
CBLOCK 0x20
SortList:10             ; List of Values to be
                       Sorted
i
  ENDC

  PAGE
; Mainline of Bubble Sort - Myke

  org      0

  clrf    i                 ; Start by Loading
                            ; SortList
  movlw   SortList          ; Setup FSR for Indirect
                            ; Addressing
  movwf   FSR
LoadLoop:                   ; Return Here Until all
                            ; Values Loaded
  movf    i, w
  call    GetValue          ; Get the Value for "i"
  movwf   INDF              ; Save in List
  incf    FSR, f            ; Point to Next Element
  incf    i, f              ; Increment Counter
  movf    i, w              ; i == 10?
  sublw   10
  btfss   STATUS, Z
   goto   LoadLoop          ; No, Do next one

  movlw   9                 ; Want to Loop Through 9x
  movwf   i
InsideLoop:
  movlw   SortList          ; Point to the Start of
                            ; the List
  movwf   FSR               ; Use FSR as Index
OutsideLoop:
  movf    INDF, w
  incf    FSR, f            ; Point to the Next List
                            ; Element
  subwf   INDF, w           ; SortList[j + 1] -
                            ; SortList[j]
  btfsc   STATUS, C         ; If Carry Reset,
                            ; SortList[j] >
```

```
   goto   OutsideLoop_Skip
                            ; SortList[j + 1]
  subwf   INDF, f           ; Second = Second -
                            ; (Second - First)
  decf    FSR, f
  addwf   INDF, f           ; First = First + (Second
                            ; - First)
  incf    FSR, f
OutsideLoop_Skip:           ; Finished, Increment "j"
                            ; to "i"
  movf    i, w              ; At the End of the List?
  addlw   SortList
  subwf   FSR, w
  btfss   STATUS, Z         ; Index at the End of the
                            ; List?
   goto   OutsideLoop
  decfsz  i, f              ; Done 9x?
   goto   InsideLoop

  goto    $                 ; Forever Loop - Done

; Return the Value for "i"
GetValue:
  clrf    PCLATH
if ((_GetValue & 0x0100) != 0)
  bsf     PCLATH, 0
endif
if ((_GetValue & 0x0200) != 0)
  bsf     PCLATH, 1
endif
if ((_GetValue & 0x0400) != 0)
  bsf     PCLATH, 2
endif
  addlw   low _GetValue ; Get Offset in 256
                            ; Address Block
  btfsc   STATUS, C
   incf   PCLATH, 0
  movwf   PCL               ; Perform the Jump into
                            ; the Table
_GetValue:
  dt      10, 100, 4, 15, 75, 150, 47, 2, 250,
175

  end
```

I'd like to note a couple of discussion points. If you work through the assembly-language code, you'll see that it is essentially a straight translation from the C language prototype with one difference. When I coded the application, I discovered that when subtracting the current array value from the next, I could use the value in the WREG (SortList[i + 1] - SortList[i]) to simplify swapping the data between the two array elements.

```
SortList[i + 1]' = SortList[i + 1] - WREG
              = SortList[i + 1] - (SortList[i
                + 1] - SortList[i])
              = SortList[i + 1] - SortList[i
                + 1] + SortList[i]
              = SortList[i]
```

Similarly, with SortList[i], the value in SortList[i + 1] could be substituted into it by simple addition:

```
SortList[i]' = SortList[i] + WREG
           = SortList[i] + (SortList[i + 1] -
             SortList[i])
           = SortList[i + 1]
```

This optimization allows for swapping the contents of the two array elements in four instructions. When I originally wrote the code, I used 10 instructions to swap the values in the elements using translated C code.

Years ago on the PICList, the question came up regarding the best way to sort a short array. My immediate response was to come up with a looping bubble-sort routine like the one above. Somebody came up with a simple macro that embodied the compare and swap code together and created a simple bubble-sort program, which ran through every possible compare and swap, as I have done in "asmbSort 2.asm," which uses the "CompareAndSwitch" macro:

```
CompareAndSwitch macro FirstValue, SecondValue
 local CASSkip
 movf FirstValue, w    ; Is SecondValue > First
                          Value?
 subwf SecondValue, w
 btfsc STATUS, C       ; If Carry Set, then
                          Second >= First
```

```
 goto CASSkip ; WREG = SecondValue - FirstValue
 addwf FirstValue, f  ; FirstValue = FirstValue
                          + (WREG)
 subwf SecondValue, f ; SecondValue =
                          SecondValue - (WREG)
CASSkip:
 endm
```

The second program runs somewhat faster than the original and is good as a demonstration tool for the squared order of operations of the algorithm. For an array of two elements, one compare/swap is required; for three elements, three compares/swaps are required; for four elements, six compares/swaps are required; and for five elements, 10 compare/swaps are required. Unfortunately, the number of instructions required by this method soon becomes the limiting factor in its adoption in using it. For sorting a maximum of three or four elements, this could be a very efficient bubble-sort method, but its usefulness diminishes as larger arrays must be sorted.

Experiment 111—Encrypt and Decrypt an ASCIIZ String Using a Simple Substitution Algorithm

PC/ Simulator

Data encryption is a fascinating topic and a number of programs are available for the PIC microcontroller that implement *data encryption standard* (DES) and other modern encryption algorithms. Simple substitution algorithms can be implemented surprisingly easily in PIC MCU assembly language, as I will show in this experiment.

The basic operation of a substitution encryption algorithm is that one character is substituted for another from a table in which the character substitutions are *symmetrical*. When the substituted character is placed back in the table, then the original character is returned. This operation is demonstrated in the following application:

```
    title  "asmEncrypt 1 - Myke - Encrypt and
Decrypt and ASCII String"
;
; This application converts a String of
;  Characters stored in the File
;  Registers and then decrypts them using
;  a symmetrical conversion table (ROT13).
;
; "C" Equivalent Code:
;
; DataString = "MYKEPREDKO";
;
; for (i = 0; i < strlen(DataString); i++)
;                     // Encrypt the String
;     DataString[i] = Encrypt(DataString[i]);
;
; for (i = 0; i < strlen(DataString); i++)
;                     // Decrypt the String
;     DataString[i] = Encrypt(DataString[i]);
;
; while(1 == 1);
;                     // Done, Loop Forever
;
;
; Hardware Notes:
;  PIC16F684 running at 4 MHz in Simulator
;
;
; Myke Predko
; 04.04.05
;
  LIST R=DEC
```

```
        INCLUDE "p16f684.inc"

; Variables
 CBLOCK 0x20
DataString:12
 ENDC

  PAGE
; Mainline of Encryption

 org    0

 movlw  DataString - 1  ; Setup FSR to Point to
                        ; the DataString
 movwf  FSR
InitLoop:
 incf   FSR, f          ; Point to the Next
                        ; Character
 movlw  DataString      ; Get Index Into String
 subwf  FSR, w
 call   InitString
 movwf  INDF            ; Store the Character
 iorlw  0               ; At End of String?
 btfss  STATUS, Z
  goto  InitLoop

 movlw  DataString      ; Now, Encrypt the
                        ; String
 movwf  FSR
EncryptLoop:
 movf   INDF, w         ; Get Character to
                        ; Encrypt
 call   Encrypt         ; Do the Encryption
 movwf  INDF            ; Put back the
                        ; Encrypted Character
 incf   FSR, f          ; Point to the Next
                        ; Character
 movf   INDF, w         ; Repeat to End of
                        ; String?
 btfss  STATUS, Z
  goto  EncryptLoop

 movlw  DataString      ; Now, Decrypt the
                        ; String
 movwf  FSR
DecryptLoop:
 movf   INDF, w         ; Get Character to
                        ; Encrypt
 call   Encrypt         ; Do the Symmetrical
                        ; Decrypt
 movwf  INDF            ; Put back the
                        ; Encrypted Character
 incf   FSR, f          ; Point to the Next
                        ; Character
 movf   INDF, w         ; Repeat to End of
                        ; String?
 btfss  STATUS, Z
  goto  DecryptLoop

 goto   $               ; Finished, Loop
                        ; Forever

; Subroutines
InitString:
 clrf   PCLATH          ; Setup PCLATH for
                        ; Table Read
 if ((_Encrypt & 0x0100) != 0)
 bsf    PCLATH, 0
 endif
 if ((_Encrypt & 0x0200) != 0)
 bsf    PCLATH, 1
 endif
 if ((_Encrypt & 0x0400) != 0)
 bsf    PCLATH, 2
```

```
 endif
 addlw  _InitString
 btfsc  STATUS, C
  incf  PCLATH, f
 movwf  PCL
_InitString:
 dt     "MYKEPREDKO", 0

Encrypt:                        ; Encrypt UPPERCASE
                                ; ASCII Character
 clrf   PCLATH                  ; Setup PCLATH for
                                ; Table Read
 if ((_Encrypt & 0x0100) != 0)
 bsf    PCLATH, 0
 endif
 if ((_Encrypt & 0x0200) != 0)
 bsf    PCLATH, 1
 endif
 if ((_Encrypt & 0x0400) != 0)
 bsf    PCLATH, 2
 endif
 addlw  (0 - 'A')
 addlw  _Encrypt                ; Zero Base the
                                ; characters
 btfsc  STATUS, C
  incf  PCLATH, f
 movwf  PCL
_Encrypt:
 dt     "NOPQRSTUVWXYZABCDEFGHIJKLM"  ; ROT 13

 end
```

This application first loads an array built with 11 file registers with the string of uppercase characters to encrypt and decrypt. The code, which is represented as the DataString initialization in the header "C" Equivalent Code, is the standard method for initializing an ASCIIZ-string variable array. Using the ASCIIZ format, the length of the string can be ignored; it is simply read until a byte with the value zero is encountered. This feature minimizes the number of file registers required to initialize, read, and write array elements.

Next, each character is passed to a subroutine that finds the character offset to the character relative to A, and the character in the Encrypt table at this offset is returned. The character returned from the Encrypt table is in the ROT 13 format in which A is changed to N, B is changed to O, and so on. This is a very basic and popular substitution algorithm used mostly for demonstration purposes.

This assignment was given to a high school student who came up with the following algorithmic version of the Encrypt subroutine:

```
if (Character >= 'N')
                // Character N-Z, Change to A-M
  Character = Character - ('N' - 'A');
else            // Character A-M, Change to N-Z
  Character = Character + ('N' - 'A');
```

This code either subtracts or adds the difference between the ASCII character code for N and the ASCII character code for A. You are probably aware that the difference between the two is 13 (assuming that there are 26 letters in the alphabet and the break is halfway through the alphabet). But by keying in the difference as an arithmetic statement that the compiler/assembler has to process makes it easier to see what's happening and minimizes the chance for keying the wrong value or changing it inadvertently.

It isn't difficult to convert to PIC MCU assembler, but the code is somewhat cumbersome, not to mention quite hard to follow. In the following code, I have indicated the value within WREG to make the operation of the function easier to follow:

```
Encrypt:
  sublw  'M'                 ;  WREG = 'M' - WREG
  btfsc  STATUS, C           ;  Skip if Carry Reset
                             ;  (WREG > 'M')
   goto  ROT13Less           ;  Carry Set, WREG is
                             ;  <= 'M'
  sublw  'M' - ('N' - 'A')   ;  WREG = 'M' - ('N' -
                             ;  'A') - ('M' - WREG)
                             ;  WREG = WREG - ('N'
                             ;  - 'A')
   goto  ROT13Done
ROT13Less:                   ;  WREG <= 'M', Add
                             ;  'N'-'A'
  sublw  'M' + ('N' - 'A')   ;  WREG = 'M' + ('N' -
                             ;  'A') - ('M' - WREG)
                             ;  WREG = WREG + ('N'
                             ;  - 'A')
ROT13Done:                   ;  WREG = ROT13(WREG);
  return
```

Interesting, if the sublw is executed twice, you end up with the original value:

```
  sublw  Number              ;  WREG = Number -
                             ;  WREG
  sublw  Number              ;  WREG = Number -
                             ;  (Number - WREG)
                             ;  WREG = WREG
```

So by using the sublw instruction a second time, the value from which the WREG was originally subtracted can be removed, and the value in WREG can be changed from a negative to a positive.

Instead of just leaving the code as is, I decided to spend a few minutes and see if I could simplify the conversion process by eliminating the two paths and looking for a value to add to the subtraction result in WREG. The subroutine I came up with follows:

```
Encrypt:                     ;  Encrypt the
                             ;  UPPERCASE ASCII
                             ;  Character
  sublw  'N'                 ;  Carry Reset if WREG
                             ;  > 'M'
  btfsc  STATUS, C
   addlw -2 * ('N' - 'A')    ;  Carry Set, Move to
                             ;  Down 'N' - 'A' Values
  sublw  'A'
  return                     ;  Return the Encoded
                             ;  Value
```

I replaced the original Encrypt subroutine in the asmEncrypt 1.c routine, with the new version above and called the application asmEncrypt 2.c. This code, although obviously very efficient, is hard to follow and understand. The best way to examine what is happening is to consider each case in which WREG contains a value from 'A' to 'M' or a value from 'N' to 'Z.' For a value between 'A' and 'M,' the code executes as follows:

```
  sublw  'N'                 ;  WREG = 'N' - WREG
  addlw  -2 * ('N' - 'A')    ;  WREG = -2'N' + 2'A'
                             ;  + ('N' - WREG)
                             ;  WREG = -'N' + 2'A' -
                             ;  WREG
  sublw  'A'                 ;  WREG = 'A' - (-'N' +
                             ;  2'A' - WREG)
                             ;  WREG = 'A' + 'N' -
                             ;  2'A' + WREG
                             ;   WREG = WREG + ('N'
                             ;  - 'A')
```

For the case in which WREG contains 'N' to 'Z,' the code is as follows:

```
  sublw  'N'                 ;   WREG = 'N' - WREG
  sublw  'A'                 ;  WREG = 'A' - ('N' -
                             ;  WREG)
                             ;  WREG = WREG + 'A' -
                             ;  'N'
                             ; WREG = WREG - ('N' -
                             ;  'A')
```

If you compare the two methods of computing the ROT13 value, you'll see that the results are the same. I don't think I would get a lot of argument that the second version is more efficient at computing the ROT13 value, although it is not obvious how I derived it. I would like to say that I have a simple process for coming up with this type of optimization, but I must confess that each case is different. To come up with the optimization for the ROT13 Encrypt function, I started with the desired end points (the final value of WREG when the contents of WREG are less than or equal to 'M' and when they are greater than 'M'). Next, I went to my test state (the sublw 'N' instruction) and worked at finding an addlw instruction that would negate the 'N' through 'A' operation.

Please do not assume that optimizations are beyond your grasp or that they come to me because I have more experience programming and working with the PIC MCU than you do. In explaining the operation of the instructions of optimized code, I have pedantically listed the register contents, listing their value as I algebraically simplify them. If you follow this methodology, you will find that you too will come up with very efficient code that meets your requirements.

Experiment 112—Generate a Fibonacci Number Sequence

PC/ Simulator

Eight hundred years ago, the Italian Leonardo Pisano Fibonacci looked at the problem of how rabbits reproduced and how the growth of their population could be plotted. Postulating that rabbits could produce a litter once a month, after they were two months old, Fibonacci came up with a sequence of numbers that is the bane of computer science students more than three-quarters of a millennium later.

Starting with one pair of new born rabbits, the population of rabbits in the first year can be listed as:

```
1, 1, 2, 3, 5, 8, 13, 21, 34, 55, 89, 144
```

This sequence is normally described as the result of a *recursive* function. As I mentioned previously, recursive functions perform part of a specified operation and then call themselves to complete the operation. I discourage the use of recursive functions in PIC MCU programming (of all types, not just assembly or C) because of the lack of an arbitrary-length program counter stack.

Without the possibility of recursion, the problem becomes quite simple, with each value in the sequence being defined by the statement:

```
Fib[i] = Fib[i - 1] + Fib[i - 2]
```

that can be checked by applying it to the sequence of the previous numbers.

The requirement for this application is to produce the Fibonacci rabbit population growth values for two years. Looking at the example sequence above, you can see that the value of the thirteenth month will be approaching 255, which means that in order to store the population values, you will have to use 16-bit variables. The general form of the addition is:

- Add high bytes
- Add low bytes
- If low-byte sum is greater than 255 (carry bit set), increment the high-byte sum

In my solution, I have set up a 24-element, 16-bit array, in which each month's element value is stored and used to produce the value for later elements. The solution to the problem uses the FSR index pointer register to keep track of which element (and which byte of the element) is currently being accessed. To change the FSR, I use incf and decf (increment and decrement instructions, respectively), and I add explicit values to the FSR.

```
    title   "asmFibonacci - Create the Fibanacci
Sequence for 24 Terms"
;
; Demonstrate working with 16 bit
;   Number Arrays to Calculate the
;   Fibonacci series for 24 Terms.
;
; "C" Equivalent Code:
;
;int Fib[24];       // 48 Byte-Fibonacci Sequence
;
; Fib[0] = Fib[1] = 1;
;                   // First Two Values are 1
;
; for (i = 2; i < 24; i++)
;    Fib[i] = Fib[i - 1] + Fig[i - 2];
;                   // Get Each Value
;
; while (1 == 1); // Finished, Loop Forever
;
; Hardware Notes:
;   PIC16F684 running at 4 MHz in Simulator
;
;
; Myke Predko
; 04.08.31
;
    LIST R=DEC
    INCLUDE "p16f684.inc"

; Variables
    CBLOCK 0x20
Fib:2*24                        ; Array Storing
                                ; Fibonacci Values
i, Temp:2
    ENDC

    PAGE

    org     0

    clrf    Fib + (0 * 2) + 1  ; Initialize 1st
                                ; Two Array Elements
    clrf    Fib + (1 * 2) + 1
    movlw   1
    movwf   Fib + (0 * 2)
    movwf   Fib + (1 * 2)
```

```
        movlw   24 - 2              ; Number of
                                    ; Elements to Add
        movwf   i

        movlw   Fib + (2 * 2)       ; Point to the
                                    ; Start of the Array
        movwf   FSR
Loop:
        decf    FSR, f              ; Point to Fib[i - 1]
        movf    INDF, w             ; and Save in Temp
        movwf   Temp + 1
        decf    FSR, f
        movf    INDF, w
        movwf   Temp
        decf    FSR, f
        movf    INDF, w             ; Add Fib[i - 2] to
                                    ; Temp
        addwf   Temp + 1, f
        decf    FSR, f
        movf    INDF, w
        addwf   Temp, f
        btfsc   STATUS, C
         incf   Temp + 1, f

        movlw   4                   ; Store Temp in
                                    ; Fib[i]
        addwf   FSR, f
        movf    Temp, w
        movwf   INDF
        incf    FSR, f
        movf    Temp + 1, w
        movwf   INDF
        incf    FSR, f

        decfsz  i, f                ; Repeat (24 - 2)x
         goto   Loop

        goto    $                   ; Finished, Loop
                                    ; Forever

        end
```

When you run this application, one of the biggest problems you will encounter is trying to determine whether or not the answers are correct. Looking at the File Register window in MPLAB-IDE (see Figure 12-5), you may have a difficult time picking out the higher values and translating them into their decimal equivalent.

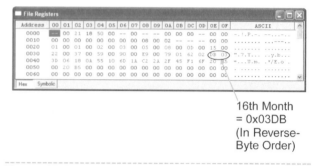

16th Month
= 0x03DB
(In Reverse-
Byte Order)

Figure 12-5   *Fibonacci value*

In Figure 12-5, I have circled the sixteenth element of the Fib array. The element's first address in the zero-based Fib array is found using the formula:

$$\text{Address} = \text{Array Start Address} + [2 \times (\text{Element} - 1)]$$

So, to find the first byte of the sixteenth element, the first address is 0x03E (assuming Fib starts at file register address 0x020). This value is saved in *Little Endian* format (low byte at low address, high byte at high address), and the value is 0x03DB—even though it looks like 0x0DB03. Converting to decimal, this value is 987 (which was found by evaluating the expression [3 x 256] + [13 x 16] + 11), exactly what is expected.

I'm going through this calculation because, like C pointer values, MPLAB IDE does not show the contents of an array (especially with elements that are 16 bits and greater), and to see what they are, you will have to employ a bit of ingenuity. For the most part, I recommend you avoid data structures with operating values that require this amount of work to understand, but in some cases calculating an address and converting the value manually cannot be avoided.

Experiment 113—Find the Largest Common Factor of Two Eight-Bit Numbers

PC/
Simulator

To round out the section, I want to leave you with an application that can be expressed reasonably crisply in the C language. However, translating the function to assembly can have a number of unique challenges that will need to be addressed. Finding the largest common factor of two eight-bit numbers is very difficult when translating between the two programming languages directly. Substituting direct translations of the C assembly language statements would have potentially made the application more difficult to write and execute efficiently.

The task of finding the largest common factor of two numbers is not particularly difficult, although it requires a number of steps. The first is to find all the factors of each of the numbers. This is done by repeatedly dividing each value with incrementing divisors until a divisor that produces a remainder of zero is found. When the remainder of zero is found, the quotient of the value and the divisor (which is a factor of the value) must be saved as the new value and the process repeated to find other factors. In high-level language programming, this is quite easy to do, but in assembly-language programming, coding the division routine requires extra work. However, this extra work can make it is possible to perform the routine only once, simplify the program, make it shorter, and make it execute faster.

Because of the need for division routines and storage of the different factors of each value, I used two unique variable arrays that are indexed using counters. Using a high-level language, this is easy to accomplish, but in the mid-range PIC MCU, it is more difficult because only one index register is available and because the address for each value needs to be recalculated. To simplify working with the index register with multiple arrays, some planning is required—planning to make sure that the index you are both reading from and writing to is the last index accessed so multiple recalculations are not required.

The assembly language result for this experiment is "asmLCFactor.asm" and its operation is best illustrated using "cLCFactor.c". Obviously, many subtleties exist in the assembly-language code that cannot be represented in the C language version of the application.

```
#include <pic.h>
/*  cLCFactor - Get the Largest Common
  Factor of 2 8 Bit Numbers"

  First get the factors for two 8-bit
  numbers and use them to find the largest
  common factor. This is a direct usage of the
  '"C" Equivalent Code' of the assembly
  language version

  Hardware Notes:
   PIC16F684 running at 4 MHz in Simulator
   Reset is tied directly to Vcc via
   Pullup/Programming Hardware


  Myke Predko
  04.09.01

*/


int Number1, Number2;
int Factor1List[16];
int Factor2List[16];
int LargestFactor;
```

```
GetFactors(int Value, int FactorList[])
{          // Find ALL the Factors of a Number

int i, j;

  FactorList[0] = 1;
  for (i = 1; i < 16; i++)
          // Initialize the FactorList
   FactorList[i] = 0;

  i = 2; j = 1;
  while ((1 != Value) && (Value != i))
          // Find the Factors
  {
   if (0 == (Value % i))
          // If Remainder or Value/i == 0
     {  //   It is a Factor
    Value = Value / i;
          // Take Away Factor
    FactorList[j++] = i;
          // Add Factor to the List
    i = 1;
          // Restart Factor Process
    }  // fi
   i++; // Look at Next Value of "i"
  }  // elihw
  if (1 != Value)
          // Remaining Prime to Save?
   FactorList[j] = Value;
          // Remainder is Final Factor

}       // end GetFactors


main()
{
int i, j;

  Number1 = 196;
  Number2 = 224;

  GetFactors(Number1, Factor1List);
          // Get Factors
  GetFactors(Number2, Factor2List);

  LargestFactor = 1;
          // Calculate Largest Factor
  for (i = 1; (i < 16) && (0 != Factor1List[i]);
     i++)
   for (j = 1; (j < 16) && (0 !=
        Factor1List[i]); j++)
   if (Factor1List[i] == Factor2List[j])
    {     // Have a Common Factor
     LargestFactor = LargestFactor *
                    Factor1List[i];
     Factor1List[i] = Factor2List[j] = 0;
    }    // fi

  while (1 == 1);
          // Finished, Loop Forever

} // End cLCFactor
```

The efficiencies of the assembly-language version over the C version became very apparent when some measurements were taken of the two applications. The assembly-language version executed in one quarter of the number of cycles of the C version (4,800 to 15,275 according to the MPLAB IDE Stopwatch) and also took up one quarter of the amount of space taken by the C language version (92 instructions to 367 instructions). The improvement in the number of file registers used was not as dramatic (the assembly-language version requires 72 file registers versus the 88 required for the C language version). From these parameters, you might argue that the assembly-language version is four times better than the C language version and is clearly the "best" of the two.

The assembly-language version would be the best if speed and size were the critical requirements for the application. If development cost was an important requirement, the C language version would win, hands down. Remember that the design of the algorithm, which I coded in the C language before attempting to write it out in assembler, has to be done in either case, and if a high-level language is used, the code can substitute for documentation. In this case, you are not developing code that implements the same function twice. And, in the case of making changes to the code, if the source code itself is used as documentation, the documentation for the application is changed automatically. Finally, the C language version can be transferred to other processors with a minimum of conversion effort. If the application was coded in assembly language, it would have to be rewritten in the appropriate assembler. This last point is known as the *portability* of the code. By definition assembly language has very poor portability across different platforms.

For Consideration

Teachers, more than anybody else, understand the importance of creating different assignment and exam problems. If a set of problems (such as the 10 in this section) were used for more than one year in programming course, chances are good that solutions would become available over the Internet in time for start of the next course. This is unfortunate because the purpose of these questions is to help the student understand problem requirements, develop an algorithm for the solution, and finally present the result as part of the project. The marks are really secondary.

The problem: It's difficult to come up with new questions. I find it useful to keep a running log of questions that I think of during the year. I write down all sorts of things that I come up in my day-to-day work. Some of the problems in my current list that could have been considered for this section include the following:

- For a set of X, Y points, find the equation for the trend line.
- Multiply two numbers together using Boothe's algorithm.

- Find the eight-bit mean and median of a list of 10 numbers.
- Produce an eight-bit random number using an LFSR and an eight-bit seed.
- Use structures to add two real numbers together.
- Sort a list of numbers using a bit-map.
- Reverse the characters in a string.
- Find the Y=0 crossing point for an equation in the format $Y = Mx^3 + B$.
- Find the day of the week for a given date.
- Compress an ASCII string (i.e., replace repeating values with a single character).
- Implement a simple transposition cipher.

Along with keeping a list of possible questions, a quick Google search for "Programming Problems" is a useful way to find additional questions to work through. In July of 2004, this search yielded over six million hits. The following selection came from the top of the list and offers many useful and appropriate questions:

ACM International Collegiate Programming Contest/European Division

www.acm.inf.ethz.ch/ProblemSetArchive.html

- Various sample problems

American Computer Science League

www.acsl.org/samples.htm

- High school level programming problems

Department of Mathematics and Computer Sciences, Western Carolina University

www.cs.wcu.edu/cscontest/²⁰⁰⁴/₂₀₀₄ problems.pdf

Programming Problems, 15th Annual Computer Science Programming Contest

Problem Set Archive

http://acm.uva.es/problemset/

- Several hundred problems, advertised as being typical to exam/project problems

University of California, 1999 Programming Contest

www.cs.berkeley.edu/~hilfingr/programming-contest/f99-contest.pdf

- Some basic programming problems

ZipZaps® Robot

PC/ Simulator

PICkit™ 1

starter kit

Required Parts

1 Radio Shack ZipZaps remote control car

1 ZipZaps Performance Booster Upgrade Kit

1 PIC16F684

1 MAX756

3 38 KHz IR TV remote control receivers

1 Two-line, 16-column LCD

1 1N5817 Schottky diode

2 LEDs

5 IR LEDs

2 Light-dependent resistors

4 10k resistors

6 1k resistors

2 4.7k resistors

2 470Ω resistors

3 100Ω resistors

  10k breadboard-mountable poten-tiometer

2 100 μF elec-trolytic capaci-

tors (rated at least 10V)

3 47 μF electrolytic capacitors

1 0.1 μF capacitor (any type)

1 0.01 μF capacitor

1 14-pin, machined pin DIP socket

1 Two-pin, right-angle straight-through connector

1 Four-pin, right-angle straight-through connector

1 Two-pin, straight-through socket

1 Four-pin, straight-through socket

1 Prototyping PCB

1 Breadboard-mountable SPDT switch (E-Switch EG1903 Recommended)

1 Breadboard-mountable momentary On push button

1 5mm heat-shrink tubing

1 Three-cell AA battery pack

3 AA alkaline batteries

2 6-32, 2-inch long nylon bolts

6 6-32 nylon nuts

Tool Box

DMM

Breadboard

Soldering iron

Solder

Rotary hand tool (Dremel tool) with carbide disk cutter

Electric drill

9/64-inch high-speed drill bit

Needle-nose pliers

20- to 24-gauge wire

Black wire

Red wire

Wire-wrap wire

Permanent marker

Wire clippers

Breadboard

Wiring kit

Scissors

Krazy Glue

Five-minute epoxy

Over the years, I have gotten at least two dozen emails from people who want to come up with their own copies of established products because what's already out there is so expensive and they feel they could produce products at a fraction of the established product's price. My only experiences with retail products are with the Tab Electronics Build Your Own Robot Kit and the Tab Electronics Sumo-Bot Kit, and they have been eye-opening experiences. In coming up with these products, we agonized over features and parts to try and meet the required price point. I very quickly learned that the product cost should ideally be one-sixth, and no more than one-fifth of the retail price. The cost differential is due to shipping, warehousing, packaging, advertising, warranty, nonrecurring expenses costs, and profits for both the manufacturer and retailer. Using this rule of thumb, I am amazed that Radio Shack can sell the ZipZaps® remote-control cars for $20.00.

The ZipZaps remote-control cars are a lot of fun to play with. If you haven't had the chance to at least try one out in the store, you should do so immediately. Figure 13-1 shows the complete package, a car and remote-control unit that doubles as a charger. The car has a 1.2 V NiMH battery that can be charged in about a minute and allows the car to run for three or four minutes at full speed. The car is capable of running forward and backward and can change direction; some earlier remote-control cars could only move forward and backwards, and turn in only one direction while in reverse. Radio Shack offers a wide range of different bodies for the kits, along with various accessories, including performance upgrades. As you play with the ZipZaps, you should be impressed with its robustness. It handles crashing into objects very well and even survives the most extreme activities (performing jumps, running off tables, and so on) without complaint. It's a pretty amazing package when you consider that they cost $3 to $4 to manufacture.

I am actually surprised to discover that nobody has explored the idea of using the ZipZaps as the basis for a mobile robot chassis, when using other larger remote-control car chassis as the basis for robots, which is very common. Looking at the ZipZaps, I can come up with a few concerns, namely:

- 1.2-volt operation
- Limited operating time
- Small size
- Minimal payload capability
- Surprisingly sophisticated motor and steering driver circuitry (and complex circuitry overall)
- Ternary steering (three states: left, right, or straight)

Taking off the body, the ZipZaps is very compactly designed. You can pop off the circuit board's plastic cover and get a better idea of the electronics that act as the radio receiver and the motor/steering drivers. You may have never seen, and have probably never worked with, the surface-mount components used in the ZipZaps' electronic board (see Figure 13-2), which may minimize your confidence in modifying the circuitry. All these concerns are legitimate, and using the ZipZaps as the basis for a PIC MCU controlled mobile robot may seem unlikely, but as I will show in the experiments that follow, the ZipZaps is designed using traditional methodologies. You can hack one into a robot over the course of a weekend for surprisingly low cost.

An important aspect of hacking the ZipZaps into a robot has been the work that has been performed in the previous chapters and the skills and knowledge that work has developed. As I work through this sec-

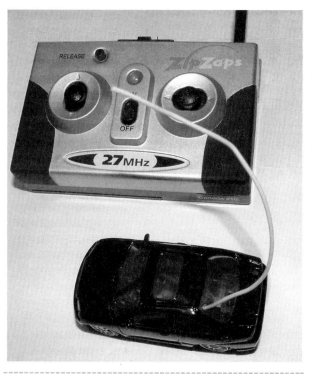

Figure 13-1   *The Radio Shack ZipZaps® car with remote control and charger*

Figure 13-2   *The ZipZaps electronics seem small and difficult to work with, but can be used as a basis for a surprisingly sophisticated robot.*

tion, I will be adding robot features to the ZipZaps chassis that have been presented and prototyped earlier in the book; in fact much of the incremental code that is added with new features was cut and pasted directly from these previous experiments. With a simple base structure for the software, using the established code base I was able to create a very sophisticated set of robot functions for the ZipZaps in just a few days.

I really had a lot of fun working through the experiments in this chapter and I think you will too. But before you start cutting up a ZipZaps for your own robot, I suggest that you spend some time reading through this section and rereading (and redoing) the various experiments that discussed and presented the control, software, and interface features required for the robot. You may not be able to find all the parts I used (although they are pretty generic) or, due to assembly differences, you may find that they work differently or have to be tuned in a different way. By reading through the section first and making sure you understand how everything is supposed to work, you will minimize the amount of work needed to complete the robot and maximize the chances for success first time at every step. Having a different perspective on the robot, you will likely come up with something even better than what's presented here!

Experiment 114—Characterizing the ZipZaps

Parts Bin

1 Radio Shack ZipZaps remote-control car

1 ZipZaps Performance Booster Upgrade Kit

Tool Box

DMM

Needle-nose pliers

Soldering iron

Solder

20- to 24-gauge wire

Before we can start designing robot hardware for a ZipZaps remote-control car, it's important to understand the characteristics of the motor drivers and the steering actuators. The information that we want to collect includes which circuits are used to drive the motors and the steering, what voltage is applied to them, and how much current they consume during operation. To do this, you will have to probe the ZipZaps very carefully, as you will be desoldering and resoldering some wires. This is somewhat fine work, and you may want to take my results on faith. Although you will be doing similar work in the next couple of experiments, now is a good time to get started and get some practice.

Just as a note of reassurance, I want you to know that during this step (and the following steps), I broke a number of wires from the car chassis to the PCB. I was able to reattach them, and the ZipZaps ran fine afterward.

The first order of business was to solder on 2- to 3-inch lengths of wire onto the red wire and black wire connections of the PCB. It is difficult to show in a photograph what was done because I attached the wire to the backside of the PCB (assuming that the side that has the wires, shown in Figure 13-3, is called the front).

These wires will be required for the next experiment. While holding the driving wheels of the ZipZaps off the table (no load condition) and with the wheels stalled against a table top (worst case with the motor essentially short circuited), I measured the voltage (which is the battery output) across the two wires as given in Table 13-1. The minimal voltage drop seen from nothing running to running and stalled is an indication that the motor drivers are well designed for the motor and that the battery has some excess capacity.

Next, I wanted to look at the motor drivers themselves. With a bit of searching with a DMM, I was able to infer that the motor driver circuit looks something like Figure 13-4. This is an innovative motor driver design with two control lines and is advantageous to us, as I will show in the next couple of experiments. To test the motor driver, I first measured the voltage across the motor (at the yellow and blue wires) for the conditions listed in Table 13-2. I then measured the current passing through the motor for the same conditions by desoldering one of the motor wires from the PCB. When I was finished, I soldered the motor wire back on.

The motor driver should be fairly easy to interface to a PIC MCU. The only problem you should be aware

Figure 13-3 *Close-up view of ZipZaps PCB's "front side"*

Table 13-2
Motor Parameters

Condition	Voltage Across Motor	Current Through Motor
No Load, Motor Running	1.20 V	25 mA
Motor Stalled	0.65 V	100+ mA

Table 13-3
Steering Driver Solenoid Parameters

Condition	Voltage Across Solenoid	Current Through Solenoid
No turn	0 V	0 mA
Turning/Solenoid Active	1.20 V	60 mA

of is the need to maintain a constant current from the PIC MCU to the transistor's base. This will require the changing the driver transistor-base current-limit resistor to a value that will provide the same amount of base current as was available to the transistors originally.

Again, the motor driver seems to be well designed with the ability to handle the current produced by a stalled motor condition. After seeing these results, I decided I wanted to use the motor driver circuit built onto the ZipZaps PCB.

Next, I wanted to characterize the steering actuator electronics. After a few minutes of probing the PCB, I came up with the circuit shown in Figure 13-5. Each of the two solenoid coils of the steering mechanism are driven by the single bipolar NPN transistors with the

parameters listed in Table 13-3. Like the motor driver, the steering gear should easily drive a PIC MCU I/O pin. The current-limiting resistance should be changed again to make sure the same amount of current passes through the base of the steering driver transistors.

While doing this characterization, I ran into two problems. First, the PCB is small and uses *surface-mount technology* (SMT) components, which are harder to rework than the pin-through-hole components that you are used to. Patience, planning, and a small-tip soldering iron are all that is required. Second, it was difficult working with the steering gear solenoid wires, as they are extremely thin magnetic wires and will pull themselves from the PCB when you are soldering adjacent components. They can be resoldered easily, but you will have to notice when they are missing, and you will find that they can be difficult to manipulate.

Table 13-1
Battery Voltage at Different Driving Motor States

Condition	Voltage
Motors, Steering Off	1.36 V
Motors Running, No Load	1.30 V
Stalled Motors	1.26 V

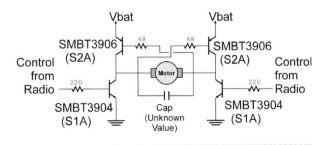

Figure 13-4 *Motor driver*

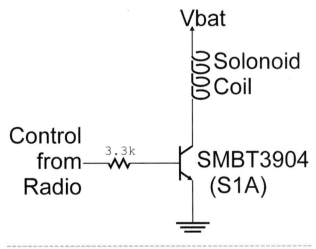

Figure 13-5 *Steering driver*

Experiment 115—PIC MCU Power Supply

PC/ Simulator

PICkit™ 1

starter kit

Parts Bin

1	Radio Shack ZipZaps remote-control car
1	PIC16F684
1	MAX756
1	1N5817 Schottky diode
2	LEDs
1	100Ω resistor
2	100 μF electrolytic capacitors (rated at least 10V)
1	0.1 μF capacitor (any type)
1	0.01 μF capacitor

Tool Box

DMM

Needle-nose pliers

Breadboard

Wiring kit

The first thing I had to do in the process of getting the ZipZaps robot together was come up with a scheme to power the PIC16F684 that would control the robot. This actually turned out to be a fairly painless process, although a couple of operational issues regarding the selected chip and its use with the PIC16F684 had to be investigated. To summarize the experiment's results, the addition of a PIC6F684 power supply went very smoothly, and I learned a bit more about the PIC MCU and had some things to think about in later robot projects like this one.

If I were to look at the ZipZaps Robot from a perspective that other people could easily understand, I would note that it is powered by a single 1.2-volt rechargeable battery. If you look at the PIC16F684 (or, indeed, any PIC MCU) datasheet, you will discover that the minimum voltage on which they run is 2.0 volts. For the ZipZaps robot PIC MCU to run at that speed, either I needed to add an additional battery or a step-up power supply. The suggestion of adding a step-up power supply may seem like overkill, but adding another battery had nothing going for it; not only did it require space that was unavailable in the chassis, but the battery used in the ZipZaps is heavy (which will decrease the performance of the robot), and I would have to come up with a special charging circuit for the two additional batteries.

To go with the step-up power supply, I had a number of requirements that it had to meet before I was comfortable using it. The requirements were:

- Noise immunity—I didn't want motor/steering electrical noise to affect the operation of the PIC MCU.

- Small size—The control chip could be no larger than an eight-pin DIP, and the number of support components had to be very minimal.

- Reasonable run time—The robot would run for a reasonable amount of time before requiring recharging.

- Able to survive PICkit programming backdriving. *Backdriving* is the term used to describe the forcing of an overvoltage condition on a chip's output. Backdriving typically becomes very wasteful in terms of power and can damage some circuits.

Using a switch mode step-up power supply eliminates many of these concerns. The capacitor filters and inductor energy storage very effectively filters out electrical noise between the motor and the circuitry powered by the step-up power supply. There are a plethora of different step-up power supplies to choose from, which made the task of choosing a small, efficient device easier, although it still took quite a bit of time for me to settle on the Maxim MAX756. This chip is designed for powering electronics using a single 1.5 V or 1.2 V battery, which is exactly the situation we have here.

To validate the operation of the MAX756 and make sure it would be appropriate for the ZipZaps robot, I built the circuit shown in Figure 13-6. This is a fairly standard application of the MAX756, using values and components recommended by the manufacturer. The first test was to see if the ZipZaps battery could run this chip and have it power a reasonable size load. The circuit in Figure 13-6 provides a simple LED (requiring about 10 mA) and the cFlash.c application, which flashes an LED between RA4 and RA5.

The application ran the first time, but did not run after that. Using my DMM, I discovered that the voltage output of the MAX756, with all its heavy capacitance, took a very long time to go below 500 mV—just enough to put the PIC microcontroller into an invalid state. To fix this problem, I enabled the brownout reset circuitry in the cFlash.c program's configuration fuses and renamed the application ZipFlash.c. As a side note, all the programs in this section start with "Zip" for easy finding, rather than the traditional c or asm. If I didn't want to enable the brownout reset circuitry, I could have put a *bleed resistor* of 4.7k or more across the PIC16F684's Vdd and Vss pins, which would drain the capacitors when power was removed.

```
#include <pic.h>
/*  ZipFlash.c - Simple C Program to Flash
    an LED on ZipZaps'
    PIC16F684

This Program is a modified version of "cFlash
2.c".  with
    Brown Out Detect Active

RA4 - LED Positive Connection
RA5 - LED Negative Connection

myke predko
04.11.25

*/

__CONFIG(INTIO & WDTDIS & PWRTEN & MCLRDIS &
UNPROTECT \
    & UNPROTECT & BOREN & IESODIS & FCMDIS);


int i, j;

main()
{

  PORTA = 0;
  CMCON0 = 7;        // Turn off Comparators
  ANSEL = 0;         // Turn off ADC
  TRISA4 = 0;        // Make RA4/RA5 Outputs
  TRISA5 = 0;

  while(1 == 1)      // Loop Forever
  {
   for (i = 0; i < 255; i++)
                     // Simple 500ms Delay
    for (j = 0; j < 129; j++);

     RA0 = RA0 ^ 1; // Toggle LED State
 }  // elihw
}  // End ZipFlash
```

The application was built on a breadboard (see Figure 13-7), and after enabling brownout rest, it ran fine and did not seem to affect the operation of the ZipZaps chassis and PCB (which, other than the addition of wires from the battery's positive and negative connections, was unmodified). The battery could be

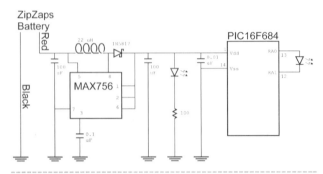

Figure 13-6   *ZipZaps power supply test circuit*

charged from a remote control while plugged into the power supply circuit. And when the battery was charged, it not only charged the ZipZaps' battery but also powered the MAX756 and the PIC16F684 to which it was connected.

With the MAX756 power supply and PIC16F684 circuit added to the ZipZaps chassis, I found that the PIC16F684's LED would flash for 10 to 11 minutes (about half the time the unmodified ZipZaps would sit on the table before loosing enough charge to the radio receiver to stop being able to move). With the radio receiver and wiring still present, I found that the motor would run, without any drag on the wheels, for about 4 minutes, about the same as the unmodified ZipZaps. The conclusion I reached here is that any task the robot is performing should be completed in 2 minutes or so, to make sure that it does not run out of power. If you were in some kind of contest, you would have to make sure that you won in 2 minutes or less. Otherwise you would loose from lack of battery power.

Essentially, all the requirements outlined here are met with the MAX756 step-up power supply. Looking at the voltage output from the MAX756 using an oscilloscope, no indication was given that motor noise was being passed to the PIC16F684. The MAX756 is an eight-pin part with five support components, which made it appropriate for use in this application. The running time of the robot did not seem to be negatively affected by the addition of the MAX756, PIC16F684, and LEDs to the load powered by the ZipZaps rechargeable battery. The only issue not addressed in this experiment is whether or not the MAX756 can tolerate the 5 volts applied to program the PIC16F684 in circuit. This is an important requirement for the application, as I will discuss below, and it is not mentioned in the MAX756 datasheet. To be on the safe side, a silicon diode will be used to isolate the PIC16F684's power supply from the MAX756's output. The following experiments will determine if this will affect the operation of the other components on the Robot's PCB.

Experiment 116—PIC MCU Electronics PCB

PC/ Simulator

PICkit™ 1

starter kit

Parts Bin

1 Radio Shack ZipZaps remote-control car

1 PIC16F684

1 MAX756

1 1N5817 Schottky diode

1 LED

2 100 µF electrolytic capacitors (rated at least 10V)

1 0.1 µF capacitor (any type)

1 0.01 µF capacitor

1 14-pin, machined pin DIP socket (see text)

1 Two-pin, right-angle straight-through connector

1 Four-pin, right-angle straight-through connector

1 Two-pin, straight-through socket

1 Four-pin, straight-through socket

1 Prototyping PCB (see text)

2 6-32, 2-inch-long nylon bolts

6 6-32 nylon nuts

Tool Box

DMM

Needle-nose pliers

Rotary hand tool (Dremel tool) with carbide disk cutter

Electric drill

9/64-inch high-speed drill bit

Wire clippers

Soldering iron

Solder

Black wire

Red wire

Wire-wrap wire

Permanent marker

Krazy Glue

Five-minute epoxy

With the power supply tested, you are now ready to start hacking into your ZipZaps car and turning it into a robot. In this experiment, you will be putting in nylon bolt attachment points for the PIC MCU PCB as well as creating the basic PCB circuit with the power supply and PIC MCU socket. This task is not going to be easy, and you will have to plan for the future experiments to make sure that enough space (and holes) is available for mounting additional hardware. As I work through this experiment, I will try to indicate the issues that you should be most aware of.

The prototyping PCB used for this experiment can really be of any type, but I used one with predefined power and ground traces and horizontally connected holes. This simplified my wiring somewhat although it restricted me in other areas for which I had to compensate when I added additional hardware. You do not

have to use a prototyping system like I used, but you should read the rest of this section to understand what prototyping PCB you want to use and how it is going to be mounted to your ZipZaps chassis.

Position the prototyping PCB over the ZipZaps chassis in such a way that the PCB will hang over the chassis evenly. Do not position the PCB so it is above the empty space in the chassis between the battery and the steering gear. The mounting bolts for the PCB will be glued into this area and as close to the steering gear as possible. When the bolts are glued in, it is important that the charging contacts and hold-down holes are not affected in any way. If they are hindered, you will not be able to charge the ZipZaps battery using the remote-control unit.

With the marks in place, find the center of the PCB at this point and drill two 9/64 -inch holes 0.250-inch away from the center. If the prototyping PCB is fully drilled (as mine was), drill out two holes, four holes apart. Remember to keep the PCB centered over the ZipZaps chassis when you are doing this.

Once the holes are drilled, run the nylon bolts through and screw them down with some of the nuts. At the end of the bolts, loosely screw on a nut and put Krazy Glue on the open face of the nuts at the end of the bolts. Carefully (making sure you do not trap any of the small steering wires) push the open face of the nuts against the bottom of the chassis. The bolts should be solid to the touch in 30 seconds or so. The Krazy Glue acts as a mild solvent to the plastic used in the ZipZaps, so leave the assembly in place for an hour or so for the Krazy Glue and chassis plastic to harden.

When the Krazy Glue has hardened, remove the bolts and cut the bolts down to 1.25 inch (3 cm). (You will probably have to loosen the nuts holding the bolts to the PCB first.) Insert the bolts into nuts, cover the nuts and the base of the bolts with 5-minute epoxy, and wait for another hour for the epoxy to harden.

When the epoxy has hardened, run two nuts down the shaft $\frac{1}{4}$ inch ($\frac{1}{2}$ cm) and Krazy Glue them in place. Once the Krazy Glue has hardened, you can place the prototype PCB on the bolts and tighten it down when you want to test the robot.

With the bolts in place, you will now have to cut down the original ZipZaps PCB. This is being done for two reasons. First, you want to avoid any kind of contention between the ZipZaps driver and the PIC MCU. And second, you want to make space for the two bolts you just installed. Before cutting the PCB with a Dremel tool and carbide wheel, I located the four resistors used for current limiting on the PCB and drew an indicator line on the PCB above them (see Figure 13-7) using an indelible marker. Cut the PCB along this line carefully, making sure you do not damage any of the marked resistors.

Once the ZipZaps PCB has been cut, solder wires to the positive (red) and negative (black) wires leading from the battery to the PCB. These wires will go to a right-angle connector to be soldered to the bottom of the prototyping PCB and will need the straight-through connector soldered to the other end. Next, solder wire-wrap wires to the ends of the four-motor driver and steering driver resistors (see Figure 13-8). These four wires will be soldered to another straight-through connector, which will go through another right angle connector and the bottom of the prototyping PCB. The ZipZaps chassis is now ready to be controlled by a PIC MCU!

When I attached the PCB to the robot, I did it with the solder traces up. This make soldering components quite a bit more difficult, but it ensures the PIC MCU power and motor/steering driver connectors are easy

Figure 13-7   *Mark cutting line above current-limiting resistors and below black blob*

to solder and their wires do not snag or rub against sharp wires on the bottom of the PCB. The machined pin socket that was specified for the PIC MCU was identified because it can be soldered to the PCB from the topside (instead of only from the bottom side, as in a regular DIP socket).

The circuit that you will be wiring to the PCB is the basic power supply with PIC16F684 (without the two LEDs) in Figure 13-9 with the bottom right-angle connector used to supply the 1.3 volts from the ZipZaps battery. Again, it's important to plan ahead; due to my wiring placement, I put the MAX756 and its related discrete components at the front of the robot, which left the rear of the PCB for the PIC16F684.

When you have completed the assembly work, your robot chassis with PCB should look something like Figure 13-10. You should still be able to put the ZipZaps chassis on the ZipZaps charger, and it should work normally (i.e., the charger's LED should start at red and change to green when the charging operation has completed). If everything is correct, after taking

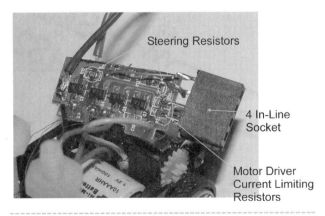

Figure 13-8   *PCB connector*

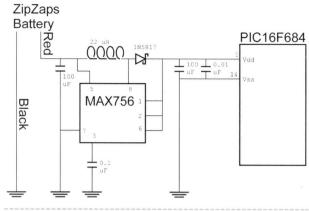

Figure 13-9   *ZipZaps power*

Figure 13-10   *ZipZaps chassis with partially assembled PCB attached*

the ZipZaps chassis and PCB off the charger, an LED placed against RA4 and RA5 (cathode) should start flashing. There is no on and off switch, the MAX756 will stop providing 3.3-volt power when the input voltage has dropped below its threshold. This should give your robot a minute or two of basic operation, which should be good for testing most applications.

I found that although the connectors were quite reliable, the solder joints and wires from the battery and motor to the ZipZaps PCB would easily break. Ideally, the prototyping PCB would be permanently attached to the ZipZaps chassis (as is the original product PCB), but this is very difficult to do and still be able to work through adding (and debugging) the different peripherals. Make sure you plan how to con-

nect and disconnect the prototype PCB to the ZipZaps chassis so that the wire solder joints receive a minimum of stress. If you don't, you will find that the wires will break periodically. You'll be left scrambling to figure out the cause of the problem and then fix it, with the wires becoming shorter over time.

This is probably the most difficult assembly experiment in the whole book. Remember to think about where everything is going next, and try to keep everything as well centered as you can. A few moments forethought will save you a lot of grief later.

Experiment 117—IR TV Remote Control

PC/ Simulator

PICkit™ 1
starter kit

Parts Bin

1 PIC16F684

1 38 KHz IR TV remote-control receivers

1 Two-line, 16-column LCD

1 10k resistor

1 100Ω resistor

1 10k breadboard-mountable potentiometer

1 47 μF electrolytic capacitor

1 0.01 μF capacitor (any type)

1 Breadboard-mountable SPDT switch (E-Switch EG1903 recommended)

1 Breadboard-mountable momentary on pushbutton

1 Three-cell AA battery pack

3 AA alkaline batteries

Tool Box

Needle-nose pliers

Breadboard

Wiring kit

I like controlling robots using an infrared TV remote control; they are cheap and the code required to read the data is surprisingly simple. Other options include RF or sound control, but these tend to be expensive and "fiddly" compared to IR control, especially considering that IR control is generally handled by the robot's controlling PIC MCU. In this experiment, I will demonstrate the code and circuitry required to read an incoming IR signal and display it on an LCD.

The IR signal produced by TV remote control comes out of the receiver module looking (see Figure 13-11). The IR signal from the remote-control code is modulated with a 38 kHz signal that is only active when a low is desired in the receiver's output. For hobby robots, I like to use the Sony TV codes, which consist of a *Packet Start* followed by 12 bits in the format shown in Figure 13-11. The timing of the different features of the waveform is based on a 550 μs T clock and consists of different multiples of this time base.

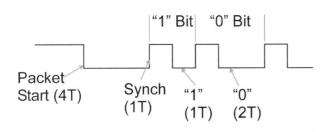

Figure 13-11 *IR remote-control codes*

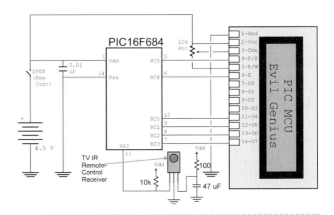

Figure 13-12 *IR Rx test circuit*

Table 13-1
Sony TV IR Remote-Control Codes Used by the ZipZaps Robot

Button	Value	Button	Value
"1"	0x010	"PIP"	0xDB0
"2"	0x810	"Enter"	0xD10
"3"	0x410	"Display"	0x5D0
"4"	0xC10	"Mute"	0x290
"5"	0x210	"Recall"	0xDD0
"6"	0xA10	"Arrow Up"	0x2F0
"7"	0x610	"Arrow Down"	0xAF0
"8"	0xE10	"Arrow Left"	0x2D0
"9"	0x110	"Arrow Right"	0xCD0
"0"	0x910	"Menu"	0x070
"Power"	0x490	"Guide"	0x764
"Vol+"	0x490	"OK"	0xA70
"Vol-"	0xC90		
"Ch+"	0x090		
"Ch-"	0x890		

To read and display the IR incoming signals, I came up with the circuit shown in Figure 13-12. This is a modification of the original LCD display experiment's circuit, which now includes an IR TV receiver module with the required filter capacitor and resistor. The C programming language code for driving the LCD was translated into assembler as asmLCD.asm, which can be found on the PICkit's CD-ROM. The actual code that reads the incoming IR TV remote-control signal is called asmLCDIR.asm. Note that to simplify the time delays, I created the Dlay macro, which will delay the operation for some set number of microseconds.

When I ran asmLCDIR.asm on the circuit shown in Figure 13-12, I came up with the codes in Table 13-1 for the different buttons on a universal remote programmed to output Sony TV codes. These different codes will be used for controlling the ZipZaps robot in the following experiments.

Experiment 118—Motor and Steering Control

PC/Simulator

PICkit™ 1

starter kit

Parts Bin

1 Radio Shack ZipZaps remote-control car chassis

1 PIC16F684

1 Prototype PCB from the previous experiment

2 4.7k resistors

2 470Ω resistors

Tool Box

DMM
Needle-nose pliers
Wire clippers
Soldering iron
Solder
Wire-wrap wire

With connections from the original ZipZaps PCB to the prototype PCB in place, the next step is to make the motor and steering connections so that the PIC MCU can control the motion of the ZipZaps chassis.

The connections require current-limiting resistors to ensure the motors and solenoids are driven with approximately the same amount of current, that they perform as designed, and that an excessive amount of current isn't flowing. This is an important point because the PIC MCU will be driving high signals at 3.3 volts instead of the 1.3 V measured on the original ZipZaps.

To calculate the new current-limiting resistors, I assumed that the base-to-collector voltage drop was 0.7 V, and using the known base current-limiting resistors, I could calculate the expected base currents.

For the steering drivers, the base current is:

```
V = isteering x 3.3k
iSteering = (1.3 - 0.7) / 3.3k A
                    = 0.18 mA
```

To maintain this base current with a 3.3-volt output:

```
V = isteering x Rsteering
Rsteering = (3.3 - 0.7) / 0.18 mA
                    = 14.4k
```

With 3.3k already in series, an additional 11.1k of resistance is required for the steering solenoid drivers.

Similarly for the motor drivers, the base current is:

$$V = i_{motor} \times 220$$

$$i_{motor} = (1.3 - 0.7) / 220 \text{ A}$$

$$= 2.73 \text{ mA}$$

To maintain this base current with a 3.3-volt output:

$$V = i_{motor} \times R_{motor}$$

$$R_{steering} = (3.3 - 0.7) / 2.73 \text{ mA}$$

$$= 950\Omega$$

With 220Ω already in series, an additional resistance of 730Ω is required for the motor drivers to be controlled by the PIC MCU.

With these values in mind, I tested an additional 10k for the steering drivers and 680Ω resistors for the motor drivers as well as no additional resistance at all points. To test the action of these circuits, I connected the transistor bases (with series resistors) directly to 3.3 volts. When the calculated additional resistors were in place, the motors and steering solenoids seemed to work exactly the same as they did with an unmodified ZipZaps. With no resistances in place, both the motors and the steering solenoids had a lot more power. In an effort to find some balance between more power and

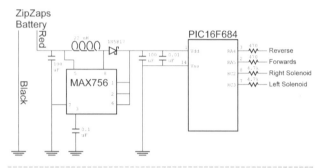

Figure 13-13   *ZipZaps motor*

Figure 13-14   *ZipZaps robot prototyping PCB with motor and steering transistor base current-limiting resistors connected from the PIC MCU socket to the backside ZipZaps PCB connector*

burning out the motors, driver transistors, or depleting the battery too quickly, I found that 4.7k and 470Ω were the optimal additional series-base current-limiting resistors for the steering and motor drivers, respectively. The ZipZaps robot prototyping PCB (see Figure 13-13) was updated appropriately and wired according to Figure 13-14, and it ran using the software presented in the next experiment.

Experiment 119—Basic Task-Control Software

PC/ Simulator

PICkit™ 1
starter kit

Parts Bin

1 Radio Shack ZipZaps remote-control car chassis

1 PIC16F684

1 Prototype PCB

Tool Box

DMM

Earlier in the book, I demonstrated how to run a motor with a fairly low frequency PWM. I'm going to take advantage of that work with the ZipZaps robot because I would like to use the PIC16F684's built-in PWM for modulating I/R output. The result is ZipBase.asm, which has a 30 Hz PWM frequency, and provides 32 PWM duty cycles, as well as up to 1 ms between PWM steps for implementing sensor and control functions.

To test out the connections to the ZipZaps chassis as well as to get an idea of how long the battery would last with a PIC MCU running a stepped-up battery voltage, I added a number of actions that take place over a 25-second repeating *life* to demonstrate running forward, forward while turning right and left, stopping, and going backward.

```
title   "ZipBase - ZipZaps Robot Base Code"
;
;   This Program Loops once every ms and
;   updates the
;   ZipZaps Motor PWM (30 Hz PWM).
;   The application also keeps track
;   of the current time for the application.
;   This application is
;   heavily based on "asmMotor 2.asm".
;
;   Hardware Notes:
;   PIC16F684 running at 4 MHz Using the
;   Internal Clock
;   RA0 - Left Light Sensor
;   RA1 - Right Light Sensor
;   RA3 - Rear IR Sensor
;   RA4 - Reverse Motor Control
#define ReversePin PORTA, 4
;   RA5 - Forwards Motor Control
```

```
#define ForwardsPin  PORTA, 5
;   RC2 - Right Steering Solenoid
#define TurnRight     PORTC, 2
;   RC3 - Left Steering Solenoid
#define TurnLeft      PORTC, 3
;   RC4 - Line Following IR PWM
;   RC5 - Object Detection IR PWM
;
;
;   Myke Predko
;   05.01.25
;
  LIST R=DEC
  INCLUDE "p16f684.inc"


  __CONFIG _FCMEN_OFF & _IESO_OFF & _BOD_OFF &
_CPD_OFF & _CP_OFF & _MCLRE_OFF & _PWRTE_ON &
_WDT_OFF & _INTOSCIO

;   Variables
  CBLOCK 0x020
RTC:2                        ; 65 Second RTC
Direction, PWMDuty, PWMCycle ; PWM Movement
                             ; Variables
Flag
  ENDC

;   Flag Bit Definitions
#define AppRun    Flag, 0    ; Set if Application
                             ; can Run
#define AppReset Flag, 1     ; Set if Application
                             ; Reset Code to Run

;   Macros
TimeCheck Macro TValue, NotVector
  movlw    HIGH TValue
  xorwf    RTC + 1, w
  btfss    STATUS, Z
  goto     NotVector
  movlw    LOW TValue
  xorwf    RTC, w
  btfss    STATUS, Z
  goto     NotVector
  endm

  PAGE
;  Mainline

  org      0

  nop                        ; For ICD Debug

  clrf     PORTA             ; Assume Everything
                             ; is Low
```

```
        clrf    PORTC
        movlw   7                       ; Turn off
                                        ; Comparators
        movwf   CMCON0
        movlw   b'00000001'             ; Enable ADC on RA0
        movwf   ADCON0
        clrf    TMR0                    ; Using TMR0 as a PWM
                                        ; Base

        bsf     STATUS, RP0
        movlw   b'11010001'             ; 1:4 Prescaler to
                                        ; TMR0
        movwf   OPTION_REG ^ 0x80;
        movlw   b'00000011'             ; RA0/RA1 (AN0/AN1)
                                        ; ADC Inputs
        movwf   ANSEL ^ 0x80
        movlw   b'00010000'             ; Select ADC Clock as
                                        ; Fosc/8
        movwf   ADCON1 ^ 0x80
        movlw   b'001111'               ; Enable Motor Bits
        movwf   TRISA ^ 0x80
        movlw   b'110011'               ; Enable Steering
                                        ; Bits
        movwf   TRISC ^ 0x80
        bcf     STATUS, RP0

        bcf     INTCON, T0IF            ; Wait for TMR0 to
                                        ; Overflow
        clrf    Flag
        clrf    PWMDuty                 ; Not Moving at First
        clrf    PWMCycle                ; Start at the
                                        ; Beginning
        clrf    Direction               ; Moving Forwards
Loop:
        btfss   INTCON, T0IF            ; Wait for Timer
                                        ; Overflow
        goto    $ - 1
        bcf     INTCON, T0IF            ; Reset and Wait for
                                        ; Next

MotorUpdate:                            ; Check to Update the
                                        ; Motor
        movf    PWMDuty, w              ; If Duty > Cycle,
                                        ; then Off
        subwf   PWMCycle, w
        movf    PORTA, w
        iorlw   1 << 5                  ; Going Forwards
        btfsc   Direction, 0            ; Forwards or
                                        ; Reverse?
        xorlw   (1 << 5) + (1 << 4) ; TRISC Reverse
        btfsc   STATUS, C
        andlw   b'001111'               ; Nothing Moving
        movwf   PORTA                   ; Save Motor Value

        incf    RTC, f                  ; Increment Real Time
                                        ; Clock
        btfsc   STATUS, Z
        incf    RTC + 1, f              ; Note, Rolls Over at
                                        ; 65 s

        incf    PWMCycle, f             ; Increment the PWM
                                        ; Cycle Count
        bcf     PWMCycle, 5             ; Maximum of 32
                                        ; Cycles
;   #### - Put Other Sensors Here
;   #### - Put in Remote Control Poll Here
;   #### - Put in Sensor State Machine
;   #### - Light Sensors (1 Full per 32 Cycles
;            - 1 Sample/8 Cycles)
;   #### - Object Sensing (1 Per 32 Cycles)
;   #### - Line Sensing (1 Per 32 Cycles)
```

```
;   #### - Put Operating Logic Here (1 Per 32
;            Cycles)/Needs Set Flag
;   #### - Check "AppReset" to See if
;            Application is to be Restarted
;   #### - Remote Control Direct (Move or
;            Stop) Resets "AppRun" Flag
;   #### - Remote Control "Recall" Sets
;            "AppRun" Flag

        goto    Loop                    ; Finished, Loop
                                        Around Again

        end
```

Test Application Software:

```
        title   "ZipMtrChk - Test Motor/Steering
Operation"
;
; This Program is a basic test of the ZipZaps
; Robot Motors and Steering.  The commands are:
; 1.  Move forwards for 5 Seconds at 10 Seconds
; 2.  Turn Right at 11 Seconds for 1 Second
; 3.  Turn Left at 13 Seconds for 1 Second
; 4.  Stop at 15 Seconds for 5 Seconds
; 5.  Go in Reverse at 20 Seconds for 5 Seconds
; 6.  Reset RTC and Repeat at 25 Seconds
;
;
;   Hardware Notes:
;   PIC16F684 running at 4 MHz Using the
;   Internal Clock
;   RA0 - Left Light Sensor
;   RA1 - Right Light Sensor
;   RA3 - Rear IR Sensor
;   RA4 - Reverse Motor Control
#define ReversePin PORTA, 4
;   RA5 - Forwards Motor Control
#define ForwardsPin PORTA, 5
;   RC2 - Right Steering Solenoid
#define TurnRight   PORTC, 2
;   RC3 - Left Steering Solenoid
#define TurnLeft    PORTC, 3
;   RC4 - Line Following IR PWM
;   RC5 - Object Detection IR PWM
;
;
;   Myke Predko
;   05.01.25
;
    LIST R=DEC
    INCLUDE "p16f684.inc"

    __CONFIG _FCMEN_OFF & _IESO_OFF & _BOD_OFF &
_CPD_OFF & _CP_OFF & _MCLRE_OFF & _PWRTE_ON &
_WDT_OFF & _INTOSCIO

; Variables
    CBLOCK 0x020
RTC:2                                   ; 65 Second RTC
Direction, PWMDuty, PWMCycle  ; PWM Movement
                                        ; Variables
Flag
    ENDC

; Flag Bit Definitions
#define AppRun   Flag, 0                ; Set if
                                        ; Application can
                                        ; Run
```

```
    #define AppReset Flag, 1    ; Set if                  subwf   PWMCycle, w
                                ; Application Reset       movf    PORTA, w
                                ; Code to be Run          iorlw   1 << 5                  ; Going Forwards
                                                          btfsc   Direction, 0            ; Forwards or
; Macros                                                                                  ; Reverse?
TimeCheck Macro TValue, NotVector                         xorlw   (1 << 5) + (1 << 4)     ; TRISC Reverse
  movlw   HIGH TValue                                     btfsc   STATUS, C
  xorwf   RTC + 1, w                                      andlw   b'001111'               ; Nothing Moving
  btfss   STATUS, Z                                       movwf   PORTA                   ; Save Motor Value
   goto   NotVector
  movlw   LOW TValue                                      incf    RTC, f                  ; Increment Real
  xorwf   RTC, w                                                                          ; Time Clock
  btfss   STATUS, Z                                       btfsc   STATUS, Z
   goto   NotVector                                       incf    RTC + 1, f              ; Note, Rolls Over
 endm                                                                                     ; at 65 s

  PAGE                                                    incf    PWMCycle, f             ; Increment the
; Mainline                                                                                ; PWM Cycle Count
                                                          bcf     PWMCycle, 5             ; Maximum of 32
  org     0                                                                               ; Cycles

  nop                           ; For ICD Debug           ;  #### - Put Other Sensors Here
                                                          ;  #### - Put in Remote Control Poll Here
  clrf    PORTA                 ; Assume                  ;  #### - Put in Sensor State Machine
                                ; Everything is Low       ;  ####   - Light Sensors (1 Full per
  clrf    PORTC                                           ;            32 Cycles - 1 Sample/8 Cycles)
  movlw   7                     ; Turn off               ;  ####   - Object Sensing (1 Per 32 Cycles)
                                ; Comparators             ;  ####   - Line Sensing (1 Per 32 Cycles)
  movwf   CMCON0
  movlw   b'00000001'           ; Enable ADC on          ;  #### - Put Operating Logic Here (1 Per
                                ; RA0                     ;          32 Cycles)/Needs Set Flag
  movwf   ADCON0                                          ;  #### - Check "AppReset" to See
  clrf    TMR0                  ; Using TMR0 as a         ;          if Application is to be Restarted
                                ; PWM Base                ;  #### - Remote Control Direct (Move
                                                          ;          or Stop) Resets "AppRun" Flag
  bsf     STATUS, RP0                                     ;  #### - Remote Control "Recall" Sets
  movlw   b'11010001'           ; 1:4 Prescaler to        ;          "AppRun" Flag
                                ; TMR0
  movwf   OPTION_REG ^ 0x80;                              ; Motor Check Test Program:
  movlw   b'00000011'           ; RA0/RA1 (AN0/AN1)      ; 1.  Move forwards for 5 Seconds at 10 Seconds
                                ; ADC Inputs             TryForwards:
  movwf   ANSEL ^ 0x80                                    TimeCheck 10000, At15sec
  movlw   b'00010000'           ; Select ADC Clock        movlw   0x1F                    ; Go Forwards at
                                ; as Fosc/8                                               ; Full Speed for 5 s
  movwf   ADCON1 ^ 0x80                                   movwf   PWMDuty
  movlw   b'001111'             ; Enable Motor            bcf     Direction, 0
                                ; Bits                    goto    Loop
  movwf   TRISA ^ 0x80
  movlw   b'110011'             ; Enable Steering        At15sec:                         ; At 15 Seconds?
                                ; Bits                    TimeCheck 15000, TryRight
  movwf   TRISC ^ 0x80                                    clrf    PWMDuty                 ; Yes, Stop Motor
  bcf     STATUS, RP0                                     goto    Loop

  bcf     INTCON, T0IF          ; Wait for TMR0 to       ; 2.  Turn Right at 11 Seconds for 1 Second
                                ; Overflow               TryRight:
  clrf    Flag                                            TimeCheck 11000, At12sec
  clrf    RTC                   ; Reset Real Time         bsf     TurnRight
                                ; Flag                    goto    Loop
  clrf    RTC + 1
  clrf    PWMDuty               ; Not Moving at          At12sec:
                                ; First                   TimeCheck 12000, TryLeft
  clrf    PWMCycle              ; Start at the            bcf     TurnRight
                                ; Beginning               goto    Loop
  clrf    Direction             ; Moving Forwards
                                                         ; 3.  Turn Left at 13 Seconds for 1 Second
Loop:                                                    TryLeft:
  btfss   INTCON, T0IF          ; Wait for Timer          TimeCheck 13000, At14sec
                                ; Overflow                bsf     TurnLeft
   goto   $ - 1                                           goto    Loop
  bcf     INTCON, T0IF          ; Reset and Wait
                                ; for Next               At14sec:
                                                          TimeCheck 14000, TryReverse
MotorUpdate:                    ; Check to Update         bcf     TurnLeft
                                ; the Motor               goto    Loop
  movf    PWMDuty, w            ; If Duty > Cycle,
                                ; then Off
```

```
; 4.  Stop at 15 Seconds for 5 Seconds
; 5.  Go in Reverse at 20 Seconds for 5 Seconds
TryReverse:
 TimeCheck 20000, At25sec
   movlw   0x1F                 ; Go Forwards at
                                ; Full Speed for 5 s
   movwf   PWMDuty
   bsf     Direction, 0
   goto    Loop

; 6.  Reset RTC and Repeat at 25 Seconds
At25sec:                        ; At 25 Seconds?
 TimeCheck 25000, Loop          ; Not there, Repeat
   clrf    PWMDuty              ; Yes, Stop Motor
   clrf    RTC + 1              ; Clear the RTC
   clrf    RTC

   goto    Loop                 ; Finished, Loop
                                ; Around Again


   end
```

As part of the development of this application, you'll see that I came up with the *TimeCheck macro*, which will execute the code following it when the *real-time clock* (RTC) variable matches the numeric value. The RTC variable is incremented once every millisecond, so fairly precise timings are possible. When you look at the operation of the ZipZaps robot in this experiment, you'll see that the motor is running for 5 seconds. And I want to warn you: The robot can go a long way over 5 seconds.

Fortunately, when the robot hits a wall or other obstacle and stalls, it does not burn out. This is why it was critical that I spent time understanding the motor and steering driver base currents and adding additional base current-limiting resistors to ensure that the drivers would not have too much current passing through them.

Experiment 120—IR Remote Control

PC/ Simulator

PICkit™ 1

starter kit

Parts Bin

1 Radio Shack ZipZaps remote-control car chassis

1 PIC16F684

1 Prototype PCB

1 38 kHz IR TV remote-control receiver

1 47 μF electrolytic capacitor

1 10k resistor

1 100Ω resistor

Tool Box

DMM

Soldering iron

Solder

Wire-wrap wire

Before you ran the ZipZaps robot in the previous experiment, you may have doubted how fast the ZipZaps robot could go. After you ran the experiment, you probably realized how difficult it would be to control the robot using just software without any type of feedback. In this experiment, you will modify the prototype PCB circuit (see Figure 13-15) so that the entire robot can be controlled by a TV remote control set to Sony TV standards (reviewed previously in this section).

The remote-control receiver was added to the rear of the robot (see Figure 13-16) as this was the easiest place I could find for the circuit, and it will be used in a later experiment. The output of the receiver drives RA3, which can be used only as an input. This was an important consideration because I wanted to make sure I left the driver values open for the remaining experiments.

The application code polls the IR receiver, waiting for a signal to come in. The code that processes the signal was taken directly from the previous IR TV remote-control experiment in this section (which also allowed me to use the codes that were recorded during this experiment). The button commands on the remote control are used to drive the ZipZaps robot forward (while optionally turning left or right) or backward (while optionally turning left or right). The PWM value of the motor driver can also be changed by using the volume controls. In the code, ZipBaseIR.asm, which is an enhancement of the previous base code, you can see that hooks have been put in for the remote control to be used as the tool that starts, stops, or suspends the execution of a test program that has been added to the application code.

As you control the ZipZaps robot, you'll probably notice a couple of things. First, running the ZipZaps robot by remote control is a lot of fun—arguably as much fun as the original ZipZaps, but different because it is just about impossible to spin it out or slide it on the floor. This change in operation seems to be due to the additional weight of the prototype PCB and its electronics. So, what was the advantage of making these modifications? After several hours of work, have you simply recreated the ZipZaps?

This is not quite true, the little robot you have is capable of a lot more as I will show in the remaining three experiments.

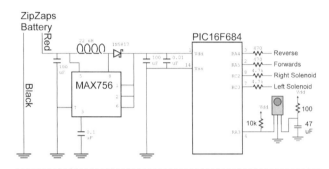

Figure 13-15   *ZipZaps IR*

Figure 13-16   *Rear-looking IR receiver mounted under the surface of the ZipZaps robot prototype PCB*

Experiment 121—Light Sensors and Light Following

PC/ Simulator

PICkit™ 1

starter kit

Parts Bin

1 Radio Shack ZipZaps remote-control car chassis

1 PIC16F684

1 Prototype PCB

2 Light-dependent resistors

2 10k resistors

Tool Box

DMM
Soldering iron
Solder
Wire-wrap wire

A basic feature of any robot is the ability to sense light and dark and, ideally, to do it from multiple points of view so that comparisons of its environment can be made. In this experiment, you are going to give your ZipZaps robot the gift of sight—that is, two light-dependent resistors that can be used to seek out or avoid light in the robot's environment. Adding the two LDRs as part of a voltage divider is quite simple (see Figure 13-17) and takes advantage of the ADC module built into the PIC16F684.

As you can see in Figure 13-18, I put the LDRs about a third of the way back from the front of the robot, with each one pointing outward at 45 degrees. This gives each LDR a unique point of view. This placement of the LDRs allows the robot to determine which side is brighter and whether or not to turn toward it or away from it.

The application code ZipBaseLDR.asm is an enhancement of ZipBaseIR.asm, as it provides an ADC operation (on RA0 and RA1, where the LDRs are connected) every 4 ms. Each LDR is sampled and the results are stored for retrieval by an application. Included in the code is a simple *light seeker* that executes when the remote control's Menu button is pressed.

The light-seeking application turns the wheels only when the darker LDR returns a value that is twice (or more) the value of the lighter LDR. This value represents quite a drastic change in light and was chosen because of the all-or-nothing operation of the ZipZaps chassis ternary steering. In a more typical robot with

proportional steering, guidance to a light source can be gentler and not so startlingly jerky (when the robot turns toward the light). The LDR ADC values are ADCValueL and ADCValueR in the application code.

I found the robot's ability to hit a flashlight on the floor (the light beam for it to follow) was largely dependent on how quickly the robot's motors turned. The faster the robot moves, the later the steering change takes place (the decision is made every 100 ms), and the more likely the robot will be to miss the flashlight.

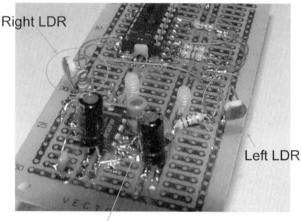

Right LDR

Left LDR

Front of Robot

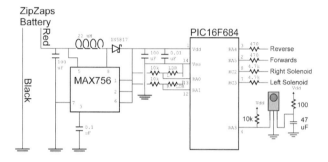

Experiment 122—IR Object-Detection Sensors

PC/ Simulator

PICkit™ 1

starter kit

Parts Bin

Tool Box

DMM

Soldering iron

Solder

Wire-wrap wire

1 Radio Shack ZipZaps remote-control car chassis

1 PIC16F684

1 Prototype PCB

2 38 kHz IR TV remote-control receivers

1 Red LED

5 IR LEDs

2 10k resistors

6 1k resistors

2 100k resistors

2 47 μF electrolytic capacitors

1 Length of 5mm heat-shrink tubing

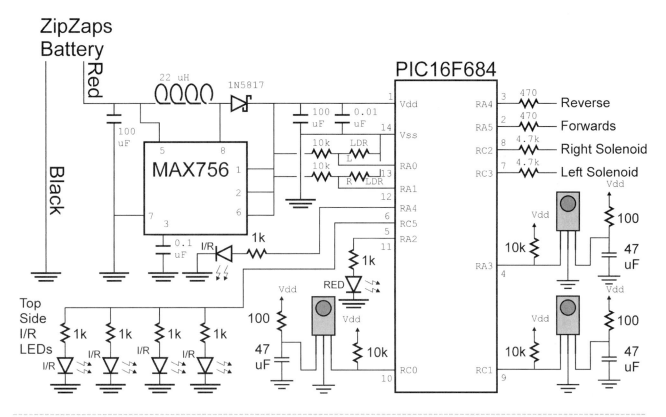

Figure 13-19   *ZipZaps complete*

In the last two experiments of this book, I will demonstrate how the IR sensors can be used to detect objects and data around them. This will be done by using the IR reflection demonstrated in other experiments. To detect objects, four LEDs are placed around the perimeter of the robot that will drive out IR pulses, which, if reflected, will be picked up by three TV IR receivers around the perimeter of the robot. By doing this, you are giving the robot a "sense of touch" with

Figure 13-20   *ZipZaps robot with IR LEDs and TV remote-control receivers*

the two front-looking IR receivers and the rear IR receiver giving the robot some idea of where the object is around it.

To modify the robot for this task, I would like you to add the six LEDs (five IR and one visible) to the prototyping PCB, as wired in Figure 3-19. Each LED should be pointing in a different direction, as shown in Figure 13-20. After you have done this, please add two IR receivers to the front of the robot. To avoid problems with conflicting IR signals, the receivers should be on the underside of the robot (see Figure 13-21).

To test the robot's sense of touch, press the Guide button on your Universal remote after building Zip-BaseObj.asm and loading it into your robot. This code continuously checks the perimeter of the robot, every four PWMCycles.

I found that I had to put small pieces (0.5 to 0.75 inch long, or 1 cm to 1.5 cm long) over the IR LEDs to get reliable operation of the sensor. But, it is amazingly reliable for each of the three IR receivers. (To test the operation of each receiver, wrap the other two in black electrical tape to prevent IR signals from reaching it.)

The state of the IR receivers and whether or not an object is around the robot can be polled by checking the PFrontRIR, PFrontLIR, and PRearIR flags.

Experiment 123—IR Line-Following Sensors

PC/Simulator

PICkit™ 1

starter kit

Parts Bin

1 Radio Shack ZipZaps
  remote-control car
  chassis

1 PIC16F684

1 Prototype PCB

This last experiment is really a follow-up to the previous one. However, rather than sensing objects, the downward-pointing IR LED at the front of the robot (see Figure 13-21) is directed toward the surface on which the robot is running and if the surface is white, it is reflected back to the IR receivers. (Use a piece of heat-shrink tubing to direct the LED toward the running surface.) The application code ZipBaseLine.asm is a modification of ZipBaseObj.asm and uses a second PWM output.

```
 title   "ZipBaseLine - ZipZaps Robot Base Code"
;
;   "ZipBaseObj.asm" with Line Sensors Added.
;
;   Hardware Notes:
;    PIC16F684 running at 4 MHz Using the
;    Internal Clock
;    RA0 - Left Light Sensor
;    RA1 - Right Light Sensor
;    RA3 - Rear IR Sensor
#define RearIR        PORTA, 3
;    RA4 - Reverse Motor Control
#define ReversePin    PORTA, 4
;    RA5 - Forwards Motor Control
#define ForwardsPin   PORTA, 5
```

Figure 13-21   *The business end of the ZipZaps robot downward-pointing IR LED used for line following*

```
;    RC2 - Right Steering Solenoid
#define TurnRight      PORTC, 2
;    RC3 - Left Steering Solenoid
#define TurnLeft       PORTC, 3
;    RC4 - Line Following IR PWM
;    RC5 - Object Detection IR PWM
;
;   IR Code Table:
Button0 EQU 0x910
Button1 EQU 0x010
Button2 EQU 0x810
Button3 EQU 0x410
Button4 EQU 0xC10
Button5 EQU 0x210
Button6 EQU 0xA10
Button7 EQU 0x610
Button8 EQU 0xE10
Button9 EQU 0x110
ArrowUp EQU 0x2F0
ArrowDn EQU 0xAF0
ArrowLt EQU 0x2D0
ArrowRt EQU 0xCD0
VolumeU EQU 0x490
VolumeD EQU 0xC90
ChanelU EQU 0x090
ChanelD EQU 0x890
OKButtn EQU 0xA70
PwrButn EQU 0x490
MenuBtn EQU 0x070
EnterBt EQU 0xD10
RecallB EQU 0xDD0
PIPButn EQU 0xDB0
DispBtn EQU 0x5D0
MuteBtn EQU 0x290
GuideBt EQU 0x764
;
;   Myke Predko
;   05.01.25
;
    LIST R=DEC
    INCLUDE "p16f684.inc"

    __CONFIG _FCMEN_OFF & _IESO_OFF & _BOD_OFF &
_CPD_OFF & _CP_OFF & _MCLRE_OFF & _PWRTE_ON &
_WDT_OFF & _INTOSCIO

; Variables
 CBLOCK 0x020
i, Dlay
RTC:2                           ; 65 Second RTC
Direction, PWMDuty, PWMCycle    ; PWM Movement
                                ; Variables
ADCCycle, ADCValueL, ADCValueR  ; ADC Variables
MoveCounter, NewPWMDuty, VolCounter
IRCode:2
ObjectPWM                       ; Set Object PWM
```

```
               ObjFrontR, ObjFrontL, ObjRear
               LinePWM
               LineFrontR, LineFrontL
               Flag:2
                 ENDC

; Flag Bit Definitions
#define AppRun      Flag, 0     ; Set if
                                ; Application
                                ; can Run
#define AppReset    Flag, 1     ; Set if
                                ; Application
                                ; Reset Code to
                                ; be Run
#define LightSeek   Flag, 2     ; Set when Light
                                ; Seeking Mode
                                ; is Enabled
#define ObjectSense Flag, 3     ; Set when Doing
                                ; an Object Sense
#define PFrontRIR   Flag, 4     ; Set When Object
                                ; to Front Right
#define PFrontLIR   Flag, 5     ; Set When Object
                                ; to Front Left
#define PRearIR     Flag, 6     ; Set When
                                ; Object to Rear
#define LineSensor  Flag, 7     ; Set when Doing
                                ; a Line Sense
#define LFrontRIR   Flag + 1, 0 ; Set When Line
                                ; to Front Right
#define LFrontLIR   Flag + 1, 1 ; Set When Line
                                ; to Front Left

; Macros
TimeCheck Macro TValue, NotVector
   movlw    HIGH TValue
   xorwf    RTC + 1, w
   btfss    STATUS, Z
    goto    NotVector
   movlw    LOW TValue
   xorwf    RTC, w
   btfss    STATUS, Z
    goto    NotVector
   endm

msDlay Macro Cycles
   movlw    HIGH ((Cycles / 5) + 256)
   movwf    Dlay
   movlw    LOW ((Cycles / 5) + 256)
   addlw    -1
   btfsc    STATUS, Z
    decfsz  Dlay, f
     goto   $ - 3
   endm

ButtonCompare Macro Button, NotVector
   movlw    HIGH Button
   xorwf    IRCode + 1, w
   btfss    STATUS, Z
    goto    NotVector
   movlw    LOW Button
   xorwf    IRCode, w
   btfss    STATUS, Z
    goto    NotVector
   endm

   PAGE
; Mainline

   org    0

   nop                      ; For ICD Debug

   clrf     PORTA           ; Assume Everything
                            ; is Low
   clrf     PORTC


   movlw    7               ; Turn off
                            ; Comparators
   movwf    CMCON0
   movlw    b'00000001'     ; Enable ADC on RA0
   movwf    ADCON0
   clrf     TMR0            ; Using TMR0 as a
                            ; PWM Base
   movlw    b'00001100'     ; Enable Basic PWM
                            ; Mode
   movwf    CCP1CON
   movlw    b'01111000'     ; Enable TMR2 with
                            ; a 16x Pre
   movwf    T2CON

   bsf      STATUS, RP0
   movlw    b'11010001'     ; 1:4 Prescaler to
                            ; TMR0
   movwf    OPTION_REG ^ 0x80;
   movlw    b'00000011'     ; RA0/RA1 (AN0/AN1)
                            ; ADC Inputs
   movwf    ANSEL ^ 0x80
   movlw    b'00010000'     ; Select ADC Clock
                            ; as Fosc/8
   movwf    ADCON1 ^ 0x80
   movlw    b'001011'       ; Enable Motor &
                            ; LED Bits
   movwf    TRISA ^ 0x80
   movlw    b'110011'       ; Enable Steering
                            ; Bits
   movwf    TRISC ^ 0x80
   bcf      STATUS, RP0

   bcf      INTCON, T0IF    ; Wait for TMR0 to
                            ; Overflow
   clrf     Flag
   clrf     PWMDuty         ; Not Moving at
                            ; First
   clrf     PWMCycle        ; Start at the
Beginning
   clrf     Direction       ; Moving Forwards
   movlw    0x1F            ; Start at Full
                            ; Speed
   movwf    NewPWMDuty
   clrf     ADCCycle
   clrf     ADCValueL
   clrf     ADCValueR

Loop:
   btfss    INTCON, T0IF    ; Wait for Timer
                            ; Overflow
    goto    $ - 1
   bcf      INTCON, T0IF    ; Reset and Wait
                            ; for Next

MotorUpdate:                ; Check to Update
                            ; the Motor
   movf     PWMDuty, w      ; If Duty > Cycle,
                            ; then Off
   subwf    PWMCycle, w
   movlw    0               ; Force Motors
   iorlw    1 << 5          ; Going Forwards
   btfsc    Direction, 0    ; Forwards or
                            ; Reverse?
    xorlw   (1 << 5) + (1 << 4) ; Reverse
   btfsc    STATUS, C
    movlw   0               ; Nothing Moving
   btfss    ObjectSense     ; Object Check to
                            ; Put LED On
    goto    SavePORTA
   btfsc    PFrontRIR       ; If Sensor Bit
                            ; Set, LED On
    iorlw   1 << 2
   btfsc    PFrontLIR
    iorlw   1 << 2
   btfsc    PRearIR
```

```
        iorlw   1 << 2                                          movlw   12                      ; Want to Get 12
SavePORTA:                                                                                      ; Bits
    movwf   PORTA              ; Save Motor Value              movwf   i

    incf    RTC, f             ; Increment Real        IRRLoop:                                  ; Loop Here for
                               ; Time Clock                                                     ; Each Bit
    btfsc   STATUS, Z                                          call    CountHigh                 ; Must be Approx
    incf    RTC + 1, f         ; Note, Rolls Over                                                ; 550 us
                               ; at 65 s                       btfsc   STATUS, C
                                                               goto    Loop                      ; Carry Set,
    incf    PWMCycle, f        ; Increment the PWM                                                ; Error/Ignore
                               ; Cycle Count                                                     ; Command
    bcf     PWMCycle, 5        ; Maximum of 32                 call    CountLow
                               ; Cycles                        movlw   80
                                                               subwf   Dlay, w                   ; "1" or "0"?
    movf    MoveCounter, f     ; Direct Control
                               ; Move?                         rlf     IRCode, f                 ; Carry Set, 1
    btfsc   STATUS, Z                                          rlf     IRCode + 1, f
    goto    NoMoveChange
    decfsz  MoveCounter, f     ; Decrement Move                decfsz  i, f                      ; Repeat?
                               ; Counter                       goto    IRRLoop
    goto    NoMoveChange
    btfsc   LightSeek          ; In Light Seeking                                                ; Forwards and Left
                               ; Mode?                  ButtonCompare Button1, CheckButton2
    goto    DoLightSeek                                        movlw   200                       ; Run for 200 ms
    clrf    PWMDuty            ; If Zero, Clear                movwf   MoveCounter
                               ; Moving Duty                   movf    NewPWMDuty, w
    bcf     TurnLeft           ; And Stop Turning              movwf   PWMDuty
    bcf     TurnRight                                          bcf     Direction, 0              ; Moving Forwards
    goto    NoMoveChange                                       bsf     TurnLeft
                                                               bcf     TurnRight
DoLightSeek:                   ; Look for Turn                 bcf     AppRun                    ; Turn Off Program
    movf    ADCValueL, w       ; Is Right > Left?              bcf     LightSeek                 ; Turn Off Light
    subwf   ADCValueR, w                                                                         ; Seeking
    btfss   STATUS, C
    goto    DoLightSeekRight                                   goto    Loop
DoLightSeekLeft:               ; Left < Right,          CheckButton2:                            ; Forwards
                               ; Turn Towards?          ButtonCompare Button2, CheckButton3
    bcf     STATUS, C                                          movlw   200                       ; Run for 200 ms
    rrf     ADCValueR, w                                       movwf   MoveCounter
    subwf   ADCValueL, w       ; Is Left <                     movf    NewPWMDuty, w
                               ; Right/2?                      movwf   PWMDuty
    movlw   0                                                  bcf     Direction, 0              ; Moving Forwards
    btfss   STATUS, C                                          bcf     TurnLeft
    movlw   1 << 3             ; Turn Towards                  bcf     TurnRight
                               ; Light                         bcf     AppRun                    ; Turn Off Program
    movwf   PORTC                                              bcf     LightSeek                 ; Turn Off Light
    movlw   100                ; Continue for                                                    ; Seeking
                               ; another 100 ms                goto    Loop
    movwf   MoveCounter
    goto    NoMoveChange                                CheckButton3:                            ; Forwards and Right
                                                        ButtonCompare Button3, CheckButton5
                                                               movlw   200                       ; Run for 200 ms
DoLightSeekRight:              ; Right < Left,                 movwf   MoveCounter
                               ; Turn Towards?                 movf    NewPWMDuty, w
    bcf     STATUS, C                                          movwf   PWMDuty
    rrf     ADCValueL, w                                       bcf     Direction, 0              ; Moving Forwards
    subwf   ADCValueR, w       ; Is Right <                    bcf     TurnLeft
                               ; Left/2?                       bsf     TurnRight
    movlw   0                                                  bcf     AppRun                    ; Turn Off Program
    btfss   STATUS, C                                          bcf     LightSeek                 ; Turn Off Light
    movlw   1 << 2             ; Turn Towards Light                                               ; Seeking
    movwf   PORTC
    movlw   100                ; Continue for                  goto    Loop
                               ; another 100 ms
    movwf   MoveCounter                                 CheckButton5:                            ; Stop
                                                        ButtonCompare Button5, CheckButton7
NoMoveChange:                                                  clrf    PWMDuty                   ; Stop Everything
    btfsc   RearIR             ; IR Value Being                bcf     TurnLeft
                               ; Received?                     bcf     TurnRight
    goto    SensorCheck                                        clrf    MoveCounter
                                                               bcf     AppRun                    ; Stop Application
IRRead:                                                                                          ; Running
    btfss   RearIR             ; Line to go High               bcf     LightSeek                 ; Turn Off Light
                               ; Again                                                           ; Seeking
    goto    $ - 1                                              goto    Loop
```

```
CheckButton7:                    ; Reverse and Left        ButtonCompare EnterBt, MenuCheck
  ButtonCompare Button7, CheckButton8                        bsf    AppRun
    movlw   200                  ; Run for 200 ms            bsf    AppReset
    movwf   MoveCounter                                      goto   Loop
    movf    NewPWMDuty, w
    movwf   PWMDuty                                      MenuCheck:                        ; Light Follow?
    bsf     Direction, 0         ; Moving Backwards        ButtonCompare MenuBtn, GuideCheck
    bsf     TurnLeft                                        bcf    AppRun           ; Stop Any Previous
    bcf     TurnRight                                                               ; Application
    bcf     AppRun               ; Turn Off Program         bsf    LightSeek        ; Enable Light
    bcf     LightSeek            ; Turn Off Light                                   ; Seeking Mode
                                 ; Seeking                  bcf    ObjectSense      ; Disable Object
    goto    Loop                                                                    ; Seeking Mode
                                                            bcf    LineSensor       ; Disable Line
CheckButton8:                    ; Reverse                                          ; Sensor
  ButtonCompare Button8, CheckButton9                       movlw  100              ; Run for 100 ms
    movlw   200                  ; Run for 200 ms           movwf  MoveCounter
    movwf   MoveCounter                                     movf   NewPWMDuty, w
    movf    NewPWMDuty, w                                   movwf  PWMDuty
    movwf   PWMDuty                                         bcf    Direction, 0     ; Moving Forwards
    bsf     Direction, 0         ; Moving Backwards         bcf    TurnLeft
    bcf     TurnLeft                                        bcf    TurnRight
    bcf     TurnRight                                       goto   Loop
    bcf     AppRun               ; Turn Off Program
    bcf     LightSeek            ; Turn Off Light       GuideCheck:                       ; Object Sense?
                                 ; Seeking                ButtonCompare MenuBtn, Loop  ; Return if Unknown
    goto    Loop                                                                     ; Button
                                                            bcf    AppRun           ; Stop Any Previous
CheckButton9:                    ; Reverse & Right                                   ; Application
  ButtonCompare Button9, CheckVolUp                         bcf    LightSeek        ; Disable Light
    movlw   200                  ; Run for 200 ms                                    ; Seeking Mode
    movwf   MoveCounter                                     bcf    LineSensor       ; Disable Line
    movf    NewPWMDuty, w                                                            ; Sensor
    movwf   PWMDuty                                         btfsc  ObjectSense      ; Toggle
    bsf     Direction, 0         ; Moving Backwards                                  ; ObjectSense
    bcf     TurnLeft                                        goto   $ + 3
    bsf     TurnRight                                       bsf    ObjectSense      ; Enable Object
    bcf     AppRun               ; Turn Off Program                                  ; Seeking Mode
    bcf     LightSeek            ; Turn Off Light           goto   Loop
                                 ; Seeking                  bcf    ObjectSense      ; Disab Object
    goto    Loop                                                                    ; Seeking Mode
                                                            goto   Loop
CheckVolUp:                      ; Increase PWM
  ButtonCompare VolumeU, CheckVolDn                     LineCheck:                        ; Line Sensor?
    movf    NewPWMDuty, w        ; At Limit?                ButtonCompare MenuBtn, LineCheck
    xorlw   0x1F                                            bcf    AppRun           ; Stop Any Previous
    btfsc   STATUS, Z                                                               ; Application
    goto    Loop                                            bsf    LightSeek        ; Enable Light
    incf    VolCounter, f        ; Delay 1x Cycle to                                 ; Seeking Mode
                                 ; Get 3s                   bcf    ObjectSense      ; Disable Object
    btfss   VolCounter, 0        ; Full Range                                        ; Seeking Mode
                                 ; Control                  bsf    LineSensor       ; Enable Line Sensor
    incf    NewPWMDuty, f                                   goto   Loop
    goto    Loop
                                                        SensorCheck:
CheckVolDn:                      ; Decrease PWM             movf   PWMCycle, w      ; Update ADC?
  ButtonCompare VolumeD, RecallCheck                        andlw  0x03             ; Every 4th Cycle
    movf    NewPWMDuty, f        ; At Limit?                btfss  STATUS, Z
    btfsc   STATUS, Z                                       goto   ObjSense
    goto    Loop
    incf    VolCounter, f        ; Delay 1x Cycle to    LightSense:                       ; Run through ADC
                                 ; Get 3s                                            ; Cycles
    btfss   VolCounter, 0        ; Full Range               incf   ADCCycle, f      ; ADCCycle =
                                 ; Control                                           ; (ADCCycle + 1) % 7
    decf    NewPWMDuty, f                                   bcf    ADCCycle, 3
    goto    Loop
                                                            movf   ADCCycle, w      ; Execute ADC State
RecallCheck:                     ; Resume                                            ; Machine
                                 ; Application?             btfss  STATUS, Z
  ButtonCompare RecallB, EnterCheck                         goto   ADCCycle1
    bsf     AppRun                                          movlw  b'00000001'      ; Enable ADC on RA0
    goto    Loop                                            movwf  ADCON0
                                                            bsf    ADCON0, GO       ; Start ADC
EnterCheck:                      ; Restart                                          ; Operation
                                 ; Application?             goto   Loop
```

```
ADCCycle1:                                          bsf     T2CON, TMR2ON     ; Turn on TMR2 for
  xorlw   1                                                                   ; Objects
  btfss   STATUS, Z                                 bsf     STATUS, RP0
   goto   ADCCycle2                                 bcf     TRISC ^ 0x80, 5   ; Enable PWM Output
  movf    ADRESH, w         ; Expect Bad ADC         bcf     STATUS, RP0
                            ; Read, Ignore Value
  goto    Loop                                      ObjLoop:                  ; Loop Here Polling
                                                                              ; Objects
ADCCycle2:                                            btfsc   PORTC, 0
  xorlw   2 ^ 1                                        incf    ObjFrontR, f
  btfss   STATUS, Z                                    btfsc   PORTC, 1
   goto   ADCCycle3                                    incf    ObjFrontL, f
  bsf     ADCON0, GO        ; Start Good Read          btfsc   PORTA, 3
  goto    Loop                                         incf    ObjRear, f
                                                       btfss   PIR1, TMR2IF
ADCCycle3:                                              goto    ObjLoop
  xorlw   3 ^ 2
  btfss   STATUS, Z                                   bsf     STATUS, RP0
   goto   ADCCycle4                                   bsf     TRISC ^ 0x80, 5   ; Disable PWM
  movf    ADRESH, w         ; Good ADC Read,                                     ; Output
                            ; Save Left Value         bcf     STATUS, RP0
  movwf   ADCValueL                                   bcf     T2CON, TMR2ON     ; Turn off TMR2
  goto    Loop
                                                      bcf     PFrontRIR         ; Clear Object Bits
ADCCycle4:                                            bcf     PFrontLIR
  xorlw   4 ^ 3                                       bcf     PRearIR
  btfss   STATUS, Z                                   movlw   30                ; Thirty Times for
   goto   ADCCycle5                                                            ; Set
  movlw   b'00000101'       ; Enable ADC on RA1       subwf   ObjFrontR, w
  movwf   ADCON0                                      btfsc   STATUS, C
  bsf     ADCON0, GO        ; Start ADC               bsf     PFrontRIR
                            ; Operation              movlw   30
  goto    Loop                                        subwf   ObjFrontL, w
                                                      btfsc   STATUS, C
ADCCycle5:                                            bsf     PFrontLIR
  xorlw   5 ^ 4                                       movlw   30
  btfss   STATUS, Z                                   subwf   ObjRear, w
   goto   ADCCycle6                                   btfsc   STATUS, C
  movf    ADRESH, w         ; Expect Bad ADC          bsf     PRearIR
                            ; Read, Ignore Value
  goto    Loop                                        goto    Loop              ; Finished, Loop
                                                                               ; Around Again
ADCCycle6:
  xorlw   6 ^ 5                                     LineSense:
  btfss   STATUS, Z                                   movf    PWMCycle, w       ; Update Line
   goto   ADCCycle7                                                            ; Sensors?
  bsf     ADCON0, GO        ; Start Good Read         andlw   0x03              ; Every 4th Cycle
  goto    Loop                                        xorlw   2
                                                      btfss   STATUS, Z
ADCCycle7:                                             goto    OperatingLogic
  movf    ADRESH, w         ; Good ADC Read,
                            ; Save Right Value        clrf    LineFrontR        ; Clear the Object
  movwf   ADCValueR                                                            ; Counters
  goto    Loop                                        clrf    LineFrontL
                                                      movf    LinePWM, w        ; Setup TMR2
ObjSense:                                             bsf     STATUS, RP0
  movf    PWMCycle, w       ; Update Object           movwf   PR2 ^ 0x80
                            ; Sensor?                 bcf     STATUS, RP0
  andlw   0x03              ; Every 4th Cycle         bcf     STATUS, C
  xorlw   1                                           rrf     LinePWM, w
  btfss   STATUS, Z                                   movwf   CCPR1L
   goto   LineSense                                   clrf    TMR2
                                                      bcf     PIR1, TMR2IF
  clrf    ObjFrontR         ; Clear the Object        bsf     T2CON, TMR2ON     ; Turn on TMR2 for
                            ; Counters                                         Line
  clrf    ObjFrontL                                   bsf     STATUS, RP0
  clrf    ObjRear                                     bcf     TRISC ^ 0x80, 4   ; Enable PWM Output
  movf    ObjectPWM, w      ; Setup TMR2              bcf     STATUS, RP0
  bsf     STATUS, RP0
  movwf   PR2 ^ 0x80                                LineLoop:                   ; Loop Here Polling
  bcf     STATUS, RP0                                                          ; Line
  bcf     STATUS, C
  rrf     ObjectPWM, w                                btfsc   PORTC, 0
  movwf   CCPR1L                                       incf    LineFrontR, f
  clrf    TMR2                                         btfsc   PORTC, 1
  bcf     PIR1, TMR2IF                                 incf    LineFrontL, f
                                                       btfss   PIR1, TMR2IF
```

```
        goto    LineLoop

        bsf     STATUS, RP0
        bsf     TRISC ^ 0x80, 4     ; Disable PWM Output
        bcf     STATUS, RP0
        bcf     T2CON, TMR2ON       ; Turn off TMR2

        bcf     LFrontRIR           ; Clear Object Bits
        bcf     LFrontLIR
        movlw   30                  ; Thirty Times for
                                    ; Set
        subwf   LineFrontR, w
        btfsc   STATUS, C
         bsf    LFrontRIR
        movlw   30
        subwf   LineFrontL, w
        btfsc   STATUS, C
         bsf    LFrontLIR

        goto    Loop                ; Finished, Loop
                                    ; Around Again

OperatingLogic:
;  #### - Put Operating Logic Here (8 Per 32
;         Cycles)/Needs Set Flag
;  #### - Check "AppReset" to See if
;         Application is to be Restarted
;  #### - Remote Control Direct (Move or Stop)
;         Resets "AppRun" Flag
;  #### - Remote Control "Enter" Sets "AppRun"
;         and "AppReset"
;  #### - Remote Control "Recall" Sets
;         "AppRun" Flag

        goto    Loop                ; Finished, Loop
                                    ; Around Again


;   Subroutines
CountLow:                           ; Count, in 10 us
                                    ; Increments
        clrf    Dlay                ; Time Signal low
CountLowLoop:
        incf    Dlay, f             ; Gone Around?
        btfsc   STATUS, Z
         goto   CountLowDone        ; Yes, Return Zero
        goto    $ + 1               ; Want 10 Cycle Loop
        goto    $ + 1
        btfss   RearIR
         goto   CountLowLoop
CountLowDone:                       ; Finished, Return
        return

CountHigh:                          ; Count, in 10 us
                                    ; Increments
        clrf    Dlay                ; Time Signal High
                                    ; and Check
CountHighLoop:
        incf    Dlay, f             ; Gone Around?
        btfsc   STATUS, Z
         goto   CountHighDone       ; Yes, Return Zero
        goto    $ + 1               ; Want 10 Cycle Loop
        goto    $ + 1
        btfsc   RearIR
         goto   CountHighLoop
CountHighDone:
        movf    Dlay, w             ; Less than 40?
        sublw   35
        btfsc   STATUS, C
         return
        movlw   80                  ; Greater than 79?
        subwf   Dlay, w
        return                      ; Carry Set if "Yes"


        end
```

I had a lot of trouble in this experiment (which figures, because it's the last one) with the IR signal being too powerful. I was able to be successful by carefully heating the heat-shrink tubing around the bottom-pointing LED to the point where very little light was let out and white and black were detected. Please check the CD-ROM and on my web page (www.myke.com) for updated code that changes the PWM period of the IR signal to decrease the sensitivity of the line sensors so they work better. Otherwise, you may wish to add a large potentiometer (say 10k) in series to see if significantly reducing the current flowing through the IR LED helps to better sense the difference between white and black surfaces.

For Consideration

Whew! Although being a lot of fun to work through, the ZipZaps hack was a pretty intensive introduction to robotics. And a pretty good introduction at that. It is possible however, that you might be questioning this from a number of perspectives, including the following:

- The fragility of the robot
- The difficulty in adding the new components precisely
- The ability of the robot to run only on flat, smooth surfaces
- The apparent complexity of the software
- What to do next

I like robots that are built like North American cars built in the 1960s—able to drive away from a serious accident with nothing more than chipped paint. I am not trying to be facetious; chances are your robot will drive itself off the table you are working on, and you'll wish for the same thing. (This is a factor because there is no on/off switch due to my relying on the ZipZaps battery for power.) So remember to chock the wheels, and you'll prevent the robot from going off on its own (it does not have the power to successfully climb over any obstacles). Yet, at some point in time, you'll probably forget to do this and accidentally lean on the remote control. Despite this danger and a few broken wires when I was taking the PCB on and off the chassis, the robot held together during the development and during about 10 hours of use without any problems. As a prototype, this robot is acceptably robust, but it is certainly not strong and resistant enough to withstand the accidents a commercial product would need to withstand.

This hack would be much easier to perform if a custom PCB were designed for it. If surface-mount technology was used, then chances are the custom PCB could be as small as the PCB that came with the original ZipZaps (although some kind of provision would have to be made for the IR TV remote-control receivers). The ease of building the circuitry could also be improved by relocating the motor and steering driver transistors from the cut-down ZipZaps PCB to the robot PCB. I did not try this because I wanted to leave the stock driver circuits intact, assuming that they have been optimized for the motors and steering solenoids used. A custom PCB could also reduce the fragility of the final robot.

The car chassis and the all-or-nothing steering limit the surfaces on which it will run and how the robot works. I prefer working with differentially driven robots that can literally turn on a dime. The very small turning radius of the ZipZaps chassis is about as good as you can expect, but still not as good as what many robots are capable of. Despite these limitations, the chassis and robot are very useful for prototyping and should help you work through the sensor and software issues of a larger, car-chassis-based robot.

The software used in this robot is not as complex as it looks. Hopefully you can see its progression as you work through the experiments and see how relatively simple changes result in substantial improvements in capability. If I had presented the final software without any of the intermediate steps, then I would agree it is very complex for somebody to understand all at once. However, by going through the individual iterations of the code, you should have some appreciation of how complex software is built from a series of small pieces.

If you can't see any way to use this robot, then I suggest that you spend some time in your local library or bookstore, or on the Internet. There are literally thousands of different applications (and methods for implementing them in hardware and software) that can be used as ideas for your robot. The ZipZaps robot will never bring you a cold drink from your refrigerator, but it can simulate the actions so you can create a larger, more capable robot. There are also many

opportunities for expanding the capabilities of the robot, including adding more or different sensors or even some kind of object-capture hardware.

On the positive side, the robot is fast and tracks as if it were on rails. I would say its speed and ability to move precisely is as good as or better than the original product. Although the car regularly spins out, this is not a desirable trait in a robot. The extra weight of the prototyping PCB and the components on it probably contribute greatly to how the robot handles. I also was pleasantly pleased by the battery life of the final product; it seems to be better than what the original product had. This doesn't seem quite possible when you consider the various electrical devices added to the chassis as well as its extra weight, which requires more power to move. It could be that the radio receiver uses a lot more power than the PIC MCU solution.

Most important, this is a robot that you can do a lot with. The software is designed to make coming up with your own application code as easy as possible; the sensors are continually polled by the base code and allow you to simply read variables to find out the latest values. Similarly, to move the robot, it is simply a matter of writing to the appropriate motor and steering variables to affect any kind of movement. Finally, the remote-control interface allows you to initiate the operation of your code and to stop it if it is not working properly. You can then move the robot to a new location to try something different or in a new environment. It's actually a nice basic robot to experiment with.

Like this book, the ZipZaps robot hack should have given you new insights into the PIC MCU, how it is interfaced, and how it is programmed. It is now up to you to take this knowledge and go forward with your own experiments. I'm looking forward to seeing what you come up with. And if at some time the book's moniker applies to you, and you use the skills and knowledge gained from this book to take over the world, please remember where those skills and knowledge came from.

For Consideration

Index

Index

Index

N

O

T

Index

working register (WREG):
    in addition instructions, 168–169
    in bitwise operations, 167–168
    loading, 164–165
    read-only array element, 208
    saving contents of, 165
    and subtract instructions, 171, 172
writing EEPROM data memory, 214–216

X

XOR (^), 35

Z

zero check module, 164
zero insertion force (ZIF) socket, 14, 51–53
ZipZaps®-based robots, 301–326
    basic task-control software, 313–316
    characterizing, 303–304
    IR line-following sensors, 320–325
    IR object-detection sensors, 318–319
    IR remote control, 316–317
    IR TV remote control, 310–311
    light sensors and light following, 317–318
    motor and steering control, 311–312
    PIC MCU electronics PCB, 307–309
    PIC MCU power supply, 305–306

Discount Coupon

20% Off Retail Price
On PICkit™ 1 Flash Start Kit (DV164101)
On Signal Analysis PICtail™ Daughter Board (AC164120)

Microchip Technology is offering a 20% discount off the retail price for the PICkit™ 1 Flash Starter Kit (DV164101) and the Signal Analysis PICtail™ Daughter Board (AC164120) low-cost development kit and daughter board. The discount may be redeemed through your local Microchip distributor, or on the Microchip e-Commerce website **http//:buy.microchip.com.** This offer cannot be used after **March 31, 2006.**

Terms and Conditions:

This offer applies only to the PICkit™ 1 Flash Starter Kit and/or the Signal Analysis PICtail™ Daughter Board. This voucher must be surrendered to authorized distributor at the time of placement of order and must include Microchip reference number shown below to allow for discount. If using Microchip's e-commerce site, select the tool and proceed to the checkout cart. Type the reference number shown below into the box labeled: "Coupon" and press the green button "apply coupon." The discount will be automatically applied to your purchase. This voucher may not be used in conjunction with any other offer, and has no cash value. Only one voucher may be used per transaction. This offer may be withdrawn without prior notice. Voucher must be used in conjunction with any other terms and conditions specified by Microchip Technology Inc.

Note To Distributor:

The reference (QTN) below must be shown on your order to gain discount.

Use this number to purchase from your local Microchip Distributor or direct from Microchip via **http//:buy.microchip.com**

PRED05
Reference Number